Carbon Neutrality and Symmetry in Power Engineering and Engineering Thermophysics

Carbon Neutrality and Symmetry in Power Engineering and Engineering Thermophysics

Editors

Zheng Xu
Guangze Li
Bin Zhang

Basel • Beijing • Wuhan • Barcelona • Belgrade • Novi Sad • Cluj • Manchester

Editors

Zheng Xu
Beihang University
Hangzhou
China

Guangze Li
Beihang University
Hangzhou
China

Bin Zhang
Beihang University
Hangzhou
China

Editorial Office
MDPI AG
Grosspeteranlage 5
4052 Basel, Switzerland

This is a reprint of articles from the Special Issue published online in the open access journal *Symmetry* (ISSN 2073-8994) (available at: https://www.mdpi.com/journal/symmetry/special_issues/carbon_neutrality_symmetry_power_engineering_engineering_thermophysics).

For citation purposes, cite each article independently as indicated on the article page online and as indicated below:

Lastname, A.A.; Lastname, B.B. Article Title. *Journal Name* **Year**, *Volume Number*, Page Range.

ISBN 978-3-7258-2565-3 (Hbk)
ISBN 978-3-7258-2566-0 (PDF)
doi.org/10.3390/books978-3-7258-2566-0

© 2024 by the authors. Articles in this book are Open Access and distributed under the Creative Commons Attribution (CC BY) license. The book as a whole is distributed by MDPI under the terms and conditions of the Creative Commons Attribution-NonCommercial-NoDerivs (CC BY-NC-ND) license.

Contents

About the Editors vii

Zheng Xu, Jinze Pei and Yue Song
The Optimization of Aviation Technologies and Design Strategies for a Carbon-Neutral Future
Reprinted from: *Symmetry* **2024**, *16*, 1226, doi:10.3390/sym16091226 1

Jiu Zhou, Jieni Zhang, Zhaoming Qiu, Zhiwen Yu, Qiong Cui and Xiangrui Tong
Two-Stage Robust Optimization for Large Logistics Parks to Participate in Grid Peak Shaving
Reprinted from: *Symmetry* **2024**, *16*, 949, doi:10.3390/sym16080949 8

Zhen Jiang, Xi Wang, Shubo Yang and Meiyin Zhu
A Deterministic Calibration Method for the Thermodynamic Model of Gas Turbines
Reprinted from: *Symmetry* **2024**, *16*, 522, doi:10.3390/sym16050522 30

Liang Chen, Quan Zhou, Guangze Li, Liuyong Chang, Longfei Chen and Yuanhao Li
Investigation into Detection Efficiency Deviations in Aviation Soot and Calibration Particles Based on Condensation Particle Counting
Reprinted from: *Symmetry* **2024**, *16*, 244, doi:10.3390/sym16020244 46

Hao Liu, Dayi Zhang, Kaicheng Liu, Jianjun Wang, Yu Liu and Yifu Long
Nonlinear Dynamic Modeling and Analysis for a Spur Gear System with Dynamic Meshing Parameters and Sliding Friction
Reprinted from: *Symmetry* **2023**, *15*, 1530, doi:10.3390/sym15081530 62

Hao Liu, Jianjun Wang, Yu Liu, Zhi Wang and Yifu Long
Stiffness Characteristics and Analytical Model of a Flange Joint with a Spigot
Reprinted from: *Symmetry* **2023**, *15*, 1221, doi:10.3390/sym15061221 90

Keqiang Miao, Xi Wang, Meiyin Zhu, Song Zhang, Zhihong Dan, Jiashuai Liu, et al.
A Multi-Cavity Iterative Modeling Method for the Exhaust Systems of Altitude Ground Test Facilities
Reprinted from: *Symmetry* **2022**, *14*, 1399, doi:10.3390/sym14071399 108

Jiashuai Liu, Xi Wang, Xitong Pei, Meiyin Zhu, Louyue Zhang, Shubo Yang and Song Zhang
Generic Modeling Method of Quasi-One-Dimensional Flow for Aeropropulsion System Test Facility
Reprinted from: *Symmetry* **2022**, *14*, 1161, doi:10.3390/sym14061161 130

Xiaojing Liu, Shuiting Ding, Longtao Shao, Shuai Zhao, Tian Qiu, Yu Zhou, et al.
Investigation on the Formation and Evolution Mechanism of Flow-Resistance-Increasing Vortex of Aero-Engine Labyrinth Based on Entropy Generation Analysis
Reprinted from: *Symmetry* **2022**, *14*, 881, doi:10.3390/sym14050881 147

Yu Zhou, Longtao Shao, Shuai Zhao, Kun Zhu, Shuiting Ding, Farong Du and Zheng Xu
An Experimental Investigation into the Thermal Characteristics of Bump Foil Journal Bearings
Reprinted from: *Symmetry* **2022**, *14*, 878, doi:10.3390/sym14050878 169

Yifu Long, Shubo Yang, Xi Wang, Zhen Jiang, Jiashuai Liu, Wenshuai Zhao, et al.
MoHydroLib: An HMU Library for Gas Turbine Control System with Modelica
Reprinted from: *Symmetry* **2022**, *14*, 851, doi:10.3390/sym14050851 189

Keqiang Miao, Xi Wang, Meiyin Zhu, Shubo Yang, Xitong Pei and Zhen Jiang
Transient Controller Design Based on Reinforcement Learning for a Turbofan Engine with Actuator Dynamics
Reprinted from: *Symmetry* **2022**, *14*, 684, doi:10.3390/sym14040684 207

Zhiping Li, Long He and Yixuan Zheng
Quasi-Analytical Solution of Optimum and Maximum Depth of Transverse V-Groove for Drag Reduction at Different Reynolds Numbers
Reprinted from: *Symmetry* **2022**, *14*, 342, doi:10.3390/sym14020342 223

About the Editors

Zheng Xu

Xu Zheng, an associate professor at Beihang University, was selected for the Young Talent Support Program of the Chinese Society of Internal Combustion Engines. His main research directions include new energy aviation power, aviation emissions and airworthiness, and the application technology and environmental impact of sustainable aviation fuels. He has been engaged in the research and prototype development of key technologies for new energy aviation power for many years. In the past five years, he has published more than 20 SCI papers and applied for 24 invention patents. He has presided over the National Natural Science Foundation of China for Young Scientists, the National Key R&D Program, sub-topics of the Civil Aviation Joint Fund, and the Zhejiang Provincial Natural Science Foundation for Young Scientists. He has won the Silver Award of the first National Postdoctoral Innovation Competition, the Excellence Award of the China Aerospace Power Innovation Competition, the Outstanding Doctoral Dissertation Award of the Chinese Society of Internal Combustion Engines, and the first prize of the National Energy Conservation and Emission Reduction Competition. He serves as a guest editor for the SCI journals *"Symmetry"* and *"Processes"*, and holds positions such as the Deputy Secretary-General of the Aviation Branch of the Chinese Society of Internal Combustion Engines, and is a member of the Plateau Branch.

Guangze Li

Guangze Li is an associate professor at the International Innovation Institute, Beihang University, China. He received a Ph.D. degree in Thermal Engineering from Beihang University in 2022, and has participated in a joint Ph.D. program at the National University of Singapore. His research focuses on the heat and mass transfer of multi-phase flow, combustion, and emissions. He has published more than 20 papers in the top SCI-indexed journals, including in *Energy*, *Fuel*, and *Environmental Science & Technology*, and has obtained sponsorship from the National Natural Science Foundation of China.

Bin Zhang

Bin Zhang, an associate professor at the Hangzhou International Innovation Institute of Beihang University, serves as a Youth Committee Member of the Ice and Anti-Ice Branch of the Chinese Aeronautical Society. He has dedicated his research to the structural strength and airworthiness safety of aircraft engines, as well as the service damage and optimization design of superhydrophobic surfaces. He has led projects funded by the National Natural Science Foundation of China, the China Postdoctoral Science Foundation, the Zhejiang Association for Science and Technology, and the Zhejiang Provincial Natural Science Foundation. He has also participated in several major projects, including the National Key Research and Development Program. Additionally, he has received the Excellence Award of the 24th China Patent Award (Rank: 3/6), the First Prize for Technical Invention from the Chinese Internal Combustion Engine Society (Rank: 10/15, 2022), and a nomination for the Early Career Award (SCI Journal—Fatigue Fract Eng M, 2022).

Editorial

The Optimization of Aviation Technologies and Design Strategies for a Carbon-Neutral Future

Zheng Xu [1], Jinze Pei [2] and Yue Song [2,3,*]

1. Hangzhou International Innovation Institute, Beihang University, Hangzhou 311115, China; zheng.xu@buaa.edu.cn
2. Research Institute of Aero-Engine, Beihang University, Beijing 102206, China; peijze@163.com
3. School of Energy and Power Engineering, Beihang University, Beijing 100083, China
* Correspondence: sy1704312@buaa.edu.cn

Citation: Xu, Z.; Pei, J.; Song, Y. The Optimization of Aviation Technologies and Design Strategies for a Carbon-Neutral Future. *Symmetry* 2024, 16, 1226. https://doi.org/10.3390/sym16091226

Received: 5 September 2024
Accepted: 10 September 2024
Published: 18 September 2024

Copyright: © 2024 by the authors. Licensee MDPI, Basel, Switzerland. This article is an open access article distributed under the terms and conditions of the Creative Commons Attribution (CC BY) license (https://creativecommons.org/licenses/by/4.0/).

This Special Issue systematically reviews and summarizes the latest research into carbon neutrality technology and symmetry principles in power engineering and engineering thermophysics. Research papers cover topics such as aero-engine experimental equipment, internal flow behavior, control system design, engine performance analysis, emission control technology, electricity system efficiency improvement, and vehicle drag reduction measures. The goal of this Special Issue is to provide new ideas and methods that can be utilized in solving current challenges by deeply exploring and optimizing key technologies. We hope that these research results can inspire more scholars and engineers to collaborate to promote the sustainable development of science, technology, and industry.

The pursuit of carbon neutrality has reached a critical juncture, placing significant pressure on the aviation industry to mitigate its environmental impact while maintaining operational efficiency. As global concern over climate change intensifies, the exploration of alternative fuel engine optimization and adaptive design has become imperative. Sustainable aviation fuel (SAF), hydrogen, and ammonia have shown promise in reducing aviation emissions. In particular, transitioning to 100% SAFs holds substantial potential in terms of achieving greater emissions reductions [1–3]. RP-3 kerosene–pentanol blends offer significant advantages in terms of enhancing the thermal efficiency of compression ignition engines and reducing smoke emissions [4,5]. The use of hydrogen and ammonia fuels in dual-fuel engines has the potential to reduce carbon emissions and enhance operational flexibility. However, attention must be given to the potential of unburned fuel and other emissions [6,7]. The use of 100% HEFA-SAF in small turbofan engines significantly reduces particulate matter emissions, offering empirical evidence that supports the aviation industry's efforts to further decrease PM emissions and informs the development of relevant airworthiness regulations [8,9]. The proportion of different fuel mixtures used impacts engine performance and emissions. By optimizing the hybrid response surface method, it is possible to identify the optimal solution, capable of effectively securing both high performances and low emissions [10–12].

Optimizing thermal cycles and components of engines is crucial in enhancing engine efficiency and reducing emissions [13–15]. While increasing spray pressure can improve spray parameters under low-temperature conditions, it can also accelerate fuel evaporation at high temperatures, potentially affecting ignition characteristics and combustion completeness [16,17]. Reducing the aerodynamic, thermal, and mechanical inertia of the turbocharger can significantly enhance its response speed, thereby improving the overall performance of the aviation piston engine [18]. By coupling the thermodynamic cycle of the piston engine and the gas turbine, the power–weight ratio and thermal efficiency of the engine under high-altitude conditions are improved [19,20]. The establishment of a theoretical model combined with the optimization of artificial intelligence technology can more accurately predict and enhance the high-altitude performance of engine gas exchange systems [21,22].

The optimized design of aircraft can further mitigate their environmental impact. However, designing to achieve emission reduction may have implications for the profitability of the aviation network [23–25]. Multi-objective optimization design can be applied to aircraft optimization, enabling the attainment of a balance between operational costs and environmental impacts. This approach offers a practical pathway to achieving ecological efficiency in aviation design [26,27].

Based on a comprehensive analysis of the literature, researchers have made significant strides in peak load regulation, alternative fuel applications, engine performance optimization, and sustainable aviation design. These advancements provide diverse technical and strategic support in order to address the complex challenges in the aviation field, laying a solid foundation for future innovation and development. All contributions are listed in the contribution list.

The integration of various new renewable energy systems is gradually disrupting the symmetry and balance between the supply side and demand side of power systems, creating opportunities to enhance power system efficiency and increase the utilization rate of renewable energy. When refrigeration units participate in power network linkage regulation, the load curve of the power network is smoothed, leading to improved power supply efficiency and reliability. Given that the cooling load during summer peak hours constitutes a significant portion of the total load in logistics parks, there is substantial potential for adjustment. Contribution 1 introduced a two-stage robust optimization method for peak shaving, establishing an optimization model aimed at maximizing the participation of cooling loads in peak shaving. This method achieves an optimal balance between operational efficiency, temperature tolerance, and peak load regulation participation, demonstrating significant advantages in enhancing power system security and stability. This work provides a framework for integrating the cold loads of logistics parks into power grid peak load regulation and includes an optimization-based case study.

The aeropropulsion propulsion system test facility (ASTF) serves not only as a critical platform for technological development and verification, but also as essential support in assessing energy-saving and emission reduction effects, setting standards and certifications, driving industrial upgrades, and fostering international cooperation and exchange. As the demands of designing and verifying advanced controllers for ASTFs continue to rise, the establishment of a high-precision mathematical model for ASTFs becomes increasingly important. Contribution 2 introduces a quasi-one-dimensional flow modeling method and constructs a high-precision quasi-one-dimensional flow model for ASTFs. The accuracy of this model in terms of predicting flow and pressure was validated through comparisons with experimental data. The experimental results demonstrate that this model has significant practical applications in the design and verification of advanced controllers for ASTF, enabling the more accurate simulation and optimization of the aero-propulsion system's operating state, thereby significantly enhancing system performance.

The leakage flow in an aero-engine can lead to structural asymmetry, negatively impacting the thrust output and fuel efficiency [28]. The labyrinth seal plays a crucial role in minimizing leakage between the blade and shroud. However, its complex flow dynamics and associated energy losses directly affect the efficiency of the engine. Contribution 3 explores the formation and evolution mechanism of drag-inducing vortices in aero-engine labyrinth seals through entropy generation analysis. The study verifies the formation mechanism and influencing factors of drag-inducing vortices using a three-dimensional simulation model and experimental data. The mechanism is clearly defined, and the impact of design parameters on vortex entropy formation is thoroughly investigated. The findings reveal that selecting appropriate design parameters can enhance vortex resistance and reduce leakage. The results indicate that leakage can be reduced by 22% by optimizing the geometric parameters of the labyrinth seal. These research outcomes hold significant practical value for the design and optimization of labyrinth seals, leading to substantial improvements in engine efficiency. Further optimization of the labyrinth seal design offers the potential for even greater performance enhancement.

An efficient hydraulic drive system ensures that the engine operates optimally under all conditions, minimizing unnecessary energy losses. Hydraulic mechanical units (HMUs) are composed of various symmetrical and asymmetrical geometries, and the complexity of existing commercial modeling software poses significant challenges. Contribution 4 addresses this issue by developing a lightweight HMU modeling library, MoHydroLib, using the Modelica language. This is designed specifically for gas turbine control systems. The library was established by studying the modeling methods of HMUs, and the detailed process of constructing the HMU model is thoroughly explained, including the design and implementation of orifice, piston, boundary, and system elements in Modelica.

The accuracy and validity of the MoHydroLib models were verified through the simulation of five typical HMU subsystems, demonstrating that the library achieves the same level of simulation accuracy as commercial modeling software. Additionally, MoHydroLib significantly reduces the number of lines of model code and enhances the modeling efficiency. The application and results indicate that this modeling method has promise for use in more complex systems, offering a more efficient alternative to traditional commercial modeling approaches.

Contribution 5 focuses on the design of turbofan transient controllers using reinforcement learning, aiming to address challenges related to uncertainty and degradation over the life cycle of the engine. The goal is to enhance the stability and performance of the turbofan system, despite these challenges. A transient controller was developed using reinforcement learning; specifically, the scholars employed the deep deterministic policy gradient algorithm. Based on a nonlinear model of the turbofan, this method successfully produced a transient controller that maintains system performance, even when dealing with uncertainties and degradation. This approach provides a valuable reference for the design of controllers in other complex systems, demonstrating the potential of reinforcement learning to improve system resilience and reliability.

The altitude ground test facility (AGTF) exhaust system is composed of symmetrical components, allowing for the circumferential variation of parameters to be ignored. Due to the growing need for accurate exhaust system modeling, Contribution 6 introduces a multi-cavity iterative modeling method to design the exhaust system of an AGTF. This method aims to address the nonlinear challenges encountered in modeling the exhaust system of a high-altitude ground test facility, as well as the difficulties in calculating the ejection coefficient.

In this study, the key components of the exhaust system, such as the exhaust diffuser and cooler, are represented as a series of volume models, which helps to overcome the limitations inherent in traditional lumped-parameter models. The accuracy and validity of this multi-cavity iterative modeling method were verified through closed-loop simulation. Further optimization of the model's parameters could enhance simulation accuracy and expand the method's applicability to other complex systems. This approach represents a significant advancement in the precise modeling of exhaust systems in high-altitude testing environments.

The application of gas foil bearings (GFBs) in aero-engines is crucial for achieving high-speed stability and reliability while also reducing the weight of small aero-engines [29,30]. The symmetry of the foil radial sliding bearing is a key aspect of its design. Contribution 7 investigates the thermal characteristics of GFBs within an aero-engine system operating at high temperatures and high speeds. The research was conducted primarily through numerical simulations and experimental methods. The temperature characteristics of the GFB were directly measured through experiments, leading to the development of a high-temperature test platform. A new GFB structure was designed and built to evaluate its load-carrying capacity under varying temperatures and rotational speeds. Several methods for reducing friction and enhancing dynamic performance were proposed and compared. The optimized GFB demonstrated superior thermal performances and stability at high temperatures and high speeds, showcasing its successful application in oil and gas turbochargers and high-speed air compressors for fuel cells.

Flanged joints are extensively used in the rotors and stators of aero-engines, and their stiffness significantly influences the dynamic characteristics of the engine. From an engineering application perspective, Contribution 8 introduces a simplified analytical model in order to simulate the deformation and slip of flange contact surfaces. This model accurately captures the stiffness characteristics of flanged joints with a spigot, including sudden changes in bending stiffness and hysteresis loop behavior. To validate the model, the author designed and constructed an experimental apparatus to study the angular bending stiffness of flanged joints, confirming the accuracy of the existing finite element model. The simplified analytical model of flange joint deformation and slip was further tested under various loading conditions, verifying its accuracy and applicability in practical scenarios. This work provides valuable insights into the mechanical behavior of flanged joints and enhances the reliability of their design in aero-engine applications.

Due to its inherent periodic symmetry, gear systems are extensively utilized in the fields of aeronautics and astronautics, where their performance directly impacts the overall effectiveness of mechanical equipment. However, existing models often neglect the influence of lateral vibration on meshing parameters and sliding friction, leading to inaccuracies in predicting the dynamic behavior of the system. Contribution 9 introduces a new sliding friction model, which forms the basis of a nonlinear dynamic model incorporating dynamic meshing parameters. The equations of motion for this model are derived using the Lagrangian method and solved via the Runge–Kutta method. The dynamic responses of the new model were compared to those of traditional models under varying input velocities and friction coefficients. The findings indicate that the new model provides more accurate predictions of the system's actual responses, particularly when accounting for lateral vibration and sliding friction. These research results are highly significant for the design, condition monitoring, and fault diagnosis of gear systems, offering improved accuracy and reliability in these critical applications.

Carbonaceous particles are the primary components of aero-engine emissions, and the reduction in particle size presents challenges for the detection efficiency of condensation particle counters (CPCs). Variations in temperature and pressure can further impact the detection efficiency of these devices, thereby affecting the accuracy and reliability of the measurement results. Contribution 10 investigates the detection efficiency of exploring aviation soot in CPCs compared to calibration particles (e.g., sodium chloride). To address this issue, the author designed and established an experimental system capable of simulating various environmental conditions. A particle generator was used to produce standard particles, which were then measured in conjunction with a Differential Electromobility Analyzer and an aerosol electrometer. The study systematically evaluates the effects of temperature and pressure on detection efficiency through experimental methods. The results demonstrate that changes in temperature and pressure significantly influence the detection efficiency of CPCs. By monitoring these environmental factors, the detection efficiency can be corrected, thereby improving the accuracy of measurement results. This research provides insights with value in terms of enhancing the reliability of emission measurements in aero-engine testing and monitoring.

Gas turbines play a critical role in the aerospace field, and the precision of their performance models is essential for effective engine control, diagnosis, and prediction [31,32]. Contribution 11 introduces a deterministic calibration method for a gas turbine thermodynamic model, which replaces traditional search algorithms with a deterministic calibration process. This approach enhances model accuracy while reducing computational complexity. The deterministic calibration method, based on an interpolation algorithm, estimates actual performance parameters through a numerical iterative process. When applied to a high-bypass-ratio turbofan model, this method reduced the prediction error from approximately 9% to around 0.6%. The author demonstrates that this method significantly improves model accuracy and provides a valuable reference for calibrating gas turbine models to accurately reflect off-design performance.

Reducing the surface friction resistance of vehicles can directly lower the energy consumption during operation and decrease greenhouse gas emissions. Contribution 12 explores the quasi-analytical solutions for the optimal and maximum depth of transverse V-shaped grooves, designed to reduce drag at different Reynolds numbers. The large eddy simulation (LES) was employed to investigate the drag reduction characteristics of grooves with varying depths for different Reynolds numbers. The study established relationships among slip velocity, viscous drag reduction rate, and the pressure increase rate. A physical model was developed to describe the relationship between dimensionless depth, dimensionless inlet velocity, and the drag reduction rate. Using this model, the optimal and maximum depths of the transverse grooves were determined. The theoretical model was found to be consistent with numerical simulations and experimental results, confirming the feasibility and practicality of this approach. This research provides valuable insights into drag reduction techniques, offering a potential pathway for reducing energy consumption and emissions in vehicle design.

Author Contributions: Conceptualization, Z.X.; writing—original draft preparation, J.P.; writing—review and editing, Y.S. All authors have read and agreed to the published version of the manuscript.

Funding: This research was funded by the National Key R&D Program of China, grant number [2022YFB2602000] and [2018YFB0104100], Basic Research Program of the National Natural Science Foundation of China (Grant number [52475503], [52206131], [U2233213] and [U2333217]), and Zhejiang Provincial Natural Science Foundation of China, grant number [LQ22E060004], Science Center of Gas Turbine Project [P2022-A-I-001-001].

Conflicts of Interest: The authors declare no conflicts of interest.

List of Contributions:

1. Zhou, J.; Zhang, J.; Qiu, Z.; Yu, Z.; Cui, Q.; Tong, X. Two-Stage Robust Optimization for Large Logistics Parks to Participate in Grid Peak Shaving. *Symmetry* **2024**, *16*, 949. https://doi.org/10.3390/sym16080949.
2. Liu, J.; Wang, X.; Pei, X.; Zhu, M.; Zhang, L.; Yang, S.; Zhang, S. Generic Modeling Method of Quasi-One-Dimensional Flow for Aeropropulsion System Test Facility. *Symmetry* **2022**, *14*, 1161. https://doi.org/10.3390/sym14061161.
3. Liu, X.; Ding, S.; Shao, L.; Zhao, S.; Qiu, T.; Zhou, Y.; Zhang, X.; Li, G. Investigation on the Formation and Evolution Mechanism of Flow-Resistance-Increasing Vortex of Aero-Engine Labyrinth Based on Entropy Generation Analysis. *Symmetry* **2022**, *14*, 881. https://doi.org/10.3390/sym14050881.
4. Long, Y.; Yang, S.; Wang, X.; Jiang, Z.; Liu, J.; Zhao, W.; Zhu, M.; Chen, H.; Miao, K.; Zhang, Y. MoHydroLib: An HMU Library for Gas Turbine Control System with Modelica. *Symmetry* **2022**, *14*, 851. https://doi.org/10.3390/sym14050851.
5. Miao, K.; Wang, X.; Zhu, M.; Yang, S.; Pei, X.; Jiang, Z. Transient Controller Design Based on Reinforcement Learning for a Turbofan Engine with Actuator Dynamics. *Symmetry* **2022**, *14*, 684. https://doi.org/10.3390/sym14040684.
6. Miao, K.; Wang, X.; Zhu, M.; Zhang, S.; Dan, Z.; Liu, J.; Yang, S.; Pei, X.; Wang, X.; Zhang, L. A Multi-Cavity Iterative Modeling Method for the Exhaust Systems of Altitude Ground Test Facilities. *Symmetry* **2022**, *14*, 1399. https://doi.org/10.3390/sym14071399.
7. Zhou, Y.; Shao, L.; Zhao, S.; Zhu, K.; Ding, S.; Du, F.; Xu, Z. An Experimental Investigation into the Thermal Characteristics of Bump Foil Journal Bearings. *Symmetry* **2022**, *14*, 878. https://doi.org/10.3390/sym14050878.
8. Liu, H.; Wang, J.; Liu, Y.; Wang, Z.; Long, Y. Stiffness Characteristics and Analytical Model of a Flange Joint with a Spigot. *Symmetry* **2023**, *15*, 1221. https://doi.org/10.3390/sym15061221.
9. Liu, H.; Zhang, D.; Liu, K.; Wang, J.; Liu, Y.; Long, Y. Nonlinear Dynamic Modeling and Analysis for a Spur Gear System with Dynamic Meshing Parameters and Sliding Friction. *Symmetry* **2023**, *15*, 1530. https://doi.org/10.3390/sym15081530.
10. Chen, L.; Zhou, Q.; Li, G.; Chang, L.; Chen, L.; Li, Y. Investigation into Detection Efficiency Deviations in Aviation Soot and Calibration Particles Based on Condensation Particle Counting. *Symmetry* **2024**, *16*, 244. https://doi.org/10.3390/sym16020244.

11. Jiang, Z.; Wang, X.; Yang, S.; Zhu, M. A Deterministic Calibration Method for the Thermodynamic Model of Gas Turbines. *Symmetry* **2024**, *16*, 522. https://doi.org/10.3390/sym16050522.
12. Li, Z.; He, L.; Zheng, Y. Quasi-Analytical Solution of Optimum and Maximum Depth of Transverse V-Groove for Drag Reduction at Different Reynolds Numbers. *Symmetry* **2022**, *14*, 342. https://doi.org/10.3390/sym14020342.

References

1. Undavalli, V.; Gbadamosi Olatunde, O.B.; Boylu, R.; Wei, C.; Haeker, J.; Hamilton, J.; Khandelwal, B. Recent advancements in sustainable aviation fuels. *Prog. Aerosp. Sci.* **2023**, *136*, 100876. [CrossRef]
2. Cabrera, E.; de Sousa, J.M.M. Use of Sustainable Fuels in Aviation—A Review. *Energies* **2022**, *15*, 2440. [CrossRef]
3. Owen, B.; Anet, J.G.; Bertier, N.; Christie, S.; Cremaschi, M.; Dellaert, S.; Edebeli, J.; Janicke, U.; Kuenen, J.; Lim, L.; et al. Review: Particulate Matter Emissions from Aircraft. *Atmosphere* **2022**, *13*, 1230. [CrossRef]
4. Chen, L.; Ding, S.; Liu, H.; Lu, Y.; Li, Y.; Roskilly, A.P. Comparative study of combustion and emissions of kerosene (RP-3), kerosene-pentanol blends and diesel in a compression ignition engine. *Appl. Energy* **2017**, *203*, 91–100. [CrossRef]
5. Wei, S.; Sun, L.; Wu, L.; Yu, Z.; Zhang, Z. Study of combustion characteristics of diesel, kerosene (RP-3) and kerosene-ethanol blends in a compression ignition engine. *Fuel* **2022**, *317*, 123468. [CrossRef]
6. Reitmayr, C.; Wiesmann, F.; Gotthard, T.; Hofmann, P. *Experimental and Numerical Investigations of a Dual-Fuel Hydrogen–Kerosene Engine for Sustainable General Aviation*; SAE International: Warrendale, PA, USA, 2024. [CrossRef]
7. Liu, J.; Liu, J. Experimental investigation of the effect of ammonia substitution ratio on an ammonia-diesel dual-fuel engine performance. *J. Clean. Prod.* **2024**, *434*, 140274. [CrossRef]
8. Xu, Z.; Wang, M.; Chang, L.; Pan, K.; Shen, X.; Zhong, S.; Xu, J.; Liu, L.; Li, G.; Chen, L. Assessing the particulate matter emission reduction characteristics of small turbofan engine fueled with 100% HEFA sustainable aviation fuel. *Sci. Total Environ.* **2024**, *945*, 174128. [CrossRef]
9. Märkl, R.S.; Voigt, C.; Sauer, D.; Dischl, R.K.; Kaufmann, S.; Harlaß, T.; Hahn, V.; Roiger, A.; Weiß-Rehm, C.; Burkhardt, U.; et al. Powering aircraft with 100% sustainable aviation fuel reduces ice crystals in contrails. *Atmos. Chem. Phys.* **2024**, *24*, 3813–3837. [CrossRef]
10. Shirneshan, A.; Mostofi, M. Optimization the Performance and Emission Parameters of a CI Engine Fueled with Aviation Fuel-Biodiesel-Diesel Blends. *J. Renew. Energy Environ.* **2020**, *7*, 33–39. [CrossRef]
11. Richter, S.; Kathrotia, T.; Naumann, C.; Scheuermann, S.; Riedel, U. Investigation of the sooting propensity of aviation fuel mixtures. *CEAS Aeronaut. J.* **2021**, *12*, 115–123. [CrossRef]
12. Betancourt, M.M.; Leon, N.F.R.; Leal, V.S.; Caranton, A.R.G.; Angarita, S.L.; Gomez, M.L. Predictive Model for the Fluidity of Jet Fuel-biofuel Mixtures for Aviation. *Chem. Eng. Trans.* **2023**, *100*, 505–510. [CrossRef]
13. Ebrahimi, R. A new comparative study on performance of engine cycles under maximum thermal efficiency condition. *Energy Rep.* **2021**, *7*, 8858–8867. [CrossRef]
14. Siswanto, E.; Darmadi, D.B.; Widodo, A.S.; Talice, M. Enhancement of combustion performances and reduction of combustible species emission using an additional of combustion-reaction of engine. *Case Stud. Therm. Eng.* **2023**, *49*, 103328. [CrossRef]
15. Qin, Z.; Liu, F.; Zhang, H.; Wang, X.; Yin, C.; Weng, W.; Han, Z. Study of hydrogen injection strategy on fuel mixing characteristics of a free-piston engine. *Case Stud. Therm. Eng.* **2024**, *56*, 104279. [CrossRef]
16. Shi, Z.; Lee, C.-f.; Wu, H.; Li, H.; Wu, Y.; Zhang, L.; Bo, Y.; Liu, F. Effect of injection pressure on the impinging spray and ignition characteristics of the heavy-duty diesel engine under low-temperature conditions. *Appl. Energy* **2020**, *262*, 114552. [CrossRef]
17. Wu, H.; Cao, W.; Li, H.; Shi, Z.; Zhao, R.; Zhang, L.; Li, X. Wall Temperature Effects on Ignition Characteristics of Liquid-phase Spray Impingement for Heavy-duty Diesel Engine at Low Temperatures. *Combust. Sci. Technol.* **2023**, *195*, 456–471. [CrossRef]
18. Zhou, Y.; Song, Y.; Zhao, S.; Li, X.; Shao, L.; Yan, H.; Xu, Z.; Ding, S. A comprehensive aerodynamic-thermal-mechanical design method for fast response turbocharger applied in aviation piston engines. *Propul. Power Res.* **2024**, *13*, 145–165. [CrossRef]
19. Song, Y.; Zhou, Y.; Zhao, S.; Du, F.-r.; Li, X.-y.; Zhu, K.; Yan, H.-s.; Xu, Z.; Ding, S.-t. Cyclic coupling and working characteristics analysis of a novel combined cycle engine concept for aviation applications. *Energy* **2024**, *301*, 131747. [CrossRef]
20. Pan, Z.; He, Q.; Pang, X. Structural Efficiency Analysis of a Piston for Aviation Engines. *Aerospace* **2022**, *9*, 718. [CrossRef]
21. Xu, Z.; Pei, J.; Ding, S.; Chen, L.; Zhao, S.; Shen, X.; Zhu, K.; Shao, L.; Zhong, Z.; Yan, H.; et al. Gas exchange optimization in aircraft engines using sustainable aviation fuel: A design of experiment and genetic algorithm approach. *Energy AI* **2024**, *17*, 100396. [CrossRef]
22. Zhou, Y.; Pei, J.; Ding, S.; Zhao, S.; Zhu, K.; Shao, L.; Zhong, Z.; Du, F.; Li, X.; Xu, Z. Theoretical model for high-altitude gas exchange process in multi-fuel poppet valves two-stroke aircraft engine. *Energy Convers. Manag.* **2024**, *301*, 118028. [CrossRef]
23. Proesmans, P.-J.; Morlupo, F.; Santos, B.F.; Vos, R. Aircraft Design Optimization Considering Network Demand and Future Aviation Fuels. In Proceedings of the AIAA AVIATION 2023 Forum, San Diego, CA, USA, 12–16 June 2023. [CrossRef]
24. Proesmans, P.-J.; Vos, R. Comparison of Future Aviation Fuels to Minimize the Climate Impact of Commercial Aircraft. In Proceedings of the AIAA AVIATION 2022 Forum, Chicago, IL, USA, 27 June–1 July 2022. [CrossRef]
25. Proesmans, P.-J.; Vos, R. Airplane Design Optimization for Minimal Global Warming Impact. *J. Aircr.* **2022**, *59*, 1363–1381. [CrossRef]

26. Mu, T.; Tao, L. Civil Aircraft Multi-Objective Design Optimization for Cost and Emission Reduction. *J. Phys. Conf. Ser.* **2023**, *2541*, 012034. [CrossRef]
27. Wang, Y.; Xing, Y.; Yu, X.; Zhang, S. Flight operation and airframe design for tradeoff between cost and environmental impact. *Proc. Inst. Mech. Eng. Part G J. Aerosp. Eng.* **2018**, *232*, 973–987. [CrossRef]
28. Liang, D.; Jin, D.; Gui, X. Investigation of Seal Cavity Leakage Flow Effect on Multistage Axial Compressor Aerodynamic Performance with a Circumferentially Averaged Method. *Appl. Sci.* **2021**, *11*, 3937. [CrossRef]
29. Hou, Y.; Zhao, Q.; Guo, Y.; Ren, X.; Lai, T.; Chen, S. Application of Gas Foil Bearings in China. *Appl. Sci.* **2021**, *11*, 6210. [CrossRef]
30. Jin, C.; Li, C.; Du, J. A Review on the Dynamic Performance Studies of Gas Foil Bearings. *Lubricants* **2024**, *12*, 262. [CrossRef]
31. Chen, Y.-Z.; Zhao, X.-D.; Xiang, H.-C.; Tsoutsanis, E. A sequential model-based approach for gas turbine performance diagnostics. *Energy* **2021**, *220*, 119657. [CrossRef]
32. Idrees, N.; Zeb, K.; Salamat, S. Model-Based Performance Analysis and Health Monitoring of a Low Bypass Ratio Turbofan Engine. In Proceedings of the 2022 19th International Bhurban Conference on Applied Sciences and Technology (IBCAST), Islamabad, Pakistan, 16–20 August 2022; pp. 736–742.

Disclaimer/Publisher's Note: The statements, opinions and data contained in all publications are solely those of the individual author(s) and contributor(s) and not of MDPI and/or the editor(s). MDPI and/or the editor(s) disclaim responsibility for any injury to people or property resulting from any ideas, methods, instructions or products referred to in the content.

Article

Two-Stage Robust Optimization for Large Logistics Parks to Participate in Grid Peak Shaving

Jiu Zhou [1], Jieni Zhang [1], Zhaoming Qiu [1], Zhiwen Yu [1], Qiong Cui [2] and Xiangrui Tong [2,*]

[1] Customer Service Center of Guangzhou Power Supply Bureau, Guangdong Power Grid Co., Ltd., Guangzhou 510620, China; dilalaj@163.com (J.Z.); zhangjieni2016@163.com (J.Z.); 13570019732@139.com (Z.Q.); yuzw@guangzhou.csg.cn (Z.Y.)

[2] Key Laboratory of Renewable Energy, Guangzhou Institute of Energy Conversion, Chinese Academy of Sciences, Guangzhou 510640, China; ciqg0716@163.com

* Correspondence: xiangruitong132@126.com

Abstract: As new energy integration increases, power grid load curves become steeper. Large logistics parks, with their substantial cooling load, show great peak shaving potential. Leveraging this load while maintaining staff comfort, product quality, and operational costs is a major challenge. This paper proposes a two-stage robust optimization method for large logistics parks to participate in grid peak shaving. First, a Cooling Load's Economic Contribution (CLEC) index is introduced, integrating the Predicted Mean Vote (PMV) and Sales Pressure Index (SPI). Then, an optimization model is established, accounting for renewable energy uncertainties and maximizing large logistics parks' participation in peak shaving. Results illustrate that the proposed method leads to a reduction in the peak shaving pressure on the distribution network. Specifically, under the scenario tolerating the maximum potential uncertainty in renewable energy output, the absolute peak-to-valley difference and fluctuation variance of the park's net load are decreased by 45.82% and 54.59%, respectively. Furthermore, the PMV and the SPI indexes are reduced by 39.12% and 26.36%, respectively. In comparison with the determined optimization method, despite a slight cost increase of 20.06%, the proposed method significantly reduces EDR load shedding by 98.1%.

Keywords: peak shaving; smart grid; demand side response; robust optimization

Citation: Zhou, J.; Zhang, J.; Qiu, Z.; Yu, Z.; Cui, Q.; Tong, X. Two-Stage Robust Optimization for Large Logistics Parks to Participate in Grid Peak Shaving. *Symmetry* **2024**, *16*, 949. https://doi.org/10.3390/sym16080949

Academic Editor: László T. Kóczy

Received: 4 June 2024
Revised: 8 July 2024
Accepted: 15 July 2024
Published: 24 July 2024

Copyright: © 2024 by the authors. Licensee MDPI, Basel, Switzerland. This article is an open access article distributed under the terms and conditions of the Creative Commons Attribution (CC BY) license (https://creativecommons.org/licenses/by/4.0/).

1. Introduction

1.1. Background

As the proportion of renewable energy integration increases, the equilibrium and symmetry between the supply side and demand side of the power system become progressively more disrupted. To maximize the utilization of renewable energy resources, it becomes imperative to fully engage the demand-side components. Also, the intensifying urban population aggregation in major cities leads to a steady surge in electricity demand, especially in core urban areas. The distinct counter-peaking attributes associated with the substantial integration of renewable energy sources further widen the peak-to-valley disparities, thus exacerbating localized supply–demand imbalances and posing significant risks to the stable operation of power grids.

1.2. Recent Works

The traditional method of managing peak loads, which involves adjustments to power generation, is economically impractical due to its consequences of inefficient plant usage and elevated pollution levels. Specifically, in recent years, the large-scale integration of intermittent and uncertain renewable energy sources, such as wind power and photovoltaics, has complicated the supply–demand balance within the electrical grid, subsequently augmenting the complexity of peak shaving. Under the above conditions, numerous contemporary studies focus on strategies to address the escalating peak shaving demands, particularly examining responses from both the supply and demand sides.

In the electricity market environment, the participation of power plants in peak shaving has significantly increased. Their active engagement in the scheduling and flexible adjustment of power generation has effectively reduced the peak-to-valley difference in the grid, alleviated operational pressures within the power system, and significantly enhanced its safety. Reference [1] proposes a cost analysis model to address the economic and environmental implications of deep peak shaving for coal-fired units. Reference [2] presents an assessment method that reflects the marginal costs of peak and off-peak electricity generation for combined heat and power plants engaged in peak shaving, discovering that the use of electric heat pumps for waste heat recovery achieves a significant depth of peak shaving. Reference [3] examines the complementary characteristics of wind power and nuclear power in peak shaving, proposing a multi-power dispatch model to optimize nuclear power output. Reference [4] addresses the serious issue of wind curtailment, proposing an optimal configuration method for power-to-heat equipment in combined heat and power plants with a focus on static payback period. Reference [5] proposes a solution combining typical peak shaving output curves with water abandonment adjustment strategies to address the challenges of short-term peak shaving operation. In reference [6], the role of carbon capture devices in peak shaving is explored. A virtual energy storage model is constructed, and a joint peak shaving strategy for carbon capture devices and virtual energy storage is proposed. Reference [7] proposes a model for pumped-storage power stations to participate in the peak shaving ancillary service market, helping to share the peak shaving pressure of thermal power units. Reference [8] explores the coordinated operation of hydropower and renewable energy to mitigate fluctuations and facilitate peak shaving, addressing the challenges posed by the intermittency and uncertainty of wind and solar energy to grid dispatch. Reference [9] presents a new peak shaving model that utilizes mixed integer linear programming without presupposing or fixing the peak shaving order of power stations. Reference [10] proposes a multi-objective unit commitment model that combines the concentration of solar power plants and wind farms for peak shaving, taking operational risks into account.

On the other hand, providing compensation to the demand side through contracting can effectively stimulate load-side resources and attract more load-side participation in the peaking process. This mechanism not only significantly reduces the peak-to-valley difference but also further enhances the safety and stability of the power system by optimizing the allocation of power resources. Reference [11] provides a comprehensive overview of peak load reduction strategies, focusing on the impact of three major strategies: demand-side management, energy storage systems, and grid-connected electric vehicles. Reference [12] constructs a demand response model considering dual uncertainty to address carbon emissions and supply–demand balance issues. In reference [13], a demand response energy management strategy adopting a peak rebate program is proposed. Reference [14] presents a joint peak shaving strategy for carbon capture equipment and virtual energy storage. It establishes a two-stage peak shaving model for day-ahead and intra-day operations, aimed at minimizing the peak-to-valley difference in the load and reducing system operating costs to ensure economically efficient peak shaving. Reference [15] explores the application of demand-side management in air conditioning and heat pumps, particularly focusing on the heat pump sector. Reference [16] proposes a community-based home energy management system that utilizes a particle swarm optimization algorithm and user-defined constraints to achieve peak load reduction. Reference [17] addresses peak load issues in district heating systems by employing a differential return water temperature adjustment strategy for peak shaving. The aforementioned research indicates that the linkage between the power generation side and the demand side can effectively coordinate, flattening the load curve. As a significant power load, the fluctuation of chillers significantly affects the safe and stable operation of the grid. Consequently, incorporating chillers into peak shaving practices is of great significance for enhancing the reliability and economy of power supply.

Large chillers, due to their inherent controllability, can effectively reduce the load pressure during peak hours through reasonable control strategies while ensuring the satisfaction of daily living and production needs. On the other hand, buildings equipped with chillers have a certain cold storage capacity, allowing for minor changes in operating status without significantly affecting indoor environmental temperature, demonstrating a high level of inertia. In large cities, air conditioning loads during summer peak hours account for a significant proportion of the total load, becoming a major contributor to peak loads, with significant adjustment potential. Reference [18] proposes a state queueing model that utilizes a bidirectional information channel to collect thermostatically controlled load information, and employs tracking curves and state prediction methods for the control of thermostatically controlled load operations. Reference [19] converts the thermodynamic equations of air conditioning loads into a finite-dimensional state-space model through the finite difference method. Additionally, reference [20] develops a stochastic energy storage charging parameter model for aggregated air conditioning loads. Reference [21] presents a method for adjusting the power of aggregated air conditioning systems through broadcasting temperature setpoints. The study in [22] proposes a comfort-constrained heat pump load management method and realizes closed-loop control.

1.3. Motivations and Contributions

While the above methods have achieved the comprehensive optimization and regulation of the chiller, most of them targeted distributed and autonomous air conditioning loads. Currently, there is no large-scale practice involving large logistics parks in peak shaving, mainly due to three reasons: (1) There is a concern that power outages or insufficient power supply during peak shaving may lead to the damage of frozen products within the logistics park. Cold storage facilities require continuous refrigeration to maintain the freshness of frozen products, and any disruption to their stable operation caused by peak shaving may result in product losses. (2) Logistics parks house a large number of office staff, whose office buildings are also cold-consuming structures with high power consumption. However, their demand for cooling loads differs significantly from that of cold storage facilities. It is necessary to establish separate mathematical models for the comfort level of office staff and the storage duration of frozen products, and incorporate them into a comprehensive consideration framework under the same dimension. (3) With the increasing penetration of renewable energy into distribution networks, the difficulty of forecasting new energy sources has increased. The traditional "forecast value as the planned value" approach is hardly applicable, and the challenge of peak shaving for cold storage facilities under such uncertainties has significantly increased, potentially exposing logistics parks to additional risks of frozen product damage. Therefore, a robust optimization approach is needed to mitigate the impact of random errors.

To solve the problems above, this paper proposes a novel optimization method for large logistics parks to participate in grid peak shaving in a two-stage robust manner. The following contributions correspond to the motivations for this study:

(1) The inertial traits inherent in the substantial cooling load within the park are exploited to engage in peak shaving, while adhering to temperature constraints. The formulated objective function factors in the comfort of the office setting within the park and the sales pressure associated with cold storage, thereby achieving an optimal balance between operational efficiency, temperature tolerance, and peak shaving engagement.

(2) A composite metric, termed Cooling Load's Economic Contribution (CLEC), is devised. This metric integrates both the Predicted Mean Vote (PMV) for human comfort and the Sales Pressure Index (SPI) for frozen goods, translating these factors into economic costs for a unified optimization process.

(3) A two-stage robust optimization model, encompassing both day-ahead and intra-day planning, is devised. This model aims to maximize the park's cooling load participation in the peak shaving efforts, while maintaining personnel comfort and ensuring the quality

of frozen products. Additionally, it accounts for the uncertainties introduced by renewable energy sources like wind and solar.

The structure of this paper is organized as follows. Section I introduces the research background and primary research focus. Section II focuses on modeling a large logistics park, establishing the CLEC index to ensure that the park participates in peak shaving while maximizing the normal operational demand for cooling load. Section III provides a two-stage robust optimization model for a large logistics park's participation in peak shaving, covering both day-ahead and intra-day periods. Section IV introduces the methodology to solve the problem using the Column and Constraint Generation (C&CG) algorithm. Section V outlines a group of case studies, creating robust optimization scenarios that consider various levels of uncertainty. Pros and cons of the proposed method are compared and summarized. Section VI presents the conclusions of this paper.

2. Modeling of Large Logistics Parks

2.1. Total Framework with Diverse Cooling Loads

Figure 1 depicts the park model constructed in this paper. The power supply consists of photovoltaic and wind power, two representative renewable energy sources, integrated into the park in a decentralized manner. The park is connected to the distribution network through a PCC node, which includes a group of gas turbines that compensates for any shortcomings in the park's renewable energy generation capacity. This gas turbine unit also has reserve capacity, providing backup when the power transmission pressure between the park and the distribution network becomes excessive. Meanwhile, the park has a significant amount of cooling load to meet the needs of frozen goods storage and the cooling demands of staff within the buildings. Traditionally, parks control the output of chillers by adjusting the temperature inside buildings to meet cooling load requirements. However, buildings possess a "cold storage" characteristic, and to avoid "wasting" this stored cold inertia, the CLEC index is employed in the park model, shown in Figure 1, to intelligently regulate the output characteristics of the chillers. In this paper, EDR is activated when the peak shaving pressure on the distribution network becomes too great, and the central optimization controller removes some of the park's electrical load.

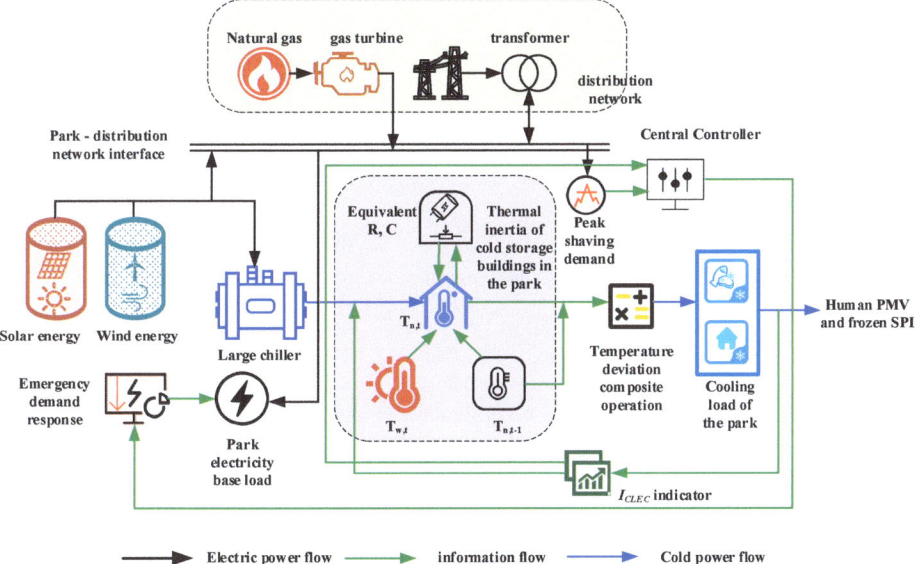

Figure 1. Configuration of the large logistics park.

In this paper, the park's central optimization controller takes into account the park's operating costs, the comfort level of cooling load users, the need to ensure the quality of frozen goods in the park, and the requirement to participate in peak shaving.

The ambient temperature of a chiller can be adjusted by modifying its cooling load (CL), thus varying the cooling capacity. Both humans and warehouses have specific acceptable temperature ranges, with warehouses favoring lower temperatures for longer food storage. As chiller-driven buildings have some heat storage capacity, CL changes do not immediately affect indoor temperature [23]. Assuming that the heat balance between the CL and the building is maintained for a short period when the outdoor temperature is constant, the electric power consumed by the chiller is

$$P_{ac} = \frac{T_{out} - T_{in}}{\lambda R_1} \quad (1)$$

where T_{in} represents the indoor temperature, T_{out} is the outdoor temperature, and λ is the energy efficiency ratio. R_1 is the equivalent heat resistance. A cold-storage building model can describe its temperature dynamic characteristics via a mathematical model that gives the relationship between indoor and outdoor cold and heat sources and room temperature changes. Its temperature dynamic characteristics can be described by an equivalent thermal parameter (ETP) model, as shown in the following equation [24]:

$$\widetilde{T}_{n,t} = \widetilde{T}_{w,t} - Q_{L,t}R - [\widetilde{T}_{w,t} - Q_{L,t}R - \widetilde{T}_{n,t-1}]e^{-\Delta t/RC} \quad (2)$$

where $Q_{L,t}$ is the total cooling load in the duration of t. R and C are the indoor equivalent thermal resistance and equivalent thermal capacity of cold-storage buildings, respectively. $\widetilde{T}_{n,t}$ and $\widetilde{T}_{W,t}$ are the indoor and outdoor temperatures, respectively. Δt is the dispatching duration. Depending on the needs of the user, the indoor temperature is usually limited to a specific range [25], as shown in the following equation:

$$T_{min} \leq T_{in}(t) \leq T_{max} \quad (3)$$

2.2. Modeling of Diverse Cooling Load's Economic Contribution

Thermal comfort, which encapsulates individuals' subjective assessments and sensations regarding their thermal environment, serves as the primary indicator of users' ambient temperature preferences. To objectively quantify the effect of temperature on user comfort, this study employs the Predicted Mean Vote (PMV) index [26], as detailed in Table 1. The mathematical form is given in Equation (4).

$$\begin{aligned}I_{PMV} =& (0.303e^{-0.036M} + 0.028)\{M - W - 3.05 \times 10^{-3} \times [5733 - 6.99(M - W) - P_a] - \\& 0.42(M - W - 58.15) - 1.7 \times 10^{-5}M \times (5867 - P_a) - 1.4 \times 10^{-3}M(34 - t_a) - \\& 3.96 \times 10^{-8}f_{cl}[(t_{cl} + 273)^4 - (t_r + 273)^4] - f_{cl}h_c(t_{cl} - t_a)\}\end{aligned} \quad (4)$$

where M and W represent the human energy metabolic rate and mechanical power, respectively. f_{cl} is the ratio between the clothed body surface area and the naked body surface area. h_c is the convective heat transfer coefficient. P_a is the water vapor pressure of the air around the human body. t_a, t_r and t_{cl} represent the air temperature around the human body, mean radiant temperature, and clothing surface temperature, respectively. This paper mainly focuses on the cooling capacity, and temperature is the most intuitive perception of indoor thermal comfort for humans. Therefore, except for the air temperature around the human body, t_a, it is assumed that other parameters are given values.

Table 1. PMV index.

PMV	−3	−2	−1	0	1	2	3
Sensation	Cold	Cool	Slightly cool	Neutral	Slightly warm	Warm	Hot

For the chiller-driven warehouses, there are upper and lower limits for the ambient temperature. For instance, frozen foods need to be stored below −18 °C, while refrigerated medicines require storage in a cold storage facility between 2 °C and 10 °C [27]. Additionally, the optimal refrigeration temperature for different frozen products should be considered. To assess the negative impact caused by deviations from the optimal refrigeration temperature, a Sales Pressure Index (SPI) is established. This index primarily focuses on the quality and freshness retention of frozen products. A lower SPI indicates that the actual temperature of the cold storage is closer to the ideal storage temperature, thereby reducing product quality loss and sales pressure. In this paper, we only focus on the value of frozen products at the end of the dispatching period, assuming no transfer of frozen products during this period. The mathematical equation for the SPI at the end of the period is as follows:

$$\lambda_{SPI} = \frac{\sum_{i=1}^{N_i} \sum_{t=1}^{N_T} a^i \left| T_{r,t}^i - T_{ideal,t}^i \right|}{N_T * T_{max}^{dev}} \quad (5)$$

where λ_{SPI} represents the comprehensive SPI of the cold storage within the microgrid at the end moment; the superscript i denotes the type of frozen product; $T_{r,t}^i$ and $T_{ideal,t}^i$ are the actual and ideal storage temperatures, respectively, for type i frozen products at time t; and a^i is the sales weighting coefficient for type i frozen products. T_{max}^{dev} is the maximum permitted temperature bias for frozen goods. The combination of PMV and SPI modeling gives rise to the definition of CLEC, denoted as I_{CLEC}, which quantifies the economic impact of the microgrid's cooling load.

$$I_{CLEC} = \frac{\omega_{pmv} \frac{\sum_{t=1}^{N_T} a_t |I_{PMV,t}|}{N_T \times |I_{PMV,max}|} + \omega_{spi} \lambda_{SPI}}{1 + e^{k(|T_{out} - T_{opt}|)}} \quad (6)$$

In Equation (6), a_t signifies the temperature sensitivity coefficient of the building's cooled occupants at a specific time t. This coefficient varies temporally, peaking during typical rest periods such as 12:00 to 14:00 and 21:00 to 24:00. The term $|I_{PMV,max}|$ represents the absolute maximum PMV index value recorded among the cooled individuals within the premises. N_T stands for the predetermined scheduling duration. Additionally, ω_{pmv} and ω_{spi} are the weighted coefficients assigned to the cooling loads associated with the building's occupants and the frozen items in cold storage, respectively, during the park's operational hours. Meanwhile, T_{out} and T_{opt} refer to the current outdoor temperature and the ideal outdoor temperature, respectively. k serves as an adjustment coefficient that regulates the sensitivity of the I_{CLEC} index to fluctuations in external temperature, thereby controlling its responsiveness to temperature variations.

The additional cost for using cooling loads is defined as

$$C_{CLEC} = F_{CLEC} I_{CLEC} \quad (7)$$

where F_{CLEC} represents the cost conversion coefficient that comprehensively considers the loss of human comfort and the sales value of frozen goods. Meanwhile, C_{CLEC} denotes the additional economic costs incurred due to the inefficient use of cooling load within the park.

3. Methodology

To alleviate the peak shaving pressure on the grid side and leverage the significant inertial cooling load present within the park, this study explores the utilization of economic incentives to motivate the park's participation in the grid's peak shaving initiatives. Specifically, the park receives economic compensation for actively engaging in peak shaving, implemented through a two-stage robust optimization framework that encompasses both day-ahead and intra-day timeframes.

During the day-ahead phase, the system computes the minimum operating costs and the associated peak shaving compensation, leveraging predicted values of photovoltaic and wind power generation. Moving into the intra-day phase, the framework incorporates backup resources from the distribution grid to mitigate any uncertainties within predefined uncertainty sets, thereby ensuring consistent, economical, and reliable system operation. The objective is to minimize daily operating costs while simultaneously reducing the peak shaving pressure to its lowest possible level. The model's objective function is articulated as follows:

$$\min C_0 + \max_U \min C_S \tag{8}$$

where C_0 and C_S denote the optimization objectives for the initial and subsequent stages, respectively, while U signifies the comprehensive uncertainty set, encompassing uncertainties inherent in photovoltaic and wind power generation.

3.1. Day-Ahead Stage
3.1.1. Objective Function

$$\min C_0 = \min[\sum_{t=1}^{N_T} C_G(t) + C_{CLEC} - C_{pls})] \tag{9}$$

In Equation (9), N_T represents the time period and $C_{G(t)}$ denotes the power distribution network's output cost during period t. C_{pls} represents the economic compensation received by the park for participating in peak shaving.

(1) Gas Turbine

The power supply from the distribution network to the park primarily relies on gas turbines. The generation cost of these turbines can be represented by a linear function, taking into account the reserve capacity. Therefore, the cost function of the gas turbine is

$$C_G(t) = \Delta t \sum_{g=1}^{N_G} [F_G P_{g,t}^0 + C_g^{R+} R_{g,t}^+ + C_g^{R-} R_{g,t}^-] \tag{10}$$

where $C_G(t)$ indicates the generation cost of the generator during period t; Δt is the scheduling step size, set to 1 h. N_G stands for the number of generators. F_G represents the cost coefficient of the gas turbine. The superscript 0 designates the day-ahead stage. $P_{g,t}^0$ is the output power of the g-th generator during period t; C_g^{R+} and C_g^{R-} are the upward and downward reserve cost coefficients of the generator, respectively; and $R_{g,t}^+$ and $R_{g,t}^-$ are the upward and downward reserve capacities provided by the generator during period t. Reserve capacity constraints and output power limitation constraints are taken into account, as shown in Equations (11)–(14):

$$R_{\min}^- \leq R_{g,t}^- \leq R_{\max}^- \quad \forall g, t \tag{11}$$

$$R_{\min}^+ \leq R_{g,t}^+ \leq R_{\max}^+ \quad \forall g, t \tag{12}$$

$$P_{g,t}^0 + R_{g,t}^+ \leq P_g^{\max} \quad \forall g, t \tag{13}$$

$$P_{g,t}^0 - R_{g,t}^- \leq P_g^{\min} \quad \forall g, t \tag{14}$$

where R_{max}^+ and R_{max}^- represent the maximum upward and downward reserve capacities that the unit can provide when transitioning from the first stage to the second stage. Meanwhile, R_{min}^+ and R_{min}^- denote the minimum upward and downward reserve capacities required by the system from the unit. P_g^{max} and P_g^{min} indicate the maximum and minimum output powers of the gas turbine, which are limited by the rated power and minimum load rate of the generator set, respectively.

(2) Evaluation of peak shaving performance

The effectiveness of peak shaving can be measured by the economic compensation C_{fls} obtained by reducing the fluctuation in the grid's net load and the peak–valley difference. C_{fls} can be expressed as

$$C_{fls} = F_{pd}\Delta P_{pd} + F_{var}\Delta P_{var} \tag{15}$$

$$\Delta P_{pd} = P_{pd}^{be} - P_{pd} \tag{16}$$

$$\Delta P_{var} = \Delta P_{var}^{be} - \Delta P_{var} \tag{17}$$

where Δp_{pd} represents the reduction in the peak–valley difference (PVD) of the net load; F_{pd} is the compensation coefficient for peak–valley difference; ΔP_{var} denotes the decrease in fluctuation variance of the net load; and F_{var} is the compensation coefficient for variance. The superscript indicates the state of the grid's net load before the park participates in peak shaving. P_{pd} and P_{var} are the PVD and variance of the microgrid's net load after peak shaving.

$$P_{pd} = P_{max} - P_{min} \tag{18}$$

In Equation (18), P_{max} is the maximum value of the net load during the scheduling period and P_{min} is the minimum value. The absolute PVD P_{pd} reflects the difference between the extreme values of the load peaks and valleys.

$$P_{var} = \sum_{t=1}^{N_T} \left(P_{l.t} - \sum_{t=1}^{N_T} P_{l.t}/N_T \right)^2 / N_T \tag{19}$$

In Equation (19), N_T is the total number of samples and $P_{l,t}$ represents the load value of the t-th sample. The fluctuation variance P_{var} of the net load reflects the degree of load dispersion. A smaller fluctuation variance of the net load indicates a lower degree of dispersion.

3.1.2. Constraints

(1) Power-balance constraints

$$\sum_{g\in G_b} P_{g,t}^0 + \sum_{g\in W_b} P_{W,t}^0 + \sum_{g\in P_b} P_{PV,t}^0 + \sum_{l=o1\_b} P_{l,t}^0 = \sum_{l=o2\_b} P_{l,t}^0 + L_{b,t} \quad \forall b, t \tag{20}$$

In Equation (20), G_b, W_b, and P_b represent the sets of generators, wind farms, and photovoltaic plants connected to node b, respectively. The notations $l = o1\_b$ and $l = o2\_b$ denote the feeder-injecting and exporting power from node b, respectively. $P_{W,t}^0$ and $P_{PV,t}^0$ refer to the actual consumption of wind and photovoltaic power at time t. $P_{l,t}^0$ represents the transmission power of line l at time t. $L_{b,t}$ signifies the load connected to node b at time t, which comprises both the base electric load and the load from the chillers.

(2) Cooling load constraints

$$Q_{AC,t} = Q_{L,t} \tag{21}$$

$$\widetilde{T}_{in}^{min} \leq \widetilde{T}_{in,t} \leq \widetilde{T}_{in}^{max} \tag{22}$$

Herein, $Q_{AC,t}$ represents the cooling power of the electric chiller in time period t, while $\widetilde{T}_{in}^{min}$ and $\widetilde{T}_{in}^{max}$ are the lower and upper limits of the indoor temperature, respectively.

The indoor cooling constraints also include Equation (2), which describes the ETP model of the thermal inertia of the cooling system.

(3) Chiller output constraints

$$Q_{AC,t} = \eta_{AC} P_{AC,t} \tag{23}$$

$$Q_{AC}^{min} \leq Q_{AC,t} \leq Q_{AC}^{max} \tag{24}$$

$P_{AC,t}$ represents the electric power consumed by the electric chiller in time period t, η_{AC} stands for the coefficient of performance (COP) of the electric chiller, and $Q_{AC}^{min}/Q_{AC}^{max}$ represent the lower/upper limits of the output of the electric chiller.

(4) Constraints on power flow

$$P_{l,t}^0 = \frac{\theta_{o1\_l}^0 - \theta_{o2\_l}^0}{x_l} \quad \forall l, t \tag{25}$$

$$-P_l^{max} \leq P_{l,t}^0 \leq P_l^{max} \quad \forall l, t \tag{26}$$

$$\theta_{min} \leq \theta_{b,t}^0 \leq \theta_{max} \quad \forall b, t \tag{27}$$

In the above equation, $\theta_{o1\_l}^0$ and $\theta_{o2\_l}^0$ represent the voltage phase angles at the ingress and egress nodes of feeder l, respectively; x_l is the reactance of transmission feeder l; P_l^{max} is the upper limit of power flow in feeder l, with positive and negative values indicating direction; and θ_{min} and θ_{max} are the minimum and maximum limits of the node voltage phase angles, respectively.

(5) Constraints of renewable energy outputs

$$0 \leq P_{PV,t}^0 \leq A_{PV,t}^0 \quad \forall PV, t \tag{28}$$

$$0 \leq P_{W,t}^0 \leq A_{W,t}^0 \quad \forall W, t \tag{29}$$

In the above equation, $A_{PV,t}^0$ and $A_{W,t}^0$ are the predicted outputs of the photovoltaic and wind power in the time period t.

3.2. Uncertainty Set

In this paper, the uncertainty set in robust optimization consists of uncertainties in the photovoltaic and wind power output, which can be expressed by Equations (30)–(35). Here, N_T represents the scheduling period of this paper, which is 24 h. $Z_{W,t}^+$ and $Z_{W,t}^-$ are the status flags for upward and downward fluctuations of wind farm W in period t, respectively. When the value is 1, it indicates the existence of fluctuations, and when it is 0, there are no fluctuations. $A_{W,t}^S$ represents the second-stage wind power prediction, which is composed of the first-stage wind power forecast superimposed with power fluctuations. $u_{t,pv}$ and $u_{t,w}$ are the robustness indicators of this paper, representing the time uncertainty limits of photovoltaic and wind power output, respectively.

$$A_{W,t}^S = A_{W,t}^0 + Z_{W,t}^+ A_{W,t}^+ - Z_{W,t}^- A_{W,t}^- \tag{30}$$

$$Z_{W,t}^+ + Z_{W,t}^- \leq 1 \quad \forall W, t \tag{31}$$

$$\sum_{t=1}^{N_T} (Z_{W,t}^+ + Z_{W,t}^-) \leq u_{t,w} \tag{32}$$

$$A_{PV,t}^S = A_{PV,t}^0 + Z_{PV,t}^+ A_{PV,t}^+ - Z_{W,t}^- A_{PV,t}^- \tag{33}$$

$$Z_{PV,t}^+ + Z_{PV,t}^- \leq 1 \quad \forall PV, t \tag{34}$$

$$\sum_{t=1}^{N_T} (Z_{W,t}^+ + Z_{W,t}^-) \leq u_{t,pv} \tag{35}$$

3.3. Intra-Day Stage

3.3.1. Objectives

$$\underset{U}{\text{maxmin}} C_S = C_{awp} \sum_{t=1}^{N_T} \sum_{W=1}^{N_{PV}} (A_{PV,t}^S - P_{PV,t}^S) + C_{aw} \sum_{t=1}^{N_T} \sum_{W=1}^{N_W} (A_{W,t}^S - P_{W,t}^S)$$
$$+ \sum_{t=1}^{N_T} \sum_{g=1}^{N_G} (C_g^{S,R+} R_{g,t}^{S+} + C_g^{S,R-} R_{g,t}^{S-}) + \sum_{b=1}^{N_B} C_{EDR} L_{b,t}^{S,EDR} \tag{36}$$

In the above equation, the superscript S represents the second intra-day stage; C_{awp} refers to the economic loss cost coefficient caused by the failure of the actual wind power output to meet the expected value in the second stage; $C_g^{S,R+}$ and $C_g^{S,R-}$ represent the upward and downward reserve dispatch rates of the unit in the second stage, respectively; and R_g^{S+} and R_g^{S-} represent the actual upward and downward reserve dispatch capacities of the unit in the second stage.

3.3.2. Remaining Constraints in the Intra-Day Stage

(1) Power balance

$$\sum_{g \in G_b} P_{g,t}^S + \sum_{g \in W_b} P_{W,t}^S + \sum_{g \in P_b} P_{P,t}^0 + \sum_{l=o1\_b} P_{l,t}^S = \sum_{l=o2\_b} P_{l,t}^S + L_{b,t} - L_{b,t}^{S,EDR} \ \forall b, t \tag{37}$$

where $P_{g,t}^S$ represents the actual output of the gas turbine in the second stage during time period t, which is composed of the actual output of the generator in the first stage during time period t and the actual reserve capacity dispatched for the unit during time period t. This can be expressed as

$$P_{g,t}^S = P_{g,t}^0 + R_{g,t}^{S+} - R_{g,t}^{S-} \ \forall g, t \tag{38}$$

(2) Feeder power flow constraints

$$P_{l,t}^S = \frac{\theta_{o1\_l}^S - \theta_{o2\_l}^S}{x_l} \ \forall l, t \tag{39}$$

$$-P_l^{\max} \leq P_{l,t}^S \leq P_l^{\max} \ \forall l, t \tag{40}$$

$$\theta_{\min} \leq \theta_{b,t}^S \leq \theta_{\max} \ \forall b, t \tag{41}$$

(3) Emergency demand response constraint

$$0 \leq L_{b,t}^{S,EDR} \leq L_{\max}^{S,EDR} \ \forall b, t \tag{42}$$

where $L_{\max}^{S,EDR}$ represents the maximum amount of emergency demand response that each node in the system can participate in during a given time period.

(4) Constraints of renewable energy outputs

$$0 \leq P_{PV,t}^S \leq A_{PV,t}^S \ \forall PV, t \tag{43}$$

$$0 \leq P_{W,t}^S \leq A_{W,t}^S \ \forall W, t \tag{44}$$

The above indicates that the actual output of photovoltaic and wind power in the second stage must not exceed the available forecast value after considering uncertainties.

The disparities between the model proposed in this paper and the existing models in the literature are outlined as follows:

(1) Modeling differences: While Reference [28] also leverages the cold load for peak shaving, its scope extends to an integrated energy system including cold, heat, electricity, and gas. Conversely, the large logistics park examined in this paper solely includes electrical and cold loads, thereby exhibiting a lower level of energy diversity. Consequently, this model is more adept at illustrating and assessing the park's peak shaving capacity in scenarios where only electrical and cold loads are present.

(2) Indicator differences: Existing models that utilize cold load for peak shaving typically consider only the costs of electrical and thermal power consumed for cooling. This paper, however, refines the usage scenarios of cold loads by incorporating the PMV and SPI, which are amalgamated into the CLEC index for peak shaving control. Additionally, the emergency demand response is introduced to evaluate and compare the resilience to uncertainties among different strategies.

(3) Method differences: Reference [29] employs fuzzy chance-constrained programming for optimized peak shaving, adopting a relatively aggressive strategy that diverges from robust optimization in terms of balancing risks and benefits. Conversely, the model in this paper establishes a two-stage robust optimization framework that accounts for uncertainties in actual wind and photovoltaic operations, while also considering the presence of gas turbines (or steam turbines). The results demonstrate that the utilization of robust optimization enhances the system's capability to manage uncertain factors, rendering it advisable to adopt this more conservative strategy when prioritizing grid security.

4. Model Solution

Similar to the Benders decomposition method, the C&CG algorithm decomposes the original problem into a master problem and a max-min subproblem [30]. It then converts the bi-level optimization subproblem into a single-level optimization form through the Karush–Kuhn–Tucker (KKT) conditions or the strong duality theory (SDT). The optimal solution of the original problem is then obtained by alternately solving the master problem and the subproblem. The key difference between the two methods lies in the fact that during the solution process of the C&CG algorithm, the subproblem continuously introduces relevant variables and constraints into the master problem, resulting in a more compact lower bound for the original objective function, improved efficiency, and effectively reduced iteration times. The model constructed from Equations (8)–(44) can be written in the following compact form:

$$\min_{x^0 \in \beta^0} (c^0)^T x^0 + \max_{z \in U}(b^T u_w + d^T u_p + \min_{x^S \in \beta^S} (c^S)^T x^S) \tag{45}$$

$$\beta^0 = \left\{ A x^0 \leq \alpha \right\} \tag{46}$$

$$\beta^S = \left\{ E x^S \leq F x^0 + G u_w^0 + H u_p^0 + I Z + j \right\} \tag{47}$$

$$U = \left\{ Z \middle| \begin{array}{l} u_w = u_w^0 + Z_w^+ \Delta U_w^+ + Z_w^- \Delta U_w^-, B_w(Z_w^+ + Z_w^-) \leq \Pi_w, \\ u_p = u_p^0 + Z_p^+ \Delta U_p^+ + Z_p^- \Delta U_p^-, B_p(Z_p^+ + Z_p^-) \leq \Pi_p \end{array} \right\} \tag{48}$$

where c^0 represents the coefficient column vector corresponding to the objective function (9); Equation (45) denotes the objective function of the two-stage robust optimization problem; x^0 and x^S are the relevant control variables in the first-stage and second-stage objective functions, respectively; and Equations (46) and (47) represent the constraint conditions for the first and second stages, including Equations (10)–(29) and Equations (37)–(44), respectively. Among them, A, E, F, G, H, I, and j are the constant column vectors corresponding to the objective functions and constraint conditions. In Equation (48), U represents the uncertainty set, where u_w^0 and u_p^0 are the available power generation from photovoltaic and wind power in the first stage, respectively, while u_w and u_p represent the available wind power in the second stage for photovoltaic and wind power, respectively.

After decomposition by the C&CG algorithm, the master problem includes the first-stage model and the worst-case operating constraints identified by the subproblem:

$$\begin{cases} \min((c^0)^T x^0 + \eta) \\ s.t. \quad Ax^0 \leq \alpha \\ \eta \geq b^T u_{w,k}^* + d^T u_{p,k}^* + (c^S)^T x_k^S \\ Ex_k^S \leq Fx^0 + Gu_w^0 + Hu_p^0 + IZ_k^* + j \\ 0 \leq k \leq m-1 \end{cases} \quad (49)$$

where m represents the current iteration number; $u_{w,k}^*$, $u_{p,k}^*$, and Z_k^* are the worst-case operating conditions solved by the lower-level problem; and η is the solution of the subproblem to be optimized. The subproblem is a two-level max-min problem. Through the SDT, it can be transformed into a max form. The transformed subproblem form in the m-th iteration is

$$\begin{cases} \max(b^T u_w + d^T u_p + (Fx_i^0 + Gu_w^0 + Hu_p^0 + IZ + j)^T \pi) \\ s.t. \quad E^T \pi \leq c^S \\ \pi \leq 0 \\ Z \in U \end{cases} \quad (50)$$

where π is the dual variable of the second-stage constraint conditions. Note that there exists a bilinear term $Z^T \pi$ in Equation (50), as Z is a 0–1 integer variable, resulting in a product form of binary and continuous variables. Therefore, the auxiliary variable and related constraints can be introduced to linearize it using the big M method [31], as shown below:

$$\begin{cases} \omega \geq -MZ \\ \omega \geq \pi \\ \omega \leq \pi - M(Z-1) \\ \omega \leq 0 \end{cases} \quad (51)$$

where ω is the introduced continuous auxiliary variable, and M is a sufficiently large constant vector. Thus, the solution process is outlined below:

(1) Initialization: Set the upper bound UB of the objective function under the final scheduling plan as $+\infty$, the lower bound LB as $-\infty$, the iteration count m as 1, and the convergence threshold as ε.

(2) Use a set of uncertain variables u_w, u_p, and Z as the initial worst-case scenario.

(3) Solve the master problem (Equation (49)) based on the worst-case scenario. The objective function of the master problem serves as the new lower bound $LB = K_m$, and the control variables x_m^0 of the first stage are obtained simultaneously.

(4) Substitute the solution x_m^0 of the master problem into Equation (50) to solve the subproblem, obtaining its objective function N_m and the corresponding uncertain variables $u_{w,k}^*$, $u_{p,k}^*$, and Z_k^* under the worst-case scenario. Update the upper bound of the objective function as $UB = \left\{ U_B, (C^0)^T x_m^0 + N_m \right\}$.

(5) If $UB - LB \leq \varepsilon$, stop the iteration and set the objective function value as UB; otherwise, continue the iteration, increment m by 1, and return to step (3).

The total methodology presented in this paper is illustrated in Figure 2.

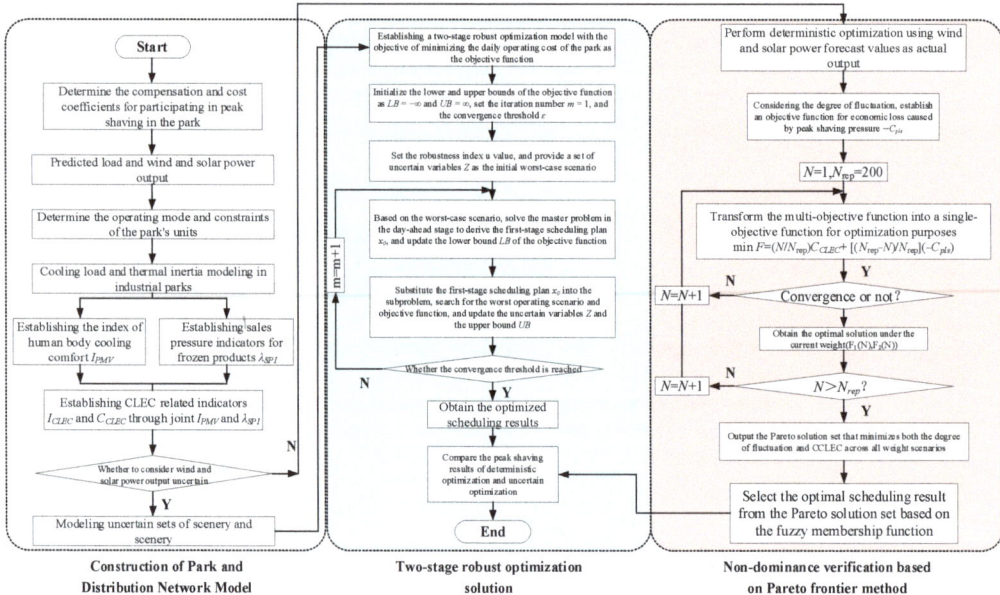

Figure 2. A flowchart for the proposed method.

5. Case Studies

Utilizing the MATLAB R2018b platform and the YALMIP toolbox, this paper establishes a load model for large-scale logistics parks and optimizes the results by invoking the Cplex solver. The case study employs the mirrored simulation of the IEEE-6 node system [32], with the system structure illustrated in Figure 3. The system comprises three gas turbine units and eleven transmission lines, with the parameters detailed in Table 2.

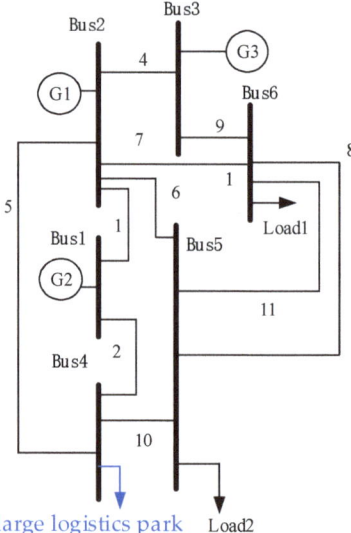

Figure 3. IEEE-6 system model.

Table 2. System parameters.

Bus-i	Pd	Qd	baseKV	Vmax	Vmin
1	0	0	230	1.05	1.05
2	0	0	230	1.05	1.05
3	0	0	230	1.07	1.05
4	70	80	230	1.05	0.95
5	80	74	230	1.05	0.95
6	90	79	230	1.05	0.95

5.1. Peak Shaving, Considering the Uncertainties

Data referencing the summer distribution network in southern China are utilized. The forward iterative search method [33] is employed to select typical days from a large number of previous scenarios as the predicted wind-power output values within a day. With a 24 h scheduling period and a 1 h scheduling unit, four scenarios are established:

Scenario 1: Considering only the uncertainty of photovoltaic power, with robustness indices $u_{t,p}$ and $u_{t,w}$ set to 8 and 0, respectively.

Scenario 2: Taking into account solely the uncertainty of wind power, where the robustness indices $u_{t,p}$ and $u_{t,w}$ are set at 0 and 8, respectively.

Scenario 3: Accounting for the uncertainties of both photovoltaic and wind power, the robustness indices $u_{t,p}$ and $u_{t,w}$ are both set to 8. In this scenario, the uncertainty set implies that the total duration of fluctuations in photovoltaic and wind power does not exceed 8 h within a scheduling period.

Scenario 4: Also addressing the uncertainties of photovoltaic and wind power, but with robustness indices $u_{t,p}$ and $u_{t,w}$ both increased to 16.

Table 3 presents the optimization results under various scenarios, as well as the indices and peak shaving performances of each scenario under their worst operating conditions. C_{CLEC} reflects the comprehensive satisfaction level of the two types of cooling loads used within the park, with a lower C_{CLEC} indicating higher satisfaction. C_{fls} reflects the peak shaving effect of the park's participation in the distribution network, with a higher C_{fls} indicating better peak shaving performance.

Table 3. Performance under different levels of uncertainties.

Case	Total Cost [USD]	Backup Volume Used [MW]	C_{CLEC}	C_{fls}	EDR [MW]
1	17,154.3	417	1725.1	5351.6	168.5
2	17,889.5	464	1811.6	4945.8	136.9
3	18,501.8	637	1904.9	4410.7	52.6
4	19,021.1	669	1991.2	4291.4	10

As evident from Table 3, there is minimal difference in various indices between Scenario 2 and Scenario 1, with the emergency demand response participation in Scenario 2 being 18.7% less than that in Scenario 1. In comparison to Scenarios 1 and 2, Scenarios 3 and 4 exhibit significant reductions in both the reserve capacity utilization and EDR participation. Figure 4 illustrates the reserve capacity utilization of power generators under Scenario 3. When compared to Scenario 2, Scenario 3 considers the uncertainties of both photovoltaic and wind power, resulting in an increase in total system cost and total reserve capacity. Specifically, the C_{CLEC} cost increased by 5%, while the C_{fls} compensation decreased by 10%. However, the reduction in EDR load shedding reached 61%.

Both Scenarios 3 and 4 take into account the uncertainties in the actual operation of photovoltaic and wind power, with the difference being that Scenario 4 incorporates greater uncertainty and a more severe worst-case scenario. It can be observed that compared to Scenario 3, Scenario 4 exhibits minimal differences in total operating cost, with slight decreases in the CLEC index and peak shaving performance. However, under the worst-case scenario, the emergency load shedding is reduced to 10 MW. Overall, Scenario 4 demonstrates highest robustness in microgrid operation, with no emergency load shedding even under the most severe conditions.

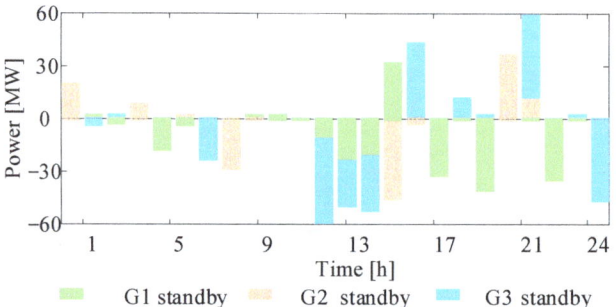

Figure 4. Reserve capacity utilization under Scenario 3.

For this study, the robustness index under Scenario 4 is selected for peak shaving, and the iteration count is set to 13 for system scheduling and control. The resulting convergence criterion is illustrated in Figure 5. In this paper, the fluctuation deviations of wind power and photovoltaic output are set based on historical maximum fluctuation deviations. By observing the final data after the convergence of iterations, a comparison curve of the actual and predicted outputs of wind power and photovoltaic is obtained, as shown in Figure 6. The results of optimization are presented in Figures 7 and 8. Figure 7 depicts the effectiveness of utilizing the CLEC index to engage chiller loads in peak shaving under the final scheme. Figure 8, on the other hand, illustrates the electrical balance histogram corresponding to this scheme.

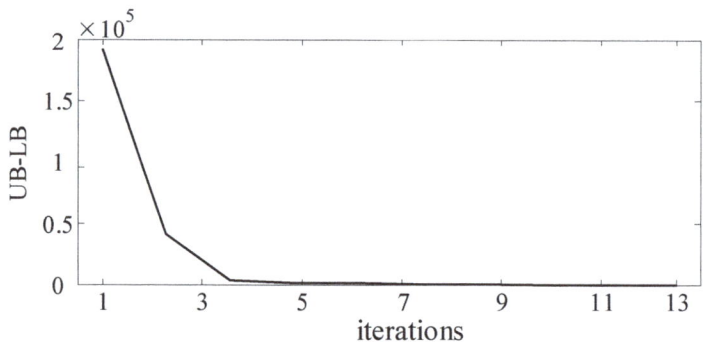

Figure 5. The convergence criterion diagram under robust optimization in scenario 4.

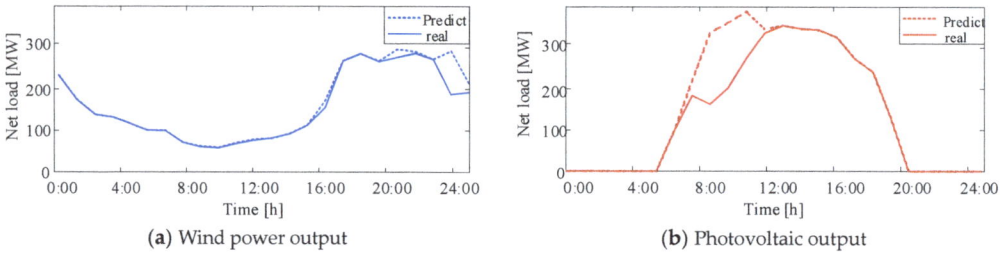

Figure 6. Photovoltaic and wind power prediction/actual output curve.

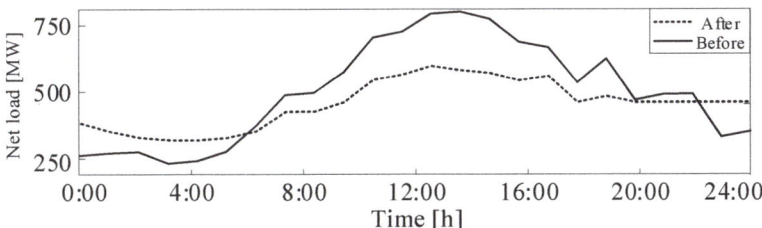

Figure 7. Peak shaving and valley-filling results in Scenario 4.

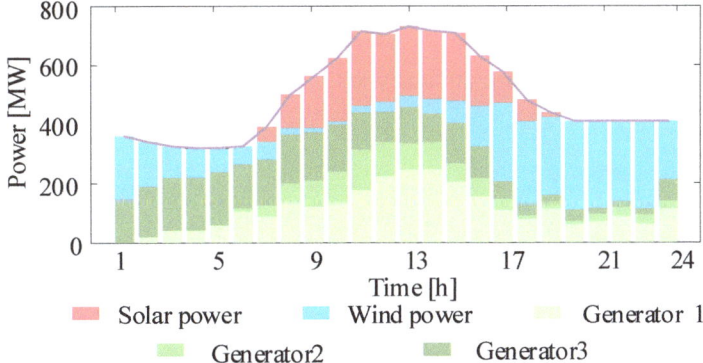

Figure 8. Electrical balance bar chart in Scenario 4.

As evident in Figure 7, prior to peak shaving by the logistics park, the net load experiences fluctuations in proximity to peak hours from 10:00 to 15:00, influenced by photovoltaic and wind turbine outputs alongside the inherent load curve. Conversely, during the early morning hours from 00:00 to 06:00, the net load remains at its nadir. However, once engaged in peak shaving activities, the net load curve undergoes smoothing, effectively achieving peak shaving. A comparative analysis of various optimization metrics before and after the process is detailed in Table 4.

Table 4. Performance of the optimization.

Indicator	Before	After	Performance [%]
PVD	547.09	296.43	45.82
FV	2890.7	1312.9	54.59
PMV	0.9445	0.5750	39.12
SPI	0.9347	0.6883	26.36

In Table 4, the PVD and Fluctuation Variance (FV) are employed as key indicators to gauge the net load variations within the park. Ideally, lower values of these indicators signify reduce peak shaving pressure on the distribution network. Furthermore, the extent of reduction in PVD and FV following optimization using the methodology outlined in this study serves as a measure of the park's effectiveness in participating in peak shaving (note that PMV and SPI values have been normalized for ease of comparison). Notably, there is a substantial decrease of 45.82% and 54.59% in PVD and FV, respectively, indicating a significant alleviation of peak shaving pressure on the distribution network. Additionally, the implementation of CLEC-based optimization has led to a 39.12% improvement in human comfort levels and a 26.36% reduction in sales pressure for frozen goods stored in cold storage facilities.

5.2. Non-Domination Verification and Comparative Validation of Indicators

While considering the uncertainties in the actual operation of wind and solar power, this paper also verifies the feasibility of the CLEC index in participating in peak shaving in deterministic optimization.

With the objectives of minimizing the target value of fluctuation degree and minimizing the economic loss caused by the inadequate utilization of the cooling load, modeling is conducted using the YALMIP toolbox in the MATLAB environment. To address the nonlinear issues in the model, the Cplex toolbox is utilized for solving the problem, inputting equality and inequality constraints to obtain the optimal boundary of the Pareto solution set. As the final decision support, the optimal compromise solution is selected from the Pareto solution set using the fuzzy membership degree method [34–36]. The specific expression of the fuzzy membership function is

$$\mu_i^j = \begin{cases} 1, & F_i \leq F_i^{min} \\ \frac{F_i^{max} - F_i^j}{F_i^{max} - F_i^{min}}, & F_i \geq F_i^{min} \\ 0, & F_i \geq F_i^{max} \end{cases} \qquad (52)$$

where μ_i^j represents the membership function of the j-th solution for the i-th objective function F_i^j, and F_i^{max} and F_i^{min} indicate the maximum and minimum values of the i-th objective function among all non-dominated solutions, respectively. The optimal compromise solution set is

$$\mu_i^{j*} = \max_{j=1,\cdots,M} \left\{ \sum_{i=1}^{n} \mu_i^j / \sum_{j=h}^{M} \sum_{i=1}^{n} \mu_i^h \right\} \qquad (53)$$

where M represents the number of non-dominated solutions and n represents the number of objective functions. Figure 9 depicts the impact of the CLEC index I_{CLEC} on the indoor cooling system in deterministic optimization. The colored dashed lines in Figure 9 represent the fluctuation range of indoor temperatures in cooling buildings under corresponding CLEC indices. As shown, the smaller the amplitude of the CLEC index, the stricter the indoor temperature requirements, resulting in a more comfortable environment for humans.

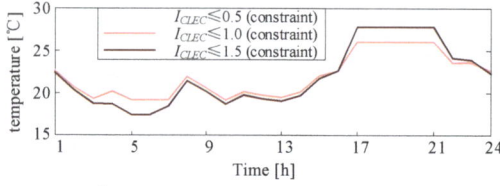
(**a**) Indoor temperature under different CLECs

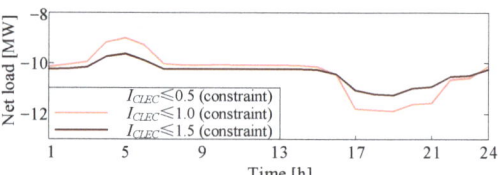
(**b**) Net load under different CLECs

Figure 9. Impact of CLECs.

The case study in this section for deterministic planning shares the same content parameters as the case study described in Section 5.1, except that it does not consider the uncertainties in the actual operation of wind and solar power. The Pareto frontier of the multi-objective optimization considering both CLEC and net load fluctuation is shown in the figure. As can be seen from Figure 10, the obtained Pareto solution set is concentrated and continuous on the frontier, which facilitates the selection of the most satisfactory solution. It can also be observed from Figure 10 that load fluctuations and the CLEC index are mutually independent, with a higher CLEC index when the fluctuation is small, and vice versa. Continuing with the fuzzy membership function method to solve Figure 10, the optimal solution with the highest satisfaction value is obtained. Further solving under this optimal solution yields the peak shaving results shown in Figure 11.

Table 5 compares the various optimization indicators under the deterministic optimization scenario with those under robust optimization Scenario 4 described in Section 5.1.

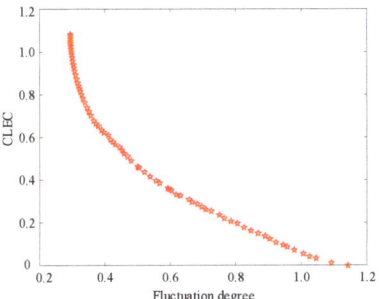

Figure 10. The Pareto frontiers between CLEC index and power fluctuation degree.

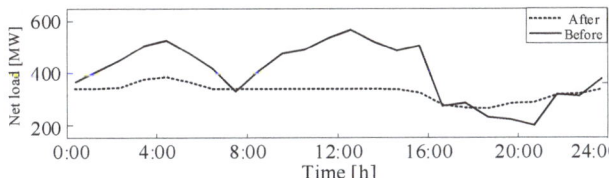

Figure 11. Peak shaving performance using a deterministic environment.

Table 5. Comparison to the deterministic optimization method.

Case	Total Cost [USD]	Reserve Capacity Used [MW]	C_{CLEC}	C_{fls}	EDR [MW]
The deterministic optimization method	15,842.6	0	1516.8	5560.1	512.9
Case 4	19,021.1	669	1991.2	4291.4	10

By integrating the data in Table 5 and comparing Figure 11 with Figure 7, it is observed that after considering the uncertainties, C_{fls} decreased from 5560.1 to 4291.4, indicating a 22.8% reduction in the park's contribution to the peak shaving. Simultaneously, C_{CLEC} increased from 1516.8 to 1991.2, representing a 31% decrease in the satisfaction level of the park's cooling load utilization. However, in Scenario 4, which incorporates robust optimization, even under the worst operating conditions, the amount of load shed through EDR was only 10 MW, significantly outperforming the 512.9 MW in deterministic optimization. This indicates that, although a scheduling scheme considering uncertainties may slightly decrease the satisfaction level of the park's cooling load utilization and diminish its role in peak shaving, the park's operation becomes safer and more robust due to the significant decrease in load shedding through EDR to 10 MW and the 98.1% reduction in emergency load shedding.

5.3. Comparison Study: Steam Turbine Replacement for Gas Turbine

Gas turbines are recognized for their swift start-up, flexible adjustment, high ramp rates, and rapid response capabilities, enabling significant energy output within short durations. Conversely, steam turbines excel in terms of efficiency, demonstrating higher energy conversion rates and offering advantages in availability and cost. In this case study, the steam turbine is utilized to replace the gas turbine for supplying electrical power to the park, facilitating a comprehensive comparison between the two technologies.

The steam turbine derives its energy from coal-fired and wood-fired boilers. The generated high-temperature and high-pressure steam is then harnessed in a waste heat boiler to drive the steam turbine for power generation. Considering the relatively slower power response rate of steam turbines in comparison to hourly scheduling, it becomes imperative to account for both their output power constraints and ramp rate constraints. When the demand for thermal load utilization is disregarded, the constraints governing the electrical power output of the steam turbine are outlined as follows:

$$P_{st,t}^0 = P_{st}^{\min} I_{st}^t + F_{st,t}^h \times \eta_{st,e} \times \eta_{st,loss} \tag{54}$$

$$F_{st,t}^h = Q_t^{coal} \times \eta_{coal} + Q_t^{wood} \times \eta_{wood} \tag{55}$$

$$P_{st}^{\min} \leq P_{st,t}^0 \leq P_{st}^{\max} \tag{56}$$

where $P_{st,t}^0$ represents the output electric power of the steam turbine at time t in the first stage; $P_{st}^{\min}$ and $P_{st}^{\max}$ are the minimum and maximum output powers of the steam turbine during operation; I_{st}^t is the operation status indicator of the steam turbine in time period t; $F_{st,t}^h$ denotes the amount of steam consumed by the steam turbine in time period t; $\eta_{se,e}$ represents the power generation efficiency of the steam turbine; $\eta_{st,loss}$ is the heat loss rate of the steam turbine; Q_t^{coal} and Q_t^{wood} are the heat generated by coal-fired and wood-fired boilers in time period t; and η_{coal} and η_{wood} are the working thermal efficiencies of the waste heat boilers for coal-fired and wood-fired boilers, respectively. Ramp constraints are as follows:

$$-R_{st}^D \leq P_{st,t+1}^0 - P_{st,t}^0 \leq R_{st}^U, \forall t \tag{57}$$

where R_{st}^D and R_{st}^U represent the downward and upward ramp rates of the steam turbine, respectively. Due to the high ramp rate and weak reserve capacity of the steam turbine, this section's case study in the second stage considers that when the power supply demand of the park cannot be met, the park can purchase electricity from the distribution network to replace the reserve capacity of the gas turbine. To avoid excessive electricity purchases from the distribution network by the park, which would weaken the peak-shaving effect of electric refrigeration air conditioning in the park, power constraints are imposed on the park's electricity purchases from the distribution network:

$$0 \leq P_{buy,t}^s \leq R_{g,t}^{\max}, \forall t \tag{58}$$

where $P_{buy,t}^s$ represents the amount of electricity purchased by the park from the distribution network in time period t of the second stage; $R_{g,t}^{\max}$ is the sum of the upper limits of the reserve capacities of all gas turbines in time period t under the gas turbine case study model in this paper. Considering the uncertainties of photovoltaic and wind power simultaneously, robustness indicators $u_{t,p}$ and $u_{t,w}$ are set to 16 and 16, respectively, for the solution. The scheduling optimization results are shown in Figures 12 and 13.

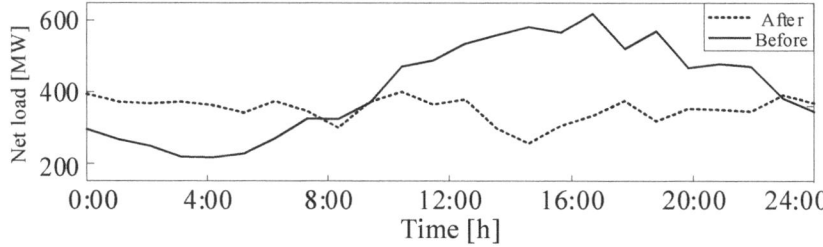

Figure 12. Peak shaving performance under steam turbine substitution output.

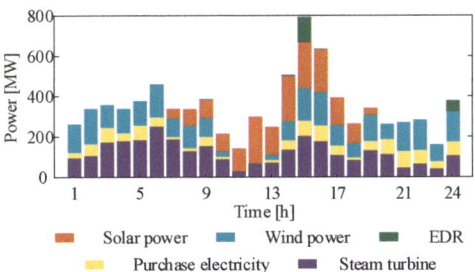

Figure 13. The 24 h electric power balance under steam turbine substitution output.

By comparing Figure 12 with Figure 7 and incorporating the optimization data from various scenarios presented in Table 3, the comparative outcomes between gas turbine and steam turbine outputs under identical uncertainty and robustness conditions are presented in Table 6.

Table 6. Comparison results between gas and steam turbines.

Case	Total Cost [USD]	C_{CLEC}	C_{fls}	EDR [MW]	PVD Enhancement [%]	FV Enhancement [%]
Gas turbine	19,021.1	1991.2	4291.4	10	45.82	54.59
Steam turbine	15,617.5	2076.9	3945.8	191.9	31.08	42.27

As Table 6 indicates, employing a steam turbine for power supply to the park results in a 17.89% reduction in the total operating cost, and this is the benefit of the steam turbine. Figure 12 demonstrates that the steam turbine still achieves satisfactory peak clipping and valley filling effects. However, when compared to scenarios utilizing gas turbines, the improvement in the PVD metric decreases by 14.74%, and the enhancement in the FV metric decreases by 12.32%. Furthermore, the C_{CLEC} indicator under the steam turbine power supply exhibits only a 4% increase compared to the gas turbine power supply, suggesting a 4% rise in economic losses due to the suboptimal utilization of the park's cooling load. It is evident that using a steam turbine for power supply meets the park's normal operational requirements and satisfies certain peak shaving demands. However, in emergency situations, the load shedding amount for demand response in the park reaches 191.9 MW, significantly higher than the 10 MW observed with gas turbines. Thus, while the steam turbine may slightly reduce the park's overall operating cost, it compromises the ability to respond effectively to emergencies.

6. Conclusions

This paper proposes a novel optimization method for large logistics parks to participate in grid peak shaving in a two-stage robust manner. The conclusions are as follows:

(1) This paper introduces a methodology for constructing the CLEC index, which comprehensively integrates the cooling comfort level, measured by PMV, of occupants within the park's buildings, and the sales pressure, indicated by SPI, of frozen goods stored in cold storage facilities. Furthermore, it takes into account the peak shaving demands of the interconnected distribution network. Validation studies confirm that the CLEC index and the park's contribution to peak shaving operate independently, affirming the rationality of this index.

(2) To address the inherent uncertainty in wind and solar power generation within the park, a two-stage robust optimization approach is employed to assess the effectiveness of the proposed method across various operational scenarios. The findings reveal that as the uncertainty surrounding renewable energy output escalates, both the satisfaction level associated with cooling load utilization and the park's peak shaving capabilities decline

by approximately 15.4% and 19.8%, respectively. Nonetheless, the amount of load shed by EDR measures drops significantly from 168.5 MW to just 10 MW, showcasing the risk mitigation capabilities of the proposed method.

(3) When juxtaposed against deterministic optimization techniques, the proposed method, despite increasing the overall system operation costs by approximately 20.06%, dramatically reduces EDR-related load shedding by 98.1%. This substantial decrease underscores the method's superior ability to withstand uncertainties and risks.

(4) One limitation of this paper is its focus solely on the suitable storage temperature range for frozen products, neglecting the diversity among them. In reality, various frozen products exhibit unique seasonality and optimal storage methods, the consideration of which would reveal further nuanced differences in optimization scheduling results. Additionally, the use of a single indicator (PMV) to assess personnel comfort within the park, without accounting for individual variations in the perception of "comfort," constitutes another limitation. These aspects require refinement in future research.

Author Contributions: Conceptualization, J.Z. (Jiu Zhou); methodology, J.Z. (Jiu Zhou), J.Z. (Jieni Zhang), and X.T.; software, Z.Q. and Z.Y.; validation, Z.Y. and Q.C.; formal analysis, X.T.; investigation, X.T.; resources, Q.C.; data curation, X.T.; writing—original draft preparation, X.T.; writing—review and editing, X.T.; visualization, X.T.; supervision, Q.C.; project administration, Q.C.; funding acquisition, Q.C. All authors have read and agreed to the published version of the manuscript.

Funding: This work was supported by the Technology Project of China Southern Power Grid (030100KC22120013 (GDKJXM20222718)).

Data Availability Statement: The datasets generated and/or analyzed during the current study are not publicly available due to privacy and confidentiality concerns, but are available from the corresponding author on reasonable request.

Conflicts of Interest: Authors Jiu Zhou, Jieni Zhang, Zhaoming Qiu, and Zhiwen Yu were employed by the Guangdong Power Grid Co., Ltd. The remaining authors declare that the research was conducted in the absence of any commercial or financial relationships that could be construed as a potential conflict of interest. The authors declare that this study received funding from the Technology Project of China Southern Power Grid. The funder was not involved in the study design, collection, analysis, interpretation of data, the writing of this article or the decision to submit it for publication.

References

1. Lin, L.; Xu, B.; Xia, S. Multi-Angle Economic Analysis of Coal-Fired Units with Plasma Ignition and Oil Injection during Deep Peak Shaving in China. *Appl. Sci.* **2019**, *9*, 5399. [CrossRef]
2. Wang, X.; Wu, Y.; Fu, L. Evaluation of combined heat and power plants with electricity regulation. *Appl. Therm. Eng.* **2023**, *227*, 120364. [CrossRef]
3. Liu, Q.; Zhao, J.; Shao, Y.; Wen, L.; Wu, J.; Liu, D.; Ma, Y. Multi-Power Joint Peak-Shaving Optimization for Power System Considering Coordinated Dispatching of Nuclear Power and Wind Power. *Sustainability* **2019**, *11*, 4801. [CrossRef]
4. Yu, Y.; Zhou, G.; Wu, K.; Chen, C.; Bian, Q. Optimal Configuration of Power-to-Heat Equipment Considering Peak-Shaving Ancillary Service Market. *Energies* **2023**, *16*, 6860. [CrossRef]
5. Liao, S.; Zhang, Y.; Liu, J.; Liu, B.; Liu, Z. Short-Term Peak-Shaving Operation of Single-Reservoir and Multicascade Hydropower Plants Serving Multiple Power Grids. *Water Resour. Manag.* **2021**, *35*, 689–705. [CrossRef]
6. Zou, Y.; Hu, Z.; Zhou, S.; Luo, Y.; Han, X.; Xiong, Y.; Li, M.E. Day-Ahead and Intraday Two-Stage Optimal Dispatch Considering Joint Peak Shaving of Carbon Capture Power Plants and Virtual Energy Storage. *Sustainability* **2024**, *16*, 2890. [CrossRef]
7. Zhang, Z.; Cong, W.; Liu, S.; Li, C.; Qi, S. Auxiliary Service Market Model Considering the Participation of Pumped-Storage Power Stations in Peak Shaving. *Front. Energy Res.* **2022**, *10*, 915125. [CrossRef]
8. Liu, B.; Lund, J.R.; Liao, S.; Jin, X.; Liu, L.; Cheng, C. Optimal power peak shaving using hydropower to complement wind and solar power uncertainty. *Energy Convers. Manag.* **2020**, *209*, 2621–2631. [CrossRef]
9. Cheng, X.; Feng, S.; Huang, Y.; Wang, J. A New Peak-Shaving Model Based on Mixed Integer Linear Programming with Variable Peak-Shaving Order. *Energies* **2021**, *14*, 887. [CrossRef]
10. Fang, Y.; Zhao, S.; Chen, Z. Multi-objective unit commitment of jointly concentrating solar power plant and wind farm for providing peak-shaving considering operational risk. *Int. J. Electr. Power Energy Syst.* **2022**, *137*, 107754. [CrossRef]
11. Uddin, M.; Romlie, M.F.; Abdullah, M.F.; Halim, S.A.; Bakar, A.H.A.; Kwang, T.C. A review on peak load shaving strategies. *Renew. Sustain. Energy Rev.* **2018**, *82*, 3323–3332. [CrossRef]

12. Yang, Y.; Wang, Y.; Gao, Y.; Gao, C. Peak Shaving Analysis of Power Demand Response with Dual Uncertainty of Unit and Demand-Side Resources under Carbon Neutral Target. *Energies* **2022**, *15*, 4588. [CrossRef]
13. Shen, J.; Jiang, C.; Liu, Y.; Qian, J. A Microgrid Energy Management System with Demand Response for Providing Grid Peak Shaving. *Electr. Mach. Power Syst.* **2014**, *44*, 843–852. [CrossRef]
14. Su, H.; Feng, D.; Zhao, Y.; Zhou, Y.; Zhou, Q.; Fang, C.; Rahman, U. Optimization of Customer-Side Battery Storage for Multiple Service Provision: Arbitrage, Peak Shaving, and Regulation. *IEEE Trans. Ind. Appl.* **2022**, *58*, 2559–2573. [CrossRef]
15. Alessia, A.; Fabio, P. Assessing the Demand Side Management Potential and the Energy Flexibility of Heat Pumps in Buildings. *Energies* **2018**, *11*, 1846. [CrossRef]
16. Abbasi, A.; Sultan, K.; Afsar, S.; Aziz, M.A.; Khalid, H.A. Optimal Demand Response Using Battery Storage Systems and Electric Vehicles in Community Home Energy Management System-Based Microgrids. *Energies* **2023**, *16*, 5024. [CrossRef]
17. Guelpa, E.; Marincioni, L. Demand side management in district heating systems by innovative control. *Energy* **2019**, *188*, 116037. [CrossRef]
18. Lu, N.; Chassin, D.P.; Widergren, S.E. Modeling uncertainties in aggregated thermostatically controlled loads using a State queueing model. *IEEE Trans. Power Syst.* **2005**, *20*, 725–733. [CrossRef]
19. Bashash, S.; Fathy, H.K. Modeling and control insights into demand-side energy management through setpoint control of thermostatic loads. In Proceedings of the American Control Conference (ACC), San Francisco, CA, USA, 29 June–1 July 2011; pp. 4546–4553.
20. Khan, S.; Shahzad, M.; Habib, U.; Gawlik, W.; Palensky, P. Stochastic Battery Model for Aggregation of Thermostatically Controlled Loads. In Proceedings of the 2016 IEEE International Conference on Industrial Technology (ICIT), Taipei, Taiwan, 14–17 March 2016; pp. 570–575.
21. Callaway, D.S. Tapping the energy storage potential in electric loads to deliver load following and regulation, with application to wind energy. *Energy Convers. Manag.* **2009**, *50*, 1389–1400. [CrossRef]
22. Parkinson, S.; Wang, D.; Crawford, C.; Djilali, N. Comfort-Constrained Distributed Heat Pump Management. In Proceedings of the International Conference on Smart Grid and Clean Energy Technologies, Chengdu, China, 27–30 September 2011; Elsevier Ltd.: Chengdu, China, 2011; pp. 849–855.
23. Ma, Y.; Xu, W.; Yang, H.; Zhang, D. Two-stage stochastic robust optimization model of microgrid day-ahead dispatching considering controllable air conditioning load. *Int. J. Electr. Power Energy Syst.* **2022**, *141*, 108174. [CrossRef]
24. Mendieta, W.; Cañizares, C.A. Primary Frequency Control in Isolated Microgrids Using Thermostatically Controllable Loads. *IEEE Trans. Smart Grid* **2021**, *12*, 93–105. [CrossRef]
25. Li, Z.; Sun, Z.; Meng, Q.; Wang, Y.; Li, Y. Reinforcement learning of room temperature set-point of thermal storage air-conditioning system with demand response. *Energy Build.* **2022**, *259*, 111903. [CrossRef]
26. Nian, F.; Wang, K. Study on indoor environmental comfort based on improved PMV index. In Proceedings of the International Conference on Computational Intelligence & Communication Technology, Ghaziabad, India, 9–10 February 2017; pp. 1–4.
27. Gormley, R.; Walshe, T.; Hussey, K.; Butler, F. The Effect of Fluctuating vs. Constant Frozen Storage Temperature Regimes on Some Quality Parameters of Selected Food Products. *LWT—Food Sci. Technol.* **2002**, *35*, 190–200. [CrossRef]
28. Yunyang, Z.; Li, Y.; Jiayong, L.I.; Lei, X.; Hao, Y.E.; Zhenzhi, L. Robust Optimal Dispatch of Micro-energy Grid with Multi-energy Complementation of Cooling Heating Power and Natural Gas. *Autom. Electr. Power Syst.* **2019**, *14*, 65–72.
29. Wang, H.; Xing, H.; Luo, Y.; Zhang, W. Optimal scheduling of micro-energy grid with integrated demand response based on chance-constrained programming. *Int. J. Electr. Power Energy Syst.* **2023**, *144*, 1086. [CrossRef]
30. Zhang, Z.; Zhou, M.; Chen, Y.; Li, G. Exploiting the operational flexibility of AA-CAES in energy and reserve optimization scheduling by a linear reserve model. *Energy* **2023**, *263*, 126084. [CrossRef]
31. Amin, G.R. A comment on modified big-M method to recognize the infeasibility of linear programming models. *Knowl. -Based Syst.* **2010**, *23*, 283–284. [CrossRef]
32. Meshram, M.S.; Sahu; Omprakash; Preeti, R. Application of ANN in Economic Generation Scheduling in IEEE 6-Bus System. *Int. J. Eng. Sci. Technol.* **2011**, *3*, 2461–2466.
33. Li, H.; Zhao, T.; Dian, S. Forward search optimization and subgoal-based hybrid path planning to shorten and smooth global path for mobile robots. *Knowl.-Based Syst.* **2022**, *258*, 110034. [CrossRef]
34. Zhang, Z.; Wang, K.; Zhu, L.; Wang, Y. A Pareto improved artificial fish swarm algorithm for solving a multi-objective fuzzy disassembly line balancing problem. *Expert Syst. Appl.* **2017**, *86*, 165–176. [CrossRef]
35. Antonsson, E.K.; Sebastian, H. Fuzzy fitness functions applied to engineering design problems. *Eur. J. Oper. Res.* **2005**, *166*, 794–811. [CrossRef]
36. Jamali, A.; Mallipeddi, R.; Salehpour, M.; Bagheri, A. Multi-objective differential evolution algorithm with fuzzy inference-based adaptive mutation factor for Pareto optimum design of suspension system. *Swarm Evol. Comput.* **2020**, *54*, 100666. [CrossRef]

Disclaimer/Publisher's Note: The statements, opinions and data contained in all publications are solely those of the individual author(s) and contributor(s) and not of MDPI and/or the editor(s). MDPI and/or the editor(s) disclaim responsibility for any injury to people or property resulting from any ideas, methods, instructions or products referred to in the content.

Article

A Deterministic Calibration Method for the Thermodynamic Model of Gas Turbines

Zhen Jiang [1], Xi Wang [1], Shubo Yang [2,*] and Meiyin Zhu [3]

[1] School of Energy and Power Engineer, Beihang University, Beijing 102206, China; jiangzhen@buaa.edu.cn (Z.J.); xwang@buaa.edu.cn (X.W.)
[2] Institute for Aero Engine, Tsinghua University, Beijing 102202, China
[3] International Innovation Institute, Beihang University, Hangzhou 310023, China; mecalzmy@buaa.edu.cn
* Correspondence: shubo_yang@tsinghua.edu.cn

Abstract: Performance adaptation is an effective way to improve the accuracy of gas turbine performance models. Although current performance adaptation methods, such as those using genetic algorithms or evolutionary computation to modify component characteristic maps, are useful for finding good solutions, they are essentially searching methods and suffer from long computation time. This paper presents a novel approach that can achieve good performance adaptation with low time complexity and without using any searching method. In this method, the actual component performance parameters are first estimated using engine measurements at different operating conditions. For each operating condition, some scaling factors are introduced and calculated to indicate the difference between the actual and predicted component performance parameters. Afterward, an interpolating algorithm is adopted to synthesize the scaling factors for modifying all major component maps. The adapted component maps are then able to make the engine model match all the gas path measurements and achieve the required accuracy of the engine performance model. The proposed approach has been tested with a model high-bypass turbofan engine using simulated data. The results show that the proposed performance adaptation approach can effectively improve the model's accuracy. Specifically, the prediction errors can be reduced from about 9% to about 0.6%. In addition, this approach has much less computational complexity compared to other optimization-based counterparts.

Keywords: performance adaptation; model calibration; characteristic shift; scaling factor; gas turbine

Citation: Jiang, Z.; Wang, X.; Yang, S.; Zhu, M. A Deterministic Calibration Method for the Thermodynamic Model of Gas Turbines. *Symmetry* **2024**, *16*, 522. https://doi.org/10.3390/sym16050522

Academic Editor: Theodore E. Simos

Received: 11 March 2024
Revised: 18 April 2024
Accepted: 24 April 2024
Published: 26 April 2024

Copyright: © 2024 by the authors. Licensee MDPI, Basel, Switzerland. This article is an open access article distributed under the terms and conditions of the Creative Commons Attribution (CC BY) license (https://creativecommons.org/licenses/by/4.0/).

1. Introduction

The gas turbine engine, a sophisticated piece of thermal equipment with symmetrical geometry, plays an important role in the aerospace industry [1]. To promote the rapid development of the gas turbine, researchers urgently need a good performance model to replace many real engine tests with numerical simulations, which could make the development process more efficient and economical [2]. Additionally, an accurate performance model is crucial for good gas engine controls, diagnostics, and prognostics [3,4].

Generally, the accuracy of a performance model mainly depends on accurate component characteristic maps. However, it is sometimes difficult to obtain the real component maps of an engine due to many different reasons, such as lacking bench testing, performance degradation, and engine-to-engine variations [5–7]. Against this limitation, multiple-point performance adaptation methods have drawn more and more attention in recent years [8]. These methods suggest a tuning of generic component maps to make an engine performance model match all measured performances. Typical published work is briefed as follows.

In the early 1990s, Stamatis et al. introduced an adaptation method that was able to calibrate existing component maps by optimizing scaling factors [9,10]. In 2009, Li et al. used a genetic algorithm method to search for an optimal set of scaling factors [11].

In the following years, such a method was further developed by applying variable scaling factors to achieve a nonlinear multiple-point adaptation [12,13]. In 2014, Alberto and Benini performed map modifications in the neighborhoods of the multiple experimental points [14]. An optimization algorithm was applied to seek optimal sets of perturbations on the map in order to minimize the deviation between experimental and predicted performance. Tsoutsanis et al. introduced a novel map representation and an optimization algorithm for the adaptation [15,16]. By employing elliptic curves to generate compressor maps, they refined gas turbine models with improved accuracy. Recently, Pang et al. proposed a joint adaptation method that added a transient performance adaptation procedure following the steady-state counterpart [17]. After a two-step optimization process, engine models can be calibrated to have more accurate transient performance predictions.

The published work for multiple-point performance adaptations involves complex numerical searching and optimizations. Facing such problems, most researchers prefer to employ metaheuristics, such as genetic algorithms [11–13,15,16] and evolutionary computation [17]. The metaheuristics can often find good solutions over a huge set of feasible solutions with less computational effort than other optimization methods and proved to be useful approaches for optimization problems [18,19]. However, some drawbacks exist, such as long computing time when the number of points included in the adaptation is large. In addition, metaheuristics do not guarantee that a globally optimal solution can be found [20]. Many heuristic methods contain stochastic procedures so that the solution found is dependent on the set of random variables generated [21]. Furthermore, it is very tricky to specify the appropriate search domain for the adaptation calculations in practice. Therefore, Li et al. explored a non-heuristic method for performance adaptation problems [22]. The method was applied to an industrial gas turbine engine and demonstrated by the comparison results. However, this approach is only suitable to the design point.

This paper proposes a deterministic approach to deal with the multiple-point performance adaptation of gas turbine engines. At first, engine measurements are employed as input information to estimate actual engine performance by a numerical iterative solution process. Next, we compare that performance with the predicted one (using the model with initial component maps) and summarize the comparison using scaling factors. Finally, an interpolation-based algorithm is developed to determine the overall modification of all major component maps. The proposed approach is applied to a model turbofan engine similar to the General Electric CF6 series. Simulation is carried out to demonstrate the capability of the proposed performance adaptation method. The results, discussions, and conclusions are made accordingly.

Compared to the heuristic methods, the novelty of this approach is that a deterministic calibration procedure replaces the searching algorithms to obtain some scaling factors and thus achieves performance adaptation with high accuracy, low computational complexity, and deterministic results.

The remainder of this paper is organized as follows: Section 2 describes the methodology of the proposed deterministic calibration method. Section 3 details the simulation results to demonstrate the effectiveness of the method. Section 4 discusses the method's advantages in three aspects. The final section provides a summary of this paper.

2. Methodology
2.1. Representation of Map Shift

A characteristic map is a chart for describing the performance behavior of an engine component, such as a compressor or a turbine at different operating conditions.

For two major categories of engine components, compressor type, and turbine type, there can be four characteristic parameters to describe the component map: the corrected relative rotational speed n, the flow capacity W, the pressure ratio π, and the isentropic efficiency η. Specifically, a map can be denoted by a group of contour lines (also called the speed lines). Each speed line may be represented by a series of points that have the same speed, n. A map with m ($m > 1$) speed lines can be defined as

$$\gamma_i : [0, 1] \to (W, \pi, \eta) \in \mathbb{R}^3, i = 1, 2, \ldots, m \tag{1}$$

where each γ_i is a continuous function from the closed real interval [0,1] into a real coordinate space of 3 dimensions. The speed value of γ_i is written as n_i. For a typical high-pressure compressor, for instance, its characteristic map may be illustrated as the solid lines ranging from γ_1 to γ_m in Figure 1.

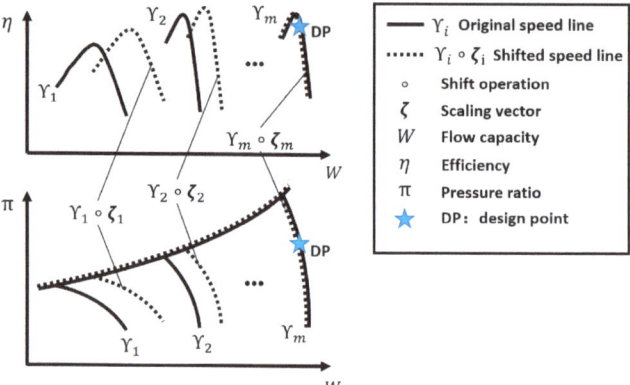

Figure 1. Shift of a compressor map.

While the dashed lines in Figure 1 represent a shift of the initial map. For each speed line in the map, an operation '∘' is defined as Equation (2) to denote its shift.

$$\gamma \circ \begin{bmatrix} \zeta_W \\ \zeta_\pi \\ \zeta_\eta \end{bmatrix} = \{(W\zeta_W, \pi\zeta_\pi - \zeta_\pi + 1, \eta\zeta_\eta) | \forall (W, \pi, \eta) \in \gamma\} \tag{2}$$

where $\gamma \circ \begin{bmatrix} \zeta_W & \zeta_\pi & \zeta_\eta \end{bmatrix}^T$ means that a speed line γ is shifted by the vector of scaling factors $\begin{bmatrix} \zeta_W & \zeta_\pi & \zeta_\eta \end{bmatrix}^T$, where the factors are three real numbers and can be written as a vector ζ for convenience. Then, scaling factor (SF) sets $\{\zeta_1, \zeta_2, \ldots, \zeta_m\}$ for all speed lines can derive the map shift.

During performance adaptation, the position of the design point (DP) on each component map is assumed to be unchanged. For instance, the DP is located on the mth speed line in Figure 1 and remains unchanged by setting $\zeta_m = [1, 1, 1]^T$ during the entire process of characteristic map shift.

2.2. Scaling Factor Determination

Characteristic maps for compressors and turbines are considered to be adapted in this study. Assume that a map has m speed lines and their speed values are $\{n_1, n_2, \ldots, n_m\}$. The objective is to determine their corresponding SF sets $\{\zeta_1, \zeta_2, \ldots, \zeta_m\}$.

Assume that measurement data contain M ($M \geq 1$) different engine conditions. For each condition, the four characteristic parameters (n, W, π, η) of each concerned component can be estimated by the method described in Section 2.2.1. Afterward, in terms of every single compressor or turbine, SF sets $\{\zeta'_1, \zeta'_2, \ldots, \zeta'_m\}$ of the M conditions can be determined according to Section 2.2.2. In Section 2.2.3, SF sets $\{\zeta_1, \zeta_2, \ldots, \zeta_m\}$ are finally obtained by using an interpolation-based algorithm. The above determination process is described in Figure 2.

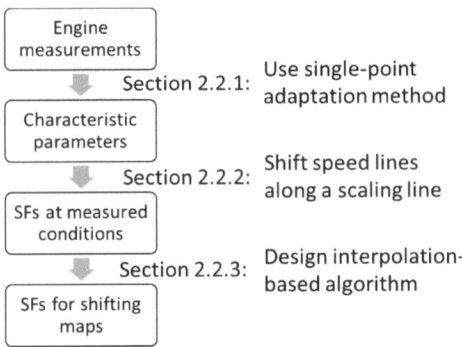

Figure 2. Scaling factor determination process.

2.2.1. Actual Performance Estimation

Based on a single-point adaptation method, the performance parameters of all components can be calculated using the given engine measurement data. The single-point adaptation method used in this subsection is developed from existing approaches [22,23] and described as follows.

Owing to the intricate thermodynamic function within the engine system, it is difficult to calculate some performance parameters through measurements directly. However, if guesses of the unknown parameters are introduced, errors can be generated to imply whether the guesses are appropriate by employing the thermodynamic function. Among the errors, the guesses, and the measurements, there exists a relationship that can be represented by Equation (3).

$$e = h(x, z) \quad (3)$$

where x is the guess vector of performance parameters, z is the vector of measurement parameters, e is the error vector representing system mismatch between measurement and calculation, and $h(\cdot)$ is a differentiable nonlinear function denoting engine thermodynamic function which can generate e from x and z.

Equation (3) is essential for estimating the performance parameters, and it can be obtained by reconstructing the thermodynamic function inside the engine. It contains all the nonlinearities of the complex system.

Remark 1. *e and x should have the same number of elements to avoid Equation (3) underdetermined or overdetermined. By omitting either some less reliable measurements in z or some less important performance parameters in x, the requirement is not difficult to meet.*

With given engine measurements z, Equation (3) can be solved by the Newton-Raphson method, which locally linearizes a nonlinear equation at the current estimated value and then uses the linearized equation to update the estimated value to the solution [24]. Iteratively, the error vector e is driven toward zero. And once a converged solution x is obtained, all the other performance parameters, including characteristic parameters of all components, can be predicted.

Remark 2. *The convergence criterion is $\|e\|_\infty < \sigma$, where σ is a set threshold.*

2.2.2. Scaling Factors at Measured Conditions

Regarding the M measured conditions, this section presents how to determine the SF sets $\{\zeta'_1, \zeta'_2, \ldots, \zeta'_m\}$ for each component.

At each measured condition, the four characteristic parameters of compressors have been obtained from Section 2.2.1. Figure 3 locates the estimated parameters using a hollow point $A(n_A, W_A, \pi_A, \eta_A)$.

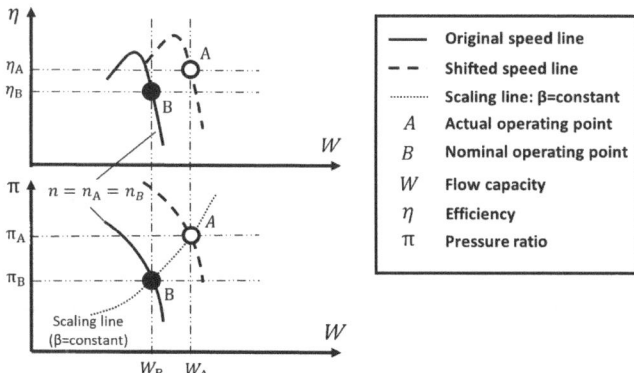

Figure 3. Determine SFs at a measured condition on a compressor map.

Normally, point A is not on the corresponding speed line of the existing compressor map. To find the matching point of the current map and the scaling factor between these two points, it is reasonable to assume that the speed line of n_A is shifted along a scaling line where the auxiliary map coordinate β is constant [25] to pass through point A. This assumption can locate a solid point B at the intersection of the speed line and the scaling line, as shown in Figure 3. By comparing points A and B, scaling factors can be calculated using Equation (4).

$$\zeta = \begin{bmatrix} \zeta_W \\ \zeta_\pi \\ \zeta_\eta \end{bmatrix} = \begin{bmatrix} W_A/W_B \\ (\pi_A - 1)/(\pi_B - 1) \\ \eta_A/\eta_B \end{bmatrix} \quad (4)$$

Remark 3. *Although there are three factors in ζ, only two of them are independent because there is a correlation between the scaling factors of corrected flow and that of pressure ratio (under the assumption β = constant).*

Remark 4. *Note that such a single vector ζ can only reflect the shift at its corresponding speed value n_A. For all M measured conditions whose speed values are $\{n'_1, n'_2, \ldots, n'_M\}$, we have to use Equation (4) for M times to obtain the SF sets $\{\zeta'_1, \zeta'_2, \ldots, \zeta'_M\}$.*

Equation (4) can also be used to determine the scaling factors of turbines. It should be mentioned that the scaling lines with a constant β are exactly the same lines with a constant pressure ratio for turbines, as shown in Figure 4. As a result, $\pi_A = \pi_B$ always holds, i.e., $\zeta_\pi = 1$ in the turbine case.

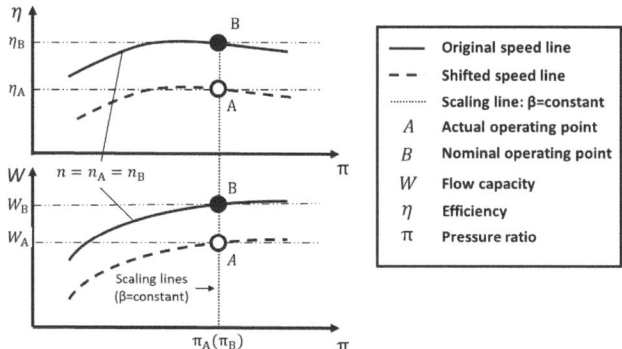

Figure 4. Determine SFs at a measured condition on a turbine map.

2.2.3. Scaling Factors for Map Shifting

For every compressor or turbine, SF sets $\{\zeta'_1, \zeta'_2, \ldots, \zeta'_M\}$ and their corresponding speed values $\{n'_1, n'_2, \ldots, n'_M\}$ can be obtained from Section 2.2.2. Meanwhile, speed values of all m initial speed lines are known as $\{n_1, n_2, \ldots, n_m\}$. Assume that speed values are written in incremental order, i.e., $n'_1 < n'_2 < \cdots < n'_M$ and $n_1 < n_2 < \cdots < n_m$.

The following part shows the procedure of how to calculate SFs $\{\zeta_1, \zeta_2, \ldots, \zeta_m\}$ for each initial map.

- Case 1. $M = 1$

In the case where there is only one available measured condition, the only scaling vector ζ'_1 should be applied to all the speed lines, i.e.,

$$\zeta_i = \zeta'_1, \forall i \in \{1, 2, \ldots, m\} \tag{5}$$

- Case 2. $M > 1$

For the case $M > 1$, if a speed value n_i is located within the interval $[n'_1, n'_M]$ or next to an endpoint of the interval, it is reasonable to calculate the corresponding scaling vector ζ_i by using a linear interpolation. In order to represent the range where interpolation can be used, a closed interval, written as "Interval A" in Figure 5, is then defined as $[n_L, n_R]$ where the subscripts "L" and "R" are determined by

$$L = \begin{cases} 1 & , n_1 \geq n'_1 \\ \max\{i \in \{1, 2, \ldots, m\} | n_i < n'_1\} & , n_1 < n'_1 \end{cases} \tag{6}$$

and

$$R = \begin{cases} m & , n_m \leq n'_M \\ \min\{i \in \{1, 2, \ldots, m\} | n_i > n'_M\} & , n_m > n'_M \end{cases} \tag{7}$$

For each $n_i \in [n_L, n_R]$, the corresponding scaling vector ζ_i can be calculated by Equation (8).

$$\zeta_i = \zeta'_{p(i)} + \frac{n_i - n'_{p(i)}}{n'_{p(i)+1} - n'_{p(i)}} (\zeta'_{p(i)+1} - \zeta'_{p(i)}) \tag{8}$$

where $i \in \{L, L+1, \ldots, R\}$ and

$$p(i) = \begin{cases} 1 & , n_i \leq n'_1 \\ \max\{k \in \{1, 2, \ldots, M-1\} | n'_k < n_i\} & , n_i > n'_1 \end{cases} \tag{9}$$

Beyond the range of Interval A, Interval B is further defined (if it exists) where the speed value n is smaller than n_L as shown in Figure 5. Since the interval covers no information on the measured conditions ζ_L, the nearest scaling vector from Interval A should be applied to all the speed lines in Interval B, i.e.,

$$\zeta_i = \zeta_L, \forall i \in \{1, 2, \ldots, L-1\} \tag{10}$$

Likewise, Interval C is defined (if it exists) and ζ_R is applied to all the speed lines in it.

$$\zeta_i = \zeta_R, \forall i \in \{R+1, R+2, \ldots, m\} \tag{11}$$

So far, all the SFs $\{\zeta_1, \zeta_2, \ldots, \zeta_m\}$ have been calculated and can be applied to shift the initial characteristic map.

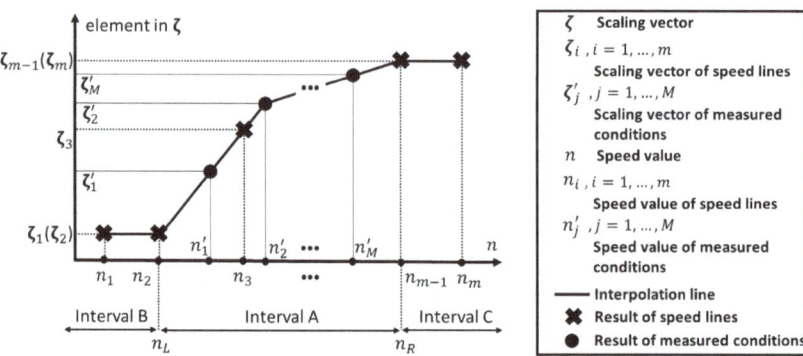

Figure 5. Interval A: the range where interpolation can be used.

3. Application and Results

3.1. Basic Information of the Model Engine

The proposed multi-point adaptation approach has been applied to a model high-bypass-ratio separate exhaust turbofan engine, commonly used in large passenger and transport aircraft. Among various types of engines, the turbofan engine is the most representative. The configuration of the model (similar to the General Electric CF6 series) is illustrated in Figure 6, with components of intake, fan, booster, high-pressure compressor (HPC), combustor, high-pressure turbine (HPT), low-pressure turbine (LPT), duct and nozzle of bypass, and core. The roles of each component can be referred to [25]. And the number in Figure 6 represents the station of the engine. For example, '13' denotes the outlet of the fan.

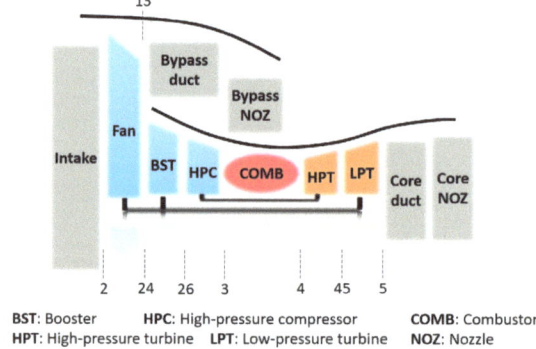

Figure 6. Model turbofan engine configuration.

The model was created using the Gas Turbine Modeling Library for Education (GTML-E), which was developed on behalf of Beihang University, and an open-source MATLAB/Simulink library for gas turbine performance modeling [26]. The Iterative Newton-Raphson Solver block used in GTML-E is based on the T-MATS package [27].

In this application, five major components are considered to be adapted: the fan, booster, high-pressure compressor, high-pressure turbine, and low-pressure turbine.

For these components, two sets of characteristic maps are prepared separately. The maps in the first set are generic component maps [28] and are regarded as the initial maps. The second set is a transformed version of the first set, where the maps were treated as "real" maps. Note that the "real" maps were only used for generating "test data".

3.2. "Test Data" Generation

The measurable parameters used in this study are summarized in Table 1, similar to those of CF6 [29]. Some of these parameters are adopted as the model input, while others are regarded as the targeted performance to judge whether the calculated performance is accurate. In the case of validation, except for ambient parameters (P_H, T_2, P_2), only the handle parameter N_L is regarded as the model input. As for the estimation of actual performance, however, more measurable parameters are assumed to be known and used as model inputs.

Table 1. CF6-80 measurable parameters.

Measurements	Symbol	For Estimation	For Validation
Ambient Pressure	P_H	Input	Input
Fan Inlet Temperature	T_2	Input	Input
Fan Inlet Pressure	P_2	Input	Input
Fuel Flow	W_f	Input	Target
Fan Speed	N_L	Input	Input
Core Speed	N_H	Input	Target
HPC Inlet Temperature	T_{26}	Target	Target
HPC Inlet Pressure	P_{26}	Target	Target
HPC Outlet Temperature	T_3	Target	Target
HPC Outlet Pressure	P_3	Target	Target
LPT Inlet Temperature	T_{45}	Target	Target
LPT Inlet Pressure	P_{45}	Target	Target
LPT Outlet Temperature	T_5	Target	Target
Fan Outlet Pressure	P_{13}	Target	Target
Net Thrust	F_N	Input	Target

With the "real" maps applied, "test data" were generated by the model at 10 steady-state operating points under ISA sea-level conditions. These operating conditions of the engine, represented by the handle parameter (normalized fan speed), are listed in Table 2. It can be noticed that 6 of them are employed to carry out the performance adaptation, and all of them are used to demonstrate the effectiveness of this method.

Table 2. Chosen conditions for generating "test data" conditions with asterisks are used for adaptation.

No.	Normalized N_L	No.	Normalized N_L
1 *	0.5	6	0.75
2	0.55	7 *	0.8
3 *	0.6	8	0.85
4	0.65	9 *	0.9
5 *	0.7	10 *	1

Among the six asterisked conditions, the one with $N_L = 1$ is assumed to be the design point (DP) of the engine. To keep the DP unchanged, this condition has to be chosen as the sticking point for performance adaptation.

3.3. Actual Performance Estimation

For each measured condition, all performance parameters can be estimated by solving Equation (3), where e should converge to zero. At the beginning, all parameters in Equation (3) should be specified. In this case, the parameters in vector z are listed in Table 1. Accordingly, the parameters in x and e are listed in Tables 3 and 4, respectively.

Table 3. Selected guess parameters.

No.	Guess	Symbol
1	Fan pressure ratio	π_{fan}
2	Fan isentropic efficiency	η_{fan}
3	Booster pressure ratio	π_{bst}
4	Booster isentropic efficiency	η_{bst}
5	HPC pressure ratio	π_{hpc}
6	HPC isentropic efficiency	η_{hpc}
7	HPT pressure ratio	π_{hpt}
8	HPT efficiency	η_{hpt}
9	LPT pressure ratio	π_{lpt}
10	LPT efficiency	η_{lpt}
11	Total air flow	W_{tot}
12	Bypass ratio	BPR

Table 4. Selected error parameters.

No.	Error	Symbol
1	Prediction error of P_{13}	$e_{P_{13}}$
2	Flow imbalance in duct nozzle	$e_{W_{duct}}$
3	Prediction error of P_{26}	$e_{P_{26}}$
4	Prediction error of T_{26}	$e_{T_{26}}$
5	Prediction error of P_3	e_{P_3}
6	Prediction error of T_3	e_{T_3}
7	Prediction error of P_{45}	$e_{P_{45}}$
8	Prediction error of T_{45}	$e_{T_{45}}$
9	Flow imbalance in core nozzle	$e_{W_{core}}$
10	Prediction error of T_5	e_{T_5}
11	Torque imbalance of HP spool	e_{TorqH}
12	Torque imbalance of LP spool	e_{TorqL}

With all the parameters in Equation (3) specified, the Newton-Raphson method was applied to calculate the true value of the guess parameters. To check whether the final solutions are converged, the infinity norm of every e, as defined in Remark 2, is listed in Table 5. It can be seen that the convergence criterion is satisfied at all six conditions when $\sigma = 1 \times 10^{-6}$.

Table 5. Convergence check.

i	$\|e_i\|_\infty$	i	$\|e_i\|_\infty$
1	5.914×10^{-7}	4	9.976×10^{-7}
2	8.952×10^{-7}	5	8.2×10^{-7}
3	8.242×10^{-7}	6	5.68×10^{-7}

As mentioned in Section 2.2.1, once a converged solution is obtained, the four characteristic parameters for all the concerned components can be calculated.

3.4. Determination of Scaling Factors

For each to-be-adapted component, scaling factor sets $\{\zeta'_1, \zeta'_2, \ldots, \zeta'_6\}$ at the six measured conditions can be calculated according to Section 2.2.2. Subsequently, SF sets $\{\zeta_1, \zeta_2, \ldots, \zeta_m\}$ of all the initial speed lines can be determined according to the procedure described in Section 2.2.3 (m represents how many speed lines the initial characteristic map has). Figure 7 plots all the scaling factors of each component, where hollow points denote the result of the six measured conditions, while asterisks represent the result of the initial speed lines whose speed values are $\{n_1, n_2, \ldots, n_m\}$. It can be seen that the number of speed lines m varies for different components. For example, $m = 12$ for the high-pressure compressor while $m = 10$ for the fan. Additionally, $\zeta_6' = [1,1,1]^T$ holds for all components referring to the DP points of the components.

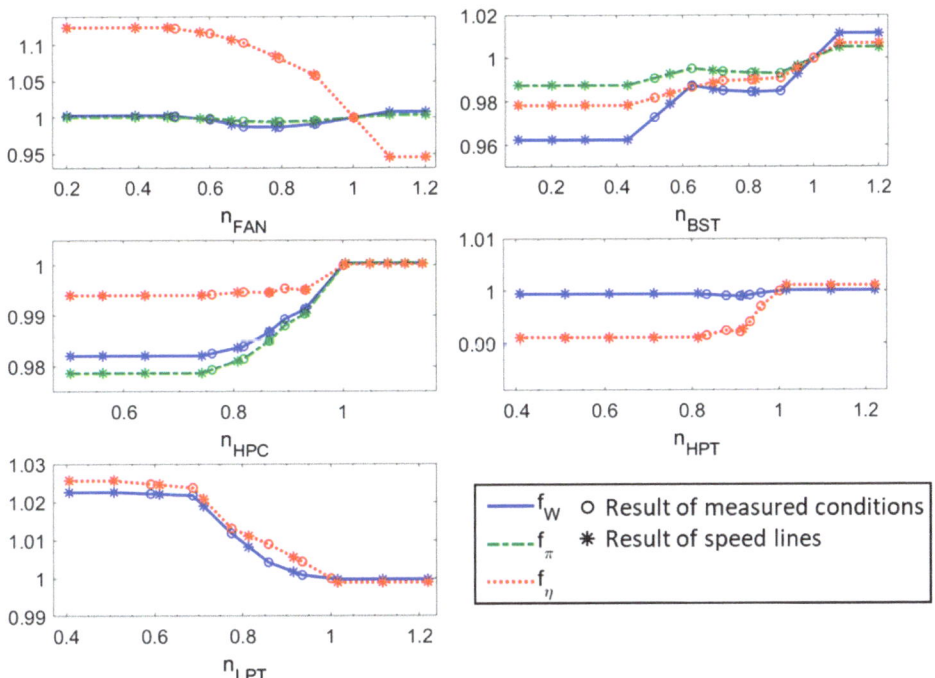

Figure 7. Scaling factors of all concerned components.

3.5. Adaptation Results

With all the above scaling factors applied to the initial maps, the shifts of the maps are displayed in Figure 8. The solid and dashed curves correspond to the maps before and after the adaptation, respectively. The hollow pentagrams shown in Figure 8 mark the design points whose positions are unchanged during the adaptation.

In order to display the effectiveness of the proposed approach, the initial and the adapted characteristic maps are employed to predict the engine performance separately. The prediction errors of the model engine using the initial set of maps covering all 10 conditions are plotted in Figure 9, while the prediction errors of the model engine using the adapted maps are shown in Figure 10. Among all 11 measurable parameters for validation, Figure 9 shows that W_f has a maximum prediction error of about 9% and F_N has a prediction error of about 4%, which greatly affects the accuracy and reliability of the model. While Figure 10 shows the maximum error of about 0.6%, and the prediction error of F_N is significantly reduced to about 0.5%, which indicates good adaptation results. The meaning of these measurable parameters can be acquired in Table 1.

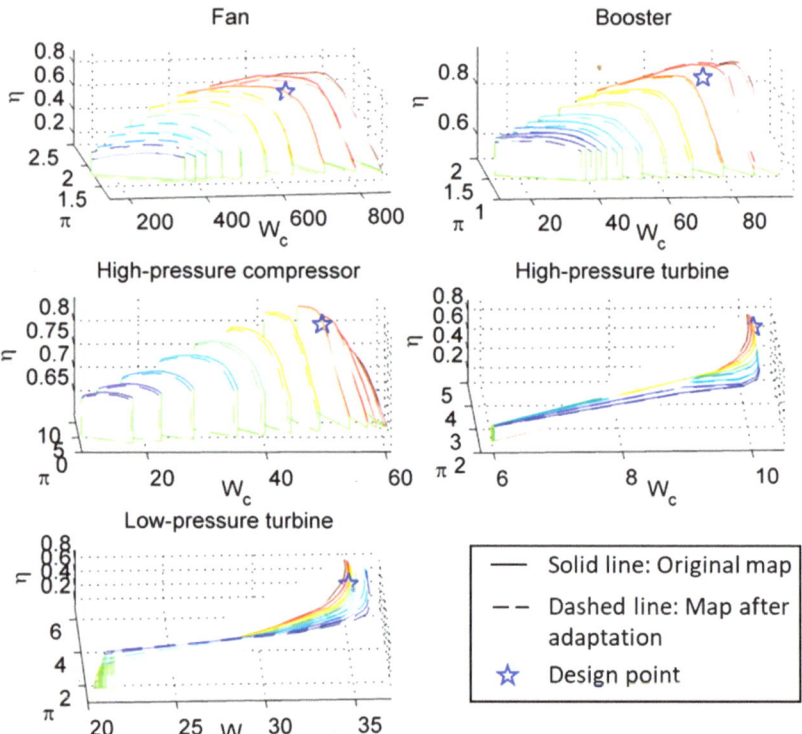

Figure 8. Map shift of all concerned components.

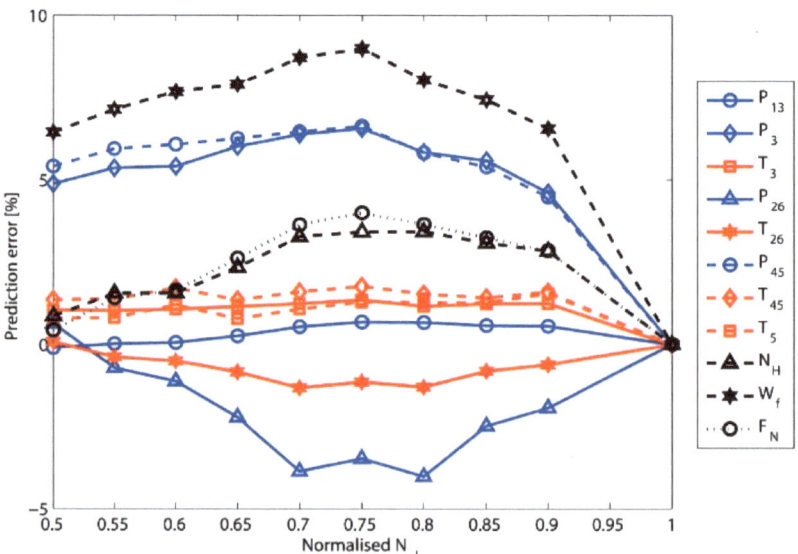

Figure 9. Prediction error of measurable parameters before the adaptation.

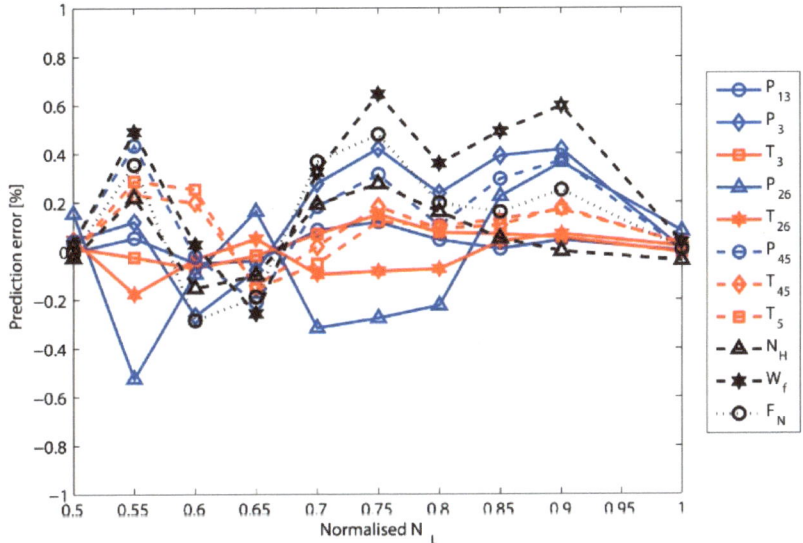

Figure 10. Prediction error of measurable parameters after the adaptation.

For further validation, five important performance parameters, i.e., bypass ratio BPR, fan pressure ratio π_{fan}, booster pressure ratio π_{bst}, HPC pressure ratio π_{hpc}, and turbine temperature ratio T_5/T_4, are selected to show the adaption results. The prediction errors before and after the adaptation are plotted in Figures 11 and 12, respectively. It can be seen that these parameters, after adaptation, reach a relatively high level of accuracy, which can support accurate prediction of engine performance.

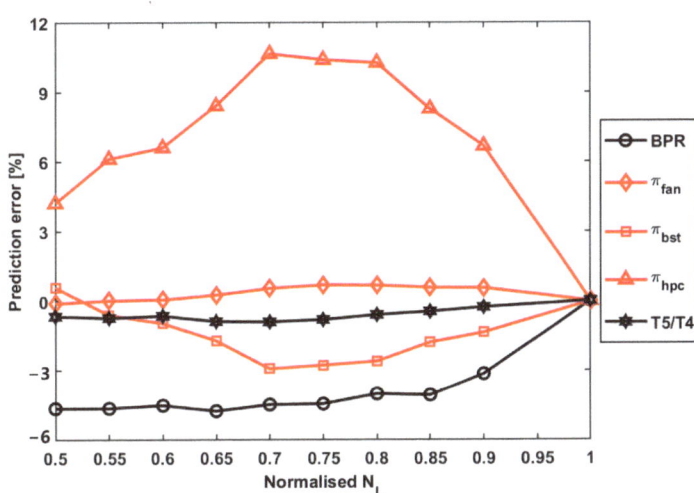

Figure 11. Prediction error of performance parameters before the adaptation.

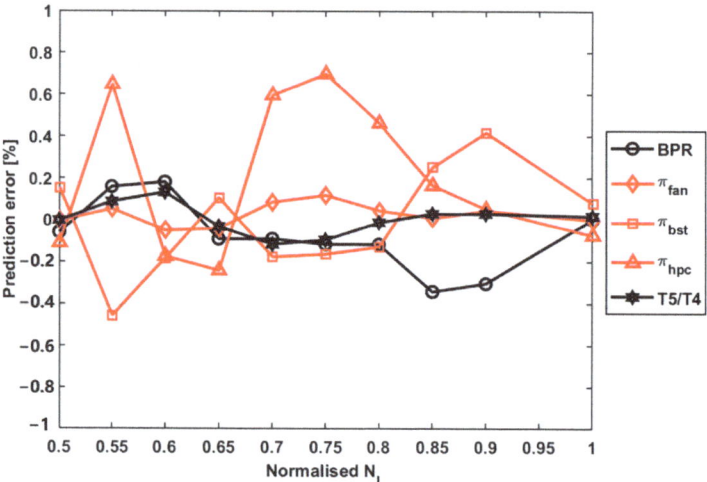

Figure 12. Prediction error of performance parameters after the adaptation.

4. Discussions

4.1. Low Time Complexity

The heuristic approaches for performance adaptation often consume plenty of time. Essentially, heuristic approaches are searching methods that need to try different solutions iteratively until a satisfactory one is found. The time complexity of the approaches mainly depends on how many attempts it takes. Unfortunately, the number of attempts is uncertain because these approaches contain stochastic processes. In practice, both the initial error and the search range may influence the number of attempts. As for the Genetic Algorithm-based adaptation addressed in [6], for example, a four-point adaptation case (20 generations with a population size of 50) took about 50 attempts, i.e., at least 50,000 runs of the model engine. Even though the method was improved by specifying the search range, the number of attempts was still more than 20 (at least 20,000 runs).

Compared to the heuristic approaches, the proposed deterministic approach has quite a lower time complexity. Apart from the actual performance estimation (at most 20 runs for each condition), even there is no need to run the model engine. The computation time of the interpolation algorithm is linearly proportional to the number of speed lines, which is very short and can be ignored in the adaptation process. For a four-point adaptation, the proposed approach needs, at most, 80 runs of the model engine.

4.2. Guaranteed Accuracy

According to Sections 2.2.1 and 2.2.2, the SF sets at each measured condition can precisely reflect the corresponding measured performance. That is, the accuracy of the model at a particular measured point only depends on measurement accuracy. There is no newly introduced error until the stage to determine SF sets for all the speed lines. At this stage, prediction error will be introduced more or less due to the interpolation algorithm. Fortunately, under the assumption that the component maps and their shifts are smooth enough, the prediction error will be maintained at a relatively low level. In extreme circumstances, we can also add extra speed lines into characteristic maps to improve prediction accuracy. As for the turbofan application, satisfactory accuracy was achieved (max error < 1%) without inserting any extra speed line. Anyway, more accuracy is meaningless for this case because 1% can be covered by measurement uncertainty.

4.3. Sure Solution

Unlike heuristic approaches that have uncertain results, the deterministic approach always has a sure result. As mentioned before, all published approaches for the multi-point adaptation are based on the heuristic algorithm. Using such an algorithm that contains a stochastic process may yield an uncertain solution. As a result, plenty of attempts are required to obtain an acceptable solution, which indeed consumes a great deal of time. Moreover, there is no guarantee that an acceptable solution will be found. In terms of the proposed approach, however, all procedures throughout this paper are deterministic. Consequently, this approach always has a sure solution.

5. Conclusions

This paper presents a deterministic approach for multiple-point performance adaptation, specifically designed to calibrate the off-design performance of gas turbine models. In contrast to the well-developed optimization-based methods, this approach advocates for determining the shift of the initial characteristic maps in a straightforward manner. This concept is realized by calculating each Scaling Factor (SF) set at its corresponding measured condition and subsequently employing an interpolation-based algorithm to ascertain the precise shift of each map.

As discussed in Section 4, this approach offers three primary advantages:

(1) The time complexity of this approach is notably low.
(2) The prediction error of the adapted model can be guaranteed at a relatively low level.
(3) In contrast to optimization-based approaches that entail stochastic procedures, the deterministic approach eliminates the need for a searching procedure and consistently delivers a reliable outcome.

The effectiveness of the proposed approach is demonstrated through its application to a model high-bypass turbofan engine. Following the adaptation process, a significant enhancement in the prediction accuracy of the model is observed, with the prediction error index decreasing from approximately 9% to about 0.6%. Owing to the modification of component maps, the method can be easily transplanted to other engine types and configurations. And with its assured results and minimal computational burden, this approach holds promise as a valuable tool for the calibration of gas turbine models.

The limitation of the proposed deterministic approach is that it is currently suitable for steady-state performance adaptation. The related dynamic performance adaptation and transient analysis need further research. Additionally, only simulated data are used in this paper to illustrate the effectiveness of the method. Applying real engine measurements for verification will be considered in future research. What is more, the impact of measurement uncertainties on model predictions is a direction worth exploring. Uncertainty analysis techniques such as Monto Carlo simulations are an optional solution for conducting future research.

Author Contributions: Conceptualization, Z.J. and S.Y.; methodology, Z.J. and S.Y.; software, S.Y.; validation, M.Z.; formal analysis, X.W.; investigation, Z.J.; resources, S.Y.; data curation, Z.J. and S.Y.; writing—original draft preparation, Z.J. and S.Y.; writing—review and editing, X.W. and M.Z.; visualization, Z.J. and M.Z.; supervision, X.W.; project administration, X.W. and S.Y. All authors have read and agreed to the published version of the manuscript.

Funding: This research received no external funding.

Data Availability Statement: The data that support the findings of this study are available from the corresponding author upon reasonable request.

Conflicts of Interest: The authors declare no conflicts of interest.

Nomenclature

DP	design point
GTML-E	Gas Turbine Modeling Library for Education
HPC	high-pressure compressor
HPT	high-pressure turbine
LPT	low-pressure turbine
SF	scaling factor
n	corrected relative rotational speed
W	flow capacity
π	pressure ratio
η	isentropic efficiency
γ	contour line of the rotational speed
ζ_W	scaling factor of flow capacity
ζ_π	scaling factor of pressure ratio
ζ_η	scaling factor of isentropic efficiency
ζ	vector of the above three scaling factors
n'	speed value at measured conditions
ζ'	scaling vector at measured conditions

References

1. Miao, K.; Wang, X.; Zhu, M.; Yang, S.; Pei, X.; Jiang, Z. Transient Controller Design Based on Reinforcement Learning for a Turbofan Engine with Actuator Dynamics. *Symmetry* **2022**, *14*, 684. [CrossRef]
2. Cao, J. Status, Challenges and Perspectives of Aero-Engine Simulation Technology. *J. Propuls. Technol.* **2018**, *39*, 961–970.
3. Hanachi, H.; Mechefske, C.; Liu, J.; Banerjee, A.; Chen, Y. Performance-based gas turbine health monitoring, diagnostics, and prognostics: A survey. *IEEE Trans. Reliab.* **2018**, *67*, 1340–1363. [CrossRef]
4. Jiang, Z.; Yang, S.; Wang, X.; Long, Y. An Onboard Adaptive Model for Aero-Engine Performance Fast Estimation. *Aerospace* **2022**, *9*, 845. [CrossRef]
5. Mamala, J.J.; Praznowski, K.; Kołodziej, S.; Ligus, G. The Use of Short-Term Compressed Air Supercharging in a Combustion Engine with Spark Ignition. *Int. J. Automot. Mech. Eng.* **2021**, *18*, 8704–8713. [CrossRef]
6. Cruz-Manzo, S.; Panov, V.; Zhang, Y. Gas path fault and degradation modelling in twin-shaft gas turbines. *Machines* **2018**, *6*, 43. [CrossRef]
7. Panov, V. Auto-tuning of real-time dynamic gas turbine models. Turbo Expo: Power for Land, Sea, and Air. In Proceedings of the ASME Turbo Expo 2014: Turbine Technical Conference and Exposition (GT2014-25606, V006T06A004), Düsseldorf, Germany, 16–20 June 2014.
8. Hayes, R.; Dwight, R.; Marques, S. Reducing parametric uncertainty in limit-cycle oscillation computational models. *Aeronaut. J.* **2017**, *121*, 940–969. [CrossRef]
9. Stamatis, A.; Mathioudakis, K.; Papailiou, K.D. Adaptive Simulation of Gas Turbine Performance. *J. Eng. Gas Turbines Power* **1990**, *112*, 168. [CrossRef]
10. Lambiris, B.; Mathioudakis, K.; Stamatis, A.; Papailiou, K. Adaptive modeling of jet engine performance with application to condition monitoring. *J. Propuls. Power* **1994**, *10*, 890–896. [CrossRef]
11. Li, Y.G.; Marinai, L.; Gatto, E.L.; Pachidis, V.; Philidis, P. Multiple-point adaptive performance simulation tuned to aeroengine test-bed data. *J. Propuls. Power* **2009**, *25*, 635–641. [CrossRef]
12. Li, Y.G.; Ghafir, M.A.; Wang, L.; Singh, R.; Huang, K.; Feng, X. Nonlinear Multiple Points Gas Turbine Off-Design Performance Adaptation Using a Genetic Algorithm. *J. Eng. Gas Turbines Power* **2011**, *133*, 42–50. [CrossRef]
13. Li, Y.G.; Abdul Ghafir, M.F.; Wang, L.; Singh, R.; Huang, K.; Feng, X.; Zhang, W. Improved Multiple Point Nonlinear Genetic Algorithm Based Performance Adaptation Using Least Square Method. *J. Eng. Gas Turbines Power* **2012**, *134*, 031701. [CrossRef]
14. Alberto Misté, G.; Benini, E. Turbojet engine performance tuning with a new map adaptation concept. *J. Eng. Gas Turbines Power* **2014**, *136*, 071202. [CrossRef]
15. Tsoutsanis, E.; Li, Y.-G.; Pilidis, P.; Newby, M. Non-linear model calibration for off-design performance prediction of gas turbines with experimental data. *Aeronaut. J.* **2017**, *121*, 1758–1777. [CrossRef]
16. Tsoutsanis, E.; Meskin, N.; Benammar, M.; Khorasani, K. A component map tuning method for performance prediction and diagnostics of gas turbine compressors. *Appl. Energy* **2014**, *135*, 572–585. [CrossRef]
17. Pang, S.; Li, Q.; Feng, H.; Zhang, H. Joint steady state and transient performance adaptation for aero engine mathematical model. *IEEE Access* **2019**, *7*, 36772–36787. [CrossRef]
18. Baklacioglu, T.; Turan, O.; Aydin, H. Dynamic modeling of exergy efficiency of turboprop engine components using hybrid genetic algorithm-artificial neural networks. *Energy* **2015**, *86*, 709–721. [CrossRef]
19. Dinc, A. Optimization of a Turboprop UAV for Maximum Loiter and Specific Power Using Genetic Algorithm. *Int. J. Turbo Jet-Engines* **2016**, *33*, 265–273. [CrossRef]

20. Gendreau, M.; Potvin, J.Y. Metaheuristics in combinatorial optimization. *Ann. Oper. Res.* **2005**, *140*, 189–213. [CrossRef]
21. Bianchi, L.; Dorigo, M.; Gambardella, L.M.; Gutjahr, W.J. A survey on metaheuristics for stochastic combinatorial optimization. *Nat. Comput.* **2009**, *8*, 239–287. [CrossRef]
22. Li, Y.G.; Pilidis, P.; Newby, M.A. An adaptation approach for gas turbine design-point performance simulation. *J. Eng. Gas Turbines Power* **2006**, *128*, 789–795. [CrossRef]
23. Roth, B.; Doel, D.L.; Mavris, D.N.; Beeson, D. High-accuracy matching of engine performance models to test data. In Proceedings of the ASME Turbo Expo 2003: Power for Land, Sea and Air (GT2003-38784), Atlanta, GA, USA, 16–19 June 2003; pp. 129–137.
24. Ortega, J.M.; Rheinboldt, W.C. Iterative Solution of Nonlinear Equations in Several Variables. *Math. Comput.* **1970**, *25*, 398.
25. Kurzke, J.; Halliwell, I. *Propulsion and Power: An Exploration of Gas Turbine Performance Modeling*; Springer International Publishing: Cham, Switzerland, 2018.
26. Yang, S. GTML-E: Gas Turbine Modeling Library for Education. 2019. Available online: https://github.com/xjysb/GTML_E (accessed on 1 January 2024).
27. Chapman, J.W.; Lavelle, T.M.; May, R.; Litt, J.S.; Guo, T.-H. Propulsion System Simulation Using the Toolbox for the Modeling and Analysis of Thermodynamic Systems (T MATS). In Proceedings of the 50th AIAA/ASME/SAE/ASEE Joint Propulsion Conference (AIAA 2014-3929), Cleveland, OH, USA, 28–30 July 2014; pp. 1–14.
28. Visser, W.P.; Broomhead, M.J. GSP, a Generic Object-Oriented Gas Turbine Simulation Environment. In Proceedings of the ASME Turbo Expo 2000: Power for Land, Sea, and Air Turbo (V001T01A002), Munich, Germany, 8–11 May 2000.
29. Dyson, R.J.; Doel, D.L. CF6-80 condition monitoring-the engine manufacturer's involvement in data acquisition and analysis. In Proceedings of the 20th AIAA//SAE/ASEE Joint Propulsion Conference, Cincinnati, OH, USA, 11–13 June 1984. [CrossRef]

Disclaimer/Publisher's Note: The statements, opinions and data contained in all publications are solely those of the individual author(s) and contributor(s) and not of MDPI and/or the editor(s). MDPI and/or the editor(s) disclaim responsibility for any injury to people or property resulting from any ideas, methods, instructions or products referred to in the content.

Article

Investigation into Detection Efficiency Deviations in Aviation Soot and Calibration Particles Based on Condensation Particle Counting

Liang Chen [1], Quan Zhou [1], Guangze Li [2], Liuyong Chang [2,*], Longfei Chen [2,*] and Yuanhao Li [3]

[1] School of Optics and Electronic Technology, China Jiliang University, Hangzhou 310018, China; 18072835032@163.com (Q.Z.)
[2] International Innovation Institute, Beihang University, Hangzhou 311115, China; liguangze@buaa.edu.cn
[3] School of Mechanical and Electrical Engineering, China Jiliang University, Hangzhou 310018, China; 1416906503@163.com
* Correspondence: changliuyong@buaa.edu.cn (L.C.); chenlongfei@buaa.edu.cn (L.C.)

Abstract: Aviation soot constitutes a significant threat to human well-being, underscoring the critical importance of accurate measurements. The condensation particle counter (CPC) is the primary instrument for quantifying aviation soot, with detection efficiency being a crucial parameter. The properties of small particles and the symmetry of their growth pathways are closely related to the detection efficiency of the CPC. In laboratory environments, sodium chloride is conventionally utilized to calibrate the CPC's detection efficiency. However, aviation soot exhibits distinctive morphological characteristics compared to the calibration particles, leading to detection efficiencies obtained from calibration particles that may not be applicable to aviation soot. To address this issue, a quantitative study was performed to explore the detection efficiency deviations between aviation soot and calibration particles. The experiment initially utilized a differential mobility analyzer to size select the two types of polydisperse particles into monodisperse particles. Subsequently, measurements of the separated particles were performed using the TSI Corporation's aerosol electrometer and a rigorously validated CPC (BH-CPC). These allowed for determining the detection efficiency deviation in the BH-CPC for the two types of particles at different particle sizes. Furthermore, the influence of the operating temperature of the BH-CPC on this detection efficiency deviation was investigated. The experimental results indicate a significant detection efficiency deviation between aviation soot and sodium chloride. In the range of 10–40 nm, the absolute detection efficiency deviation can reach a maximum of 0.15, and the relative deviation can reach a maximum of 0.75. And this detection efficiency deviation can be reduced by establishing a relevant relationship between the detection efficiency of the operating temperature and the calibration temperature. Compared to the saturated segment calibration temperature of 50 °C, the aviation soot detection efficiency is closer to the sodium chloride detection efficiency at the calibration temperature of 50 °C when the saturated segment operates at a temperature of 45 °C. These studies provide crucial theoretical guidance for enhancing the precision of aviation soot emission detection and establish a foundation for future research in monitoring and controlling soot emissions within the aviation sector.

Keywords: aviation soot; condensation particle counting; particle measurement; detection efficiency; operating temperature

Citation: Chen, L.; Zhou, Q.; Li, G.; Chang, L.; Chen, L.; Li, Y. Investigation into Detection Efficiency Deviations in Aviation Soot and Calibration Particles Based on Condensation Particle Counting. *Symmetry* **2024**, *16*, 244. https://doi.org/10.3390/sym16020244

Academic Editor: Sergei D. Odintsov

Received: 9 January 2024
Revised: 10 February 2024
Accepted: 14 February 2024
Published: 16 February 2024

Copyright: © 2024 by the authors. Licensee MDPI, Basel, Switzerland. This article is an open access article distributed under the terms and conditions of the Creative Commons Attribution (CC BY) license (https://creativecommons.org/licenses/by/4.0/).

1. Introduction

Carbonaceous particles constitute the predominant constituents of pollutant emissions from aviation engines [1]. Given the predominantly high-altitude operation of aviation power systems, aviation soot plays a pivotal role as a major contributor to carbonaceous particles present in the upper troposphere and lower stratosphere [2]. These particles, through their capacity to absorb and scatter sunlight, disrupt the radiation balance of the atmosphere, thus exerting a discernible influence on the Earth's climate [3,4]. Furthermore, significant particle emissions are generated during aircraft takeoff and landing, exacerbating pollution in the vicinity of airports and posing potential health hazards to human populations [5]. Projections for the next two decades suggest a possible threefold increase in air transport, intensifying these environmental problems [6]. In light of these circumstances, precise quantification of the number concentration of carbonaceous particles emitted by aviation is crucial for guiding regulatory and control measures.

Currently, the detection of aviation soot emissions is predominantly carried out using condensation particle counters (CPCs) [7]. CPCs measure carbon soot particles, which have undergone condensation-induced growth, through the utilization of light scattering. The detection efficiency stands as a crucial parameter for CPCs, and its calibration is performed within a laboratory environment. In a laboratory setting, Sakurai et al. [8] calibrated the condensation particle counter using polystyrene spheres and extended the traceability of the calibration concentration to 1 P/cm^3. Krasa et al. [9] calibrated the condensation particle counter using atomized inorganic salt particles and simplified the calibration method through analysis. Hermann et al. [10] utilized sodium chloride particles for calibrating the condensation particle counter under conditions involving both water and butanol as working fluids. Wang X et al. [11] calibrated the condensation particle counter using sodium chloride particles and assessed its performance in vehicle emission detection. Liu P. S. K. and Deshler T. [12] calibrated the condensation particle counter using sodium chloride particles and conducted a comparative analysis with a scanning mobility particle sizer. In summary, CPC is commonly calibrated using sodium chloride in laboratory experiments.

As the particle size decreases, the challenge of particle growth through condensation to a detectable range becomes more pronounced, underscoring the critical importance of detection efficiency for small-sized particles in CPCs. The symmetry of the condensation section affects the flow field distribution of sheath gas and sample gas within the condensation section, subsequently influencing the condensation and growth of small particles. This indicates a highly significant relationship between the design of the condensation section's symmetry and the particle detection efficiency of the CPC. Therefore, the research findings presented in this paper, regarding the CPC detection efficiency and its influencing factors, hold crucial guiding implications for the symmetry design of the CPC's condensation section. The detection efficiency of small particles is also profoundly influenced by the inherent characteristics, including their morphology and chemical properties [13]. Aviation soot exhibits distinct disparities compared with calibration particles (sodium chloride) due to their irregular shapes and rough surfaces. This would impact the detection efficiency of the CPC, resulting in a reduction in the accuracy of aviation soot measurements. Research on the detection efficiency bias between these two types of particles is rare. In addressing the aforementioned issues, this paper quantitatively determined the detection efficiency deviation in different particles through experimental methods. Subsequently, the factors influencing this deviation were investigated, leading to the identification of a method to mitigate such discrepancies. These studies offer crucial theoretical guidance for enhancing the accuracy of detecting aviation soot emissions, laying the foundation for future research focused on monitoring and controlling aviation emissions. This contributes to the development of sustainable aviation practices.

2. Experimental System and Data Processing
2.1. Fundaments of Condensation Particle Counting

The condensation particle counter (CPC) functions on the principle of heterogeneous condensation, which involves the enlargement of nanoscale particles to the micrometer range before quantification. It consists of the optical particle counter and the particle growth chamber, as shown in Figure 1. In the course of its operation, aerosol particles are introduced into the particle growth chamber, where supersaturated vapor undergoes heterogeneous condensation on their surfaces, resulting in an augmentation of the particle size. These enlarged particles are then quantified using the optical particle counter.

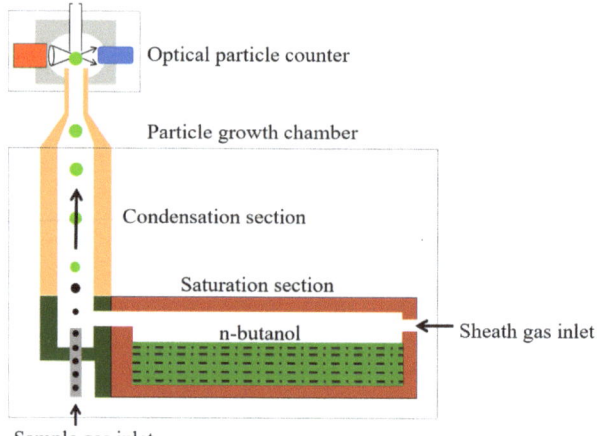

Figure 1. Schematic diagram of the structure of a condensation particle counter.

The optical particle counter is a device used for counting individual particles, operating on the principle of light scattering. Figure 2 presents a schematic depiction of the optical particle counter's structure, primarily comprising an incident light unit, a scattered light collection unit, and a signal processing circuit. During its functioning, a laser emits incident light, which is precisely directed and focused by a lens to create a narrow laser beam. As aerosol particles traverse this laser beam, the scatter light emits from the particles. A fraction of the scattered light is captured by a reflector and reflected towards a photodetector. The photodetector translates the light signal into an electrical current signal, which is subsequently processed by the signal processing circuit. This circuit generates pulse signals, which are then subjected to statistical analysis to ascertain the particle concentration [14].

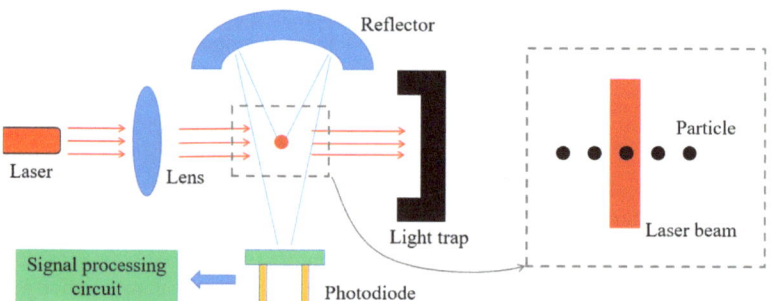

Figure 2. Schematic diagram of the structure of an optical particle counter.

Within a CPC, particle growth relies on the principle of heterogeneous condensation, entailing nucleation and subsequent growth of particles through the utilization of dissimilar substances as nuclei. This process transpires in a non-equilibrium state, with the presence of mass transport effects [15]. When gradients exist within thermodynamic systems, such as temperature or concentration gradients, the system departs from equilibrium. Important factors contributing to this non-equilibrium state include particle surface roughness, surface free energy, and vapor saturation.

In a vapor-containing system, its non-equilibrium state can be characterized by the saturation ratio (S), expressed as the ratio of partial pressure of vapor (P_V) to the saturation vapor pressure (P_{sat}) at the system's temperature. When S exceeds 1, the system enters a supersaturated state. Conversely, when S falls below 1, the system transitions to a subsaturated state, indicating a shortage of vapor [16–18]. The calculation formula for S is given by Equation (1) [19]:

$$S = \frac{P_V}{P_{sat}(T)} \quad (1)$$

The critical nucleation radius of the particle can be calculated using Equation (2) [20]:

$$r^* = \frac{2\sigma_s}{(M_l v_l / N_A) k_B \cdot T \cdot \ln(S)} \quad (2)$$

where r^* is the critical nucleation radius of the particle, σ_s represents the surface tension (N/m), M_l is the molar mass of the liquid (kg/mol), v_l is the specific volume of the liquid (m^3/kg), N_A is Avogadro's constant (approximately 6×10^{23} /mol), k_B is the Boltzmann constant (1.38×10^{-23} J/K), T signifies the temperature in Kelvin (K), and S indicates the saturation ratio. According to Equation (2), it is evident that the presence of surface curvature on aerosol particles necessitates a higher equilibrium saturation vapor pressure to maintain surface equilibrium. Smaller particles exhibiting greater surface curvature require a greater equilibrium saturation vapor pressure for condensation to occur. Under the same saturation conditions, heterogeneous condensation may not transpire on the surface of small-sized particles. This occurs when the vapor pressure within the system is lower than the saturation vapor pressure on the surface of small-sized particles, thereby impeding condensation growth on the particle surface. Consequently, the detection efficiency for small-sized particles is diminished.

$$\cos\theta = \phi_s \cdot m_\infty - \frac{\tau \cdot \phi_l}{\sigma_s \cdot R_p \cdot \tan\phi} \quad (3)$$

Equation (3) [21] delineates the functional relationship between particle surface roughness and the contact angle of condensation droplets on the surface. In this equation, θ represents the microscopic contact angle between the droplet nucleus and the particle, m_∞ represents the cosine of the macroscopic contact angle, ϕ_s represents the surface roughness factor of the particle, ϕ_l represents the line roughness factor of the particle, τ represents the three-phase line tension, R_P represents the particle radius, and ϕ represents the angle between the line connecting the droplet nucleus center to the particle center, and the line connecting the particle center to a three-phase contact point. Equation (3) indicates that as the particle roughness increases, the contact angle decreases, which promotes condensation growth, and subsequently affects the detection efficiency of various particles.

2.2. Experimental Setup

Figure 3 illustrates a schematic diagram of the experimental system used in this study. Firstly, the particle generator produces the desired polydisperse aerosol particles, which is then carried by the clean air into the drying tube. The drying tube is filled with silica, leveraging the principle of diffusion capture to effectively remove moisture. Silica exhibits a strong affinity for water molecules, ensuring thorough drying of the polydisperse aerosol particles. Subsequently, the particles pass through a dilution bridge and enter the differential mobility analyzer (DMA). The dilution bridge is composed a filter and a valve. The filter prevents particles while maintaining a supply of clean air. The dilution ratio can be regulated by adjusting the valve resistance, thus controlling the particle source concentration. The mixing chamber promotes turbulent mixing, ensuring a uniform distribution of particles. Additionally, a bypass fitted with an additional filter is connected to maintain flow balance. The DMA selectively separates particles of specific sizes, transforming the polydisperse aerosol particles into monodisperse aerosol particles.

Ultimately, the monodisperse aerosol particles are simultaneously directed into the self-developed condensation particle counter (BH-CPC), the TSI aerosol electrometer (TSI-AE3068B, TSI Incorporated, Minneapolis, MN, USA), and the TSI condensation particle counter (TSI-CPC3775) for measurement. The connecting pipes of the entire experimental system are constructed from polytetrafluoroethylene (PTFE) with a carbon impregnation base [22].

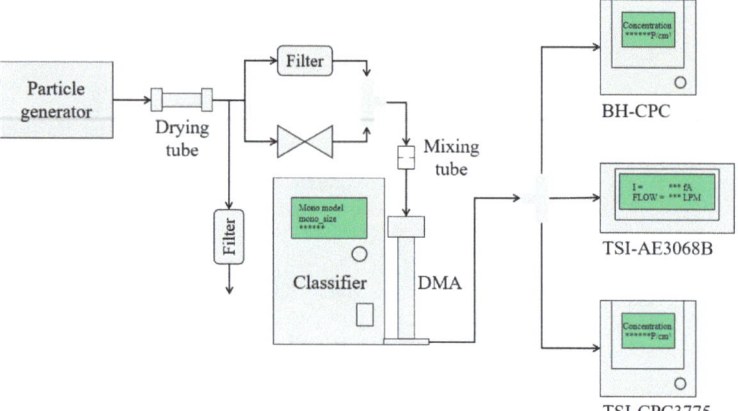

Figure 3. Schematic diagram of the experimental system.

2.2.1. Particle Generator

The experiment employed two types of particle generators: an atomizing aerosol generator and a micro-combustion aerosol generator (MiniCAST).

The atomizing aerosol generator (model: AG-21, Shanghai Aoxuan Measurement Technology Co., Ltd., Shanghai, China) demonstrates the capability to produce diverse particles using various solutions. Initially, compressed air is introduced into the generator through micro-holes, generating a jet-like airflow used to atomize the liquid for aerosol generation. Coarse particles are redirected upon collision with the generator walls and recirculated, while fine particles are expelled through the top outlet of the generator. This process results in the production of sub-micron-sized aerosol particles [23].

The MiniCAST (model: MOD6204C, Jing Ltd., Zollikofen, Switzerland) generates carbonaceous particles through diffusion flames in the pyrolysis process. Within the generator, the chemical reaction between the fuel and oxidizer is deliberately left incomplete. The quenching airflow at a specific flame height extinguishes the upper flame, preventing further participation of the carbonaceous particles produced by the diffusion flame in the combustion process. Consequently, a high concentration of particles is generated within the combustion chamber. Furthermore, the quenching airflow ensures the stability of the particles production process and suppresses the condensation of carbonaceous particles in the particle stream [24]. The MiniCAST offers flexible operational parameters, enabling control over the particle size range, morphology, and organic composition of the carbonaceous particles. It also exhibits high stability and reproducibility [25]. The morphology, organic carbon composition, density, and Raman spectra of the carbonaceous particles generated by the MiniCAST closely resemble those of aviation soot [26], making it suitable as an aviation soot generator [27]. Moreover, collecting carbonaceous particles from aircraft engine exhaust necessitates complex and costly facilities [28–31], making it practically unfeasible to replicate in a laboratory setting. Therefore, the present study utilized MiniCAST-generated carbonaceous particles as a viable substitute for aviation soot. In the experiment, Minicast generates polydisperse carbonaceous particles with a center diameter of 26 nm, as illustrated in the probability distribution shown in Figure 4. These polydisperse particles will undergo size selection through DMA to monodisperse particles.

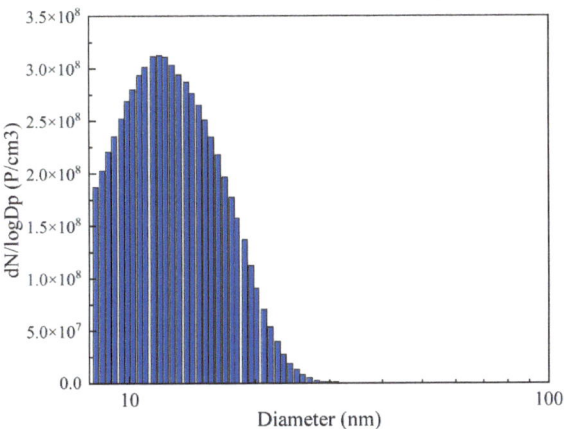

Figure 4. Probability distribution of carbon soot.

2.2.2. The Differential Mobility Analyzer

The differential mobility analyzer (DMA, model: BMI-DMA2101C, Brechtel Manufacturing Incorporated, Hayward, CA, USA) exhibits the capability to selectively segregate particles according to their specific sizes. The process commences with the polydisperse aerosol being charged to equilibrium by a radiation source before entering the DMA through inlet ports. Sheath gas is introduced from the top and mixed with the aerosol particles. During operation, a negative voltage is applied to the long central column of the DMA, while the outer shell is grounded, thus creating an electric field within the DMA. As the aerosol particles descend through the column, they experience a force resulting from the electric field. This force leads to accumulation of particles with higher electric mobility near the upper section of the column, while smaller particles gather toward the lower portion. Particles that successfully traverse the narrow gap at the bottom of the DMA are effectively screened and separated, while the remaining particles are carried out from the DMA alongside the sheath gas [32].

The size selection of the DMA is not perfectly monodisperse, but characterized by a narrow size range, as dictated by the DMA transfer function. Rose et al., demonstrated that when the sheath gas-to-sample flow ratio exceeds 5:1, the particles screened by the DMA exhibit perfect monodispersity [33]. Therefore, during the experiment, the sheath gas-to-sample flow ratio was set to 8:1 to ensure the production of highly monodisperse particles.

2.2.3. The Aerosol Electrometer

The aerosol electrometer is internationally recognized as the foremost standard for particle counting (number concentration) [34], and serves as the primary instrument for quantifying aerosol number concentration. The aerosol electrometer consists of a Faraday cup and a low-current measurement module. When aerosol particles carrying a single charge are introduced into the sampling inlet, they are effectively captured by a high-efficiency filter located within the Faraday cup. The interaction with the filter leads to the release of the particle's charge due to the space charge effect, consequently generating a quantifiable electric current. In recent years, notable contributions to the field of aerosol electrometer calibration have been made by national metrology institutes in several countries, including China, the United States, Japan [35]. These studies have played a vital role in advancing the accuracy and reliability of particle concentration measurement, providing valuable technical support of accurately measuring particle concentration using an aerosol electrometer.

2.3. Data Analysis

In the experiment, monodisperse particles of varying sizes were employed, with diameters of 10 nm, 12 nm, 14 nm, 18 nm, 22 nm, 26 nm, 30 nm, 35 nm and 40 nm. Each individual data point was recorded at a data acquisition rate of 1 Hz, spanning a duration of 2 min. The particle number concentration was continuously monitored using a BH-CPC, while the electric current was recorded using the aerosol electrometer of TSI-AE3068B (TSI Incorporated, Minneapolis, MN, USA). The particle number concentration can be obtained from the Equation (4) [36]:

$$N_E = \frac{I}{neQ_E} \quad (4)$$

where N_E represents the number concentration of particles measured by the aerosol electrometer, I denotes the electric current measured by the electrometer, n represents the elementary charge of each particle, e represents the fundamental charge of an electron (i.e., 1.6×10^{-19} Cb), and Q_E represents the flow rate passing through the aerosol electrometer. It is worth noting that the elementary charge of each particle is assumed to be equal to 1. This assumption is based on the fact that in this study the particles with the diameter smaller than 50 nm are primarily capable of acquiring a single charge within the aerosol neutralizer [37,38].

The determination of the CPC's detection efficiency is computed as the ratio of the number concentration measured by the CPC to the number concentration measured by the aerosol electrometer, as illustrated in Equation (5):

$$DE = \frac{N_C}{N_E} \quad (5)$$

where DE represents the detection efficiency, and N_C represents the number concentration measured by the CPC.

The evaluation of the CPC's detection efficiency involves the computation of the mean and standard deviation in the detection efficiencies for each data point. This process is carried out through the utilization of Equations (6) and (7) as follows:

$$\overline{DE} = \frac{DE_1 + DE_2 + \cdots\cdots + DE_t}{T_{DE}} \quad (6)$$

$$S_{DE} = \sqrt{\frac{(DE_1 - \overline{DE})^2 + (DE_2 - \overline{DE})^2 + \cdots\cdots + (DE_t - \overline{DE})^2}{T_{DE}}} \quad (7)$$

where $\overline{DE}$ represents the average detection efficiency, DE_t represents the detection efficiency at time t, S_{DE} represents the standard deviation in the detection efficiencies, and T_{DE} represents the recording duration in seconds. Given the sampling frequency of 1 Hz, the recording duration T_{DE} is equal to the total number of data points for the detection efficiencies.

The comparison assessment of the CPC detection efficiency for aviation soot and calibration particles is conducted by employing both absolute deviation and relative deviation. This procedure is accomplished through the application of Equations (8) and (9) as follows:

$$\Delta = |DE_m - DE_s| \quad (8)$$

$$\varepsilon = \frac{\Delta}{DE_s} \quad (9)$$

where Δ represents the absolute deviation, DE_m represents the detection efficiency for aviation soot, DE_s represents the detection efficiency for calibration particles, and ε represents the relative deviation.

3. Results and Discussion

3.1. Verification of Detection Performance of BH-CPC

Due to uncontrollable parameters such as the working temperature of commercial CPCs, a self-developed condensation particle counter (BH-CPC) was utilized in this study. Before studying detection efficiency deviations in aviation soot and calibration particles, the response performance and counting accuracy of the BH-CPC underwent thorough validation. Figure 5 illustrates the results of the BH-CPC response performance test. In Figure 5a, the raw voltage signals of particles measured by the BH-CPC are displayed, while Figure 5b presents the normalized particle number concentration. Before the experiment, the particle number concentration measured by the BH-CPC was reduced to zero using a filter. Subsequently, at a specific time (0 time in Figure 5), the filter was removed to allow particles to enter. For this experiment, sodium chloride particles with a diameter of 50 nm, were chosen as the test particles. Figure 5a reveals that after the introduction of particles, the first particle voltage signal appeared approximately 2.2 s later, followed by a gradual increase in the density of particle voltage signals. Figure 5b presents the particle number concentration with an interval of 0.1 s calculated using the particle voltage signals in Figure 5a. The results indicate that the particle number concentration initially remained close to zero, gradually increased after 2.5 s, and reached a steady state after 4 s. The particle number concentration was fitted using the Boltzmann function, revealing that the detection efficiency of the particle number concentration by the BH-CPC reached 95% at 3.7 s. This indicates that the response time of BH-CPC used in this study is 3.7 s. Notably, the response time is shorter than the 4 s response time of the TSI-CPC3775.

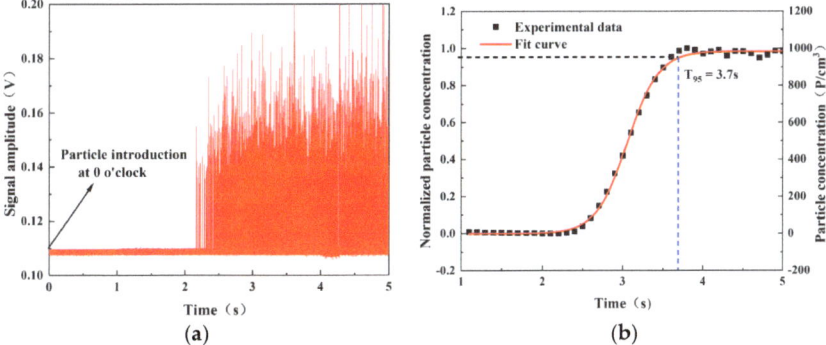

Figure 5. Results of BH-CPC responsiveness performance test: (**a**) raw particle signals; (**b**) transient particle number concentration.

Figure 6 illustrates the counting accuracy test of the BH-CPC. In Figure 6a, the real-time measurement results of both the BH-CPC and TSI-CPC3775 are presented, while Figure 6b provides the counting fitting results between the two instruments. Prior to conducting the experiment, a short conductive hose was used to connect the sampling inlet of the BH-CPC and TSI-CPC3775 in parallel, and a filter was employed to ensure that both instruments exhibited zero counts. Subsequently, the filter was removed to introduce particles. For this experiment, carbon soot particles generated from the burning of mosquito coils were selected. These particles originate from an unstable combustion source, resulting in a wide range of particle concentrations. Figure 6a indicates that the initial particle concentration was relatively high. After approximately 10 min, the mosquito coil was extinguished, resulting in a decrease in particle concentration. At approximately the 30 min mark, the mosquito coil was reignited, leading to fluctuations in the particle number concentration, ranging between 10,000 and 50,000 P/cm^3. Overall, the counting variations in the BH-CPC and TSI-CPC3775 exhibited consistent trends. Figure 6b demonstrates a linear fitting of the counting results between the BH-CPC and TSI-CPC3775, and the red line from Figure 6b represents a regression line. The slope of this red line is 1.03. The closer the slope of this red line is to 1, the smaller the counting error between the two instruments. The R^2 of this red line is 0.9859. The closer the R^2 of this red line is to 1, the higher the similarity between the regression line and the original data.

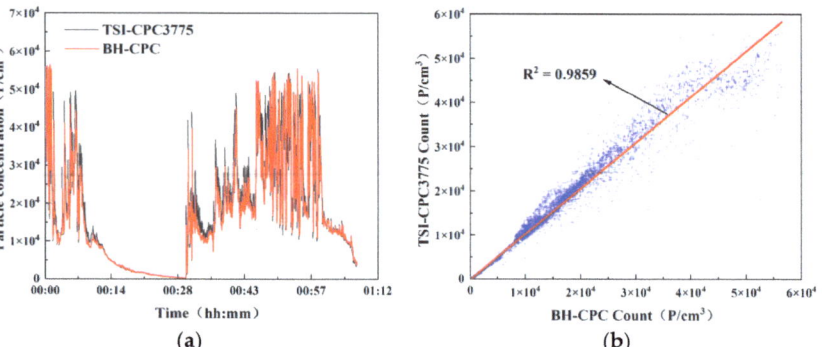

Figure 6. Results of BH-CPC counting accuracy test: (**a**) measurement results; (**b**) linearity analysis.

3.2. Comparison of Detection Efficiency between Aviation Soot and Calibration Particles

The morphology of aviation soot and sodium chloride particles was investigated using a scanning electron microscope (SEM, model: Quattro S, Thermo Fisher Scientific, Waltham, MA, USA). In the experiment, carbon soot particles with the median diameter of 26 nm were generated with the MiniCAST. Sodium chloride particles were generated by aerosolizing a sodium chloride solution with an aerosol generator. To collect these two types of particles for observation, a atmospheric single-particle sampler (model DKL-2, Qingdao Genxing Electronic Technology Co., Ltd., Qingdao, China) was employed to deposited them onto separate 3 × 3 mm single crystal silicon wafers. The sampling process lasted approximately 20 s for each sample. To ensure the reproducibility of the experimental results, we initiated particle collection roughly 2 min after the particle generation device had stabilized. Figure 7 presents the SEM observations of the aviation soot and sodium chloride particles collected during the experiment. In Figure 7a, the SEM image of sodium chloride particles is displayed, while Figure 7b presents the SEM image of aviation soot particles. The results indicate that sodium chloride particles exhibit a nearly square shape with a relatively smooth surface. In contrast, aviation soot particles are composed of several or hundreds of primary spherical-like particles, forming agglomerates, chains, or branched structures, with a relatively rough surface.

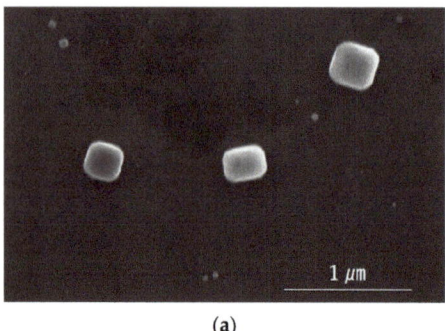
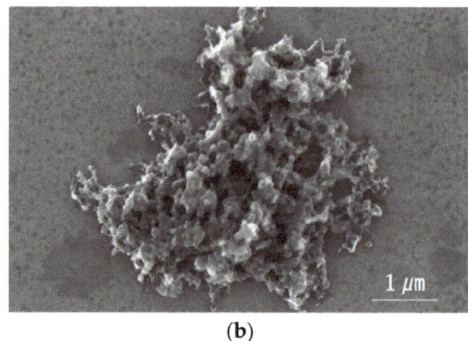

Figure 7. Particle material characterization image: (**a**) SEM image of sodium chloride; (**b**) SEM image of aviation soot.

The number concentrations of aviation soot and laboratory-calibrated sodium chloride particles were determined using the experimental system illustrated in Figure 3. During the experiment, the particle number concentration was maintained below 10,000 P/cm^3 by adjusting the dilution ratio of the dilution bridge, thus mitigating the influence of high concentrations on particle detection efficiency. The DMA controller was set to maintain a constant sheath flow rate of 12 L per minute (LPM), resulting in a sheath-to-aerosol flow rate ratio of approximately 8:1, surpassing the threshold of 5:1. This ensured a narrow size distribution of the classified particles. Subsequently, particles with sizes ranging from 10 nm to 40 nm (10 nm, 12 nm, 14 nm, 18 nm, 22 nm, 26 nm, 30 nm, 35 nm, and 40 nm) were sequentially classified by the DMA and introduced in parallel into the BH-CPC and TSI-AE3068B instruments. The BH-CPC operated at a flow rate of 0.36 LPM, while the TSI-AE3068B maintained a flow rate of 1 LPM. The particle number concentration measured by the BH-CPC served as the measured value, and the particle number concentration obtained through charge inversion in the TSI-AE3068B was regarded as the standard value for comparison.

Figure 8 presents the experimental results depicting the detection efficiency of BH-CPC for different particle sizes of aviation soot and sodium chloride particles. Within the size range of 10–40 nm, the detection efficiency for both aviation soot and sodium chloride particles exhibited a gradual increase with larger particle sizes, indicating a consistent overall trend. However, it is noteworthy that the detection efficiency of sodium chloride particles displayed a noticeable deceleration in the rate of increase beyond a particle size of 18 nm, whereas the detection efficiency of aviation soot continued to exhibit a more pronounced upward trend. Furthermore, in the particle size range of 10–14 nm, the detection efficiency of aviation soot was generally lower compared to that of sodium chloride. In the particle size range of 18–22 nm, the detection efficiency of aviation soot was similar to that of sodium chloride. In the particle size range of 26–40 nm, the detection efficiency of aviation soot was generally higher than that of sodium chloride. In the particle size range of 10–40 nm, the measurement errors for sodium chloride vary, with a maximum of 3.6% and a minimum of 1.3%. In the case of aviation soot, the measurement errors range from a maximum of 3.9% to a minimum of 1.1%. Larger particle sizes demonstrate smaller measurement errors when compared to smaller particle sizes.

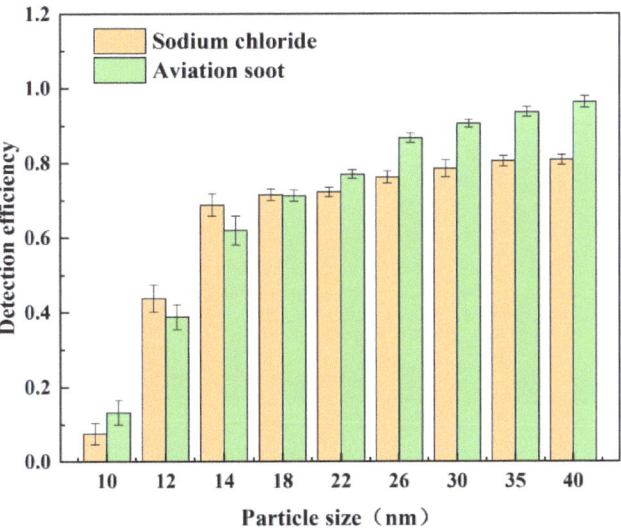

Figure 8. Comparison of detection efficiency between aviation soot and sodium chloride.

Figure 9 presents the statistical analysis of the detection efficiency deviation between the two types of particles shown in Figure 8, namely aviation soot and calibration particles (sodium chloride). In terms of relative deviation, the detection efficiency of aviation soot and sodium chloride particles showed the most substantial difference at approximately 10 nm, while the smallest difference was observed at approximately 18 nm. However, in terms of absolute deviation, the detection efficiency of aviation soot and sodium chloride particles exhibited the greatest difference at approximately 40 nm, while the smallest difference occurring at approximately 18 nm. These results indicate that within the particle size range of 10–40 nm, both aviation soot and sodium chloride particles exhibit notable deviations in detection efficiency at both small and large particle sizes.

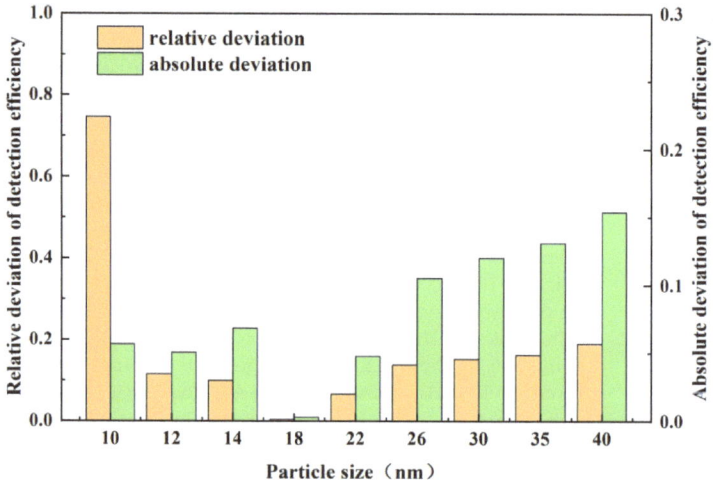

Figure 9. Deviation in detection efficiency of aviation soot compared to sodium chloride.

3.3. Impact of the Operating Temperature on Detection Efficiency for Aviation Soot

Section 3.2 highlights significant deviations in the detection efficiency between aviation soot and laboratory-calibrated particles. To mitigate the impact of this deviation on practical measurements, this section investigates the influence of the operating temperature of BH-CPC on the detection efficiency of aviation soot.

The operating temperature of BH-CPC primarily refers to the temperatures of its saturation section (T_S) and condensation section (T_C). Figure 10 illustrates the results of the detection efficiency for aviation soot by BH-CPC at varying T_S values, ranging from the calibrated temperature of 50 °C to a non-calibrated temperature of 40 °C, while T_C remains constant at the calibrated temperature of 10 °C. Across various T_S settings, the detection efficiency for aviation soot gradually increases with the particle size. However, as T_S decreases, the detection efficiency for aviation soot also decreases, particularly for smaller particle sizes. In the particle size of 18 nm, the detection efficiency decreased by 4.3% from 50 °C to 45 °C, and by 59.5% from 45 °C to 40 °C. When T_S is set to 40 °C, the detection efficiency of aviation soot within the range of 10–14 nm is almost zero. Within the size range of 10–40 nm, the measurement errors for aviation soot vary from a maximum of 3.9% to a minimum of 1.1% at T_S of 50 °C. When T_S is set to 45 °C, the measurement errors for aviation soot range from a maximum of 3.0% to a minimum of 1.0%. When T_S is set to 40 °C, the measurement errors for aviation soot range peak at 4.2%, with a minimum of 1.0%.

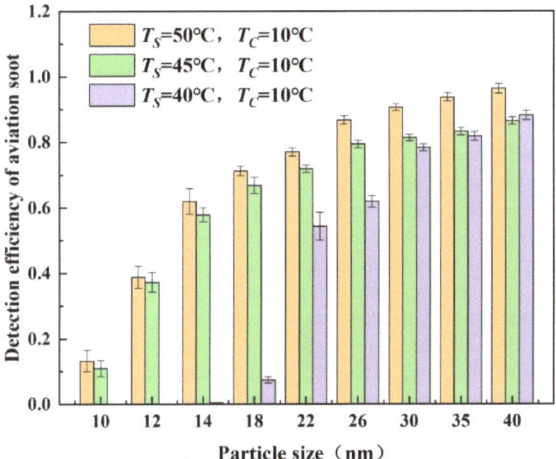

Figure 10. Detection efficiency of aviation soot at different saturation section temperatures (T_S).

Figure 11 presents the results of detection efficiency for aviation soot at different T_C values, ranging from the calibrated temperature of 10 °C to a non-calibrated temperature of 20 °C, while T_S remains constant at the calibrated temperature of 50 °C. Across various T_C settings, the detection efficiency of aviation soot gradually increases with the particle size. However, as T_C increases, the detection efficiency of aviation soot decreases, particularly for smaller particle sizes. In the particle size of 18 nm, the detection efficiency decreased by 7.7% from 10 °C to 15 °C, and by 60.7% from 15 °C to 20 °C. When T_C is set to 20 °C, the detection efficiency of aviation soot within the range of 10–14 nm is nearly zero. Within the size range of 10–40 nm, the measurement errors for aviation soot vary from a maximum of 3.9% to a minimum of 1.1% at T_C of 10 °C. When T_C is set to 15 °C, the measurement errors for aviation soot range from a maximum of 3.7% to a minimum of 1.1%. When T_C is set to 20 °C, the measurement errors for aviation soot range peak at 2.4%, with a minimum of 0.3%.

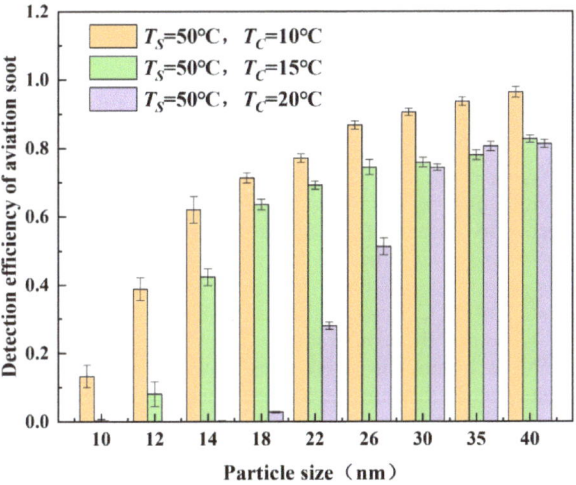

Figure 11. Detection efficiency for aviation soot at varying condensation section temperatures (T_C).

3.4. Discussion

Based on the experimental results from Sections 3.2 and 3.3, it is evident that there is a significant disparity in the detection efficiency between aviation soot and calibration particles (sodium chloride). This disparity can lead to a performance gap between industrial field measurements using CPC and experimental calibration. Additionally, the variation in the CPC operating temperature has a noticeable impact on its detection efficiency. When the saturation section temperature decreases from 45 °C to 40 °C or the condensation section temperature increases from 15 °C to 20 °C, the detection efficiency of small-sized aviation soot can decrease by up to 60%. To investigate whether adjusting the operating temperature can mitigate the detection efficiency deviation caused by particle characteristics, a comparison was conducted between aviation soot and calibration particles (sodium chloride) at different operating temperatures. Figure 12 illustrates the results of detection efficiency deviation for aviation soot at varying operating temperatures, specifically highlighting the temperature conditions where the detection efficiency is non-zero within the 10–40 nm range.

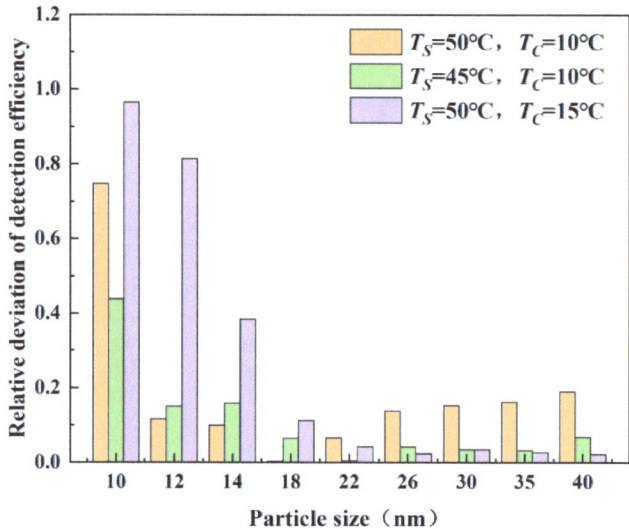

Figure 12. Relative deviation in detection efficiency of aviation soot at varying operating temperatures.

Within the particle size range of 18–40 nm, the relative deviation in the detection efficiency of aviation soot is relatively small for all three temperature combinations. The combination with a saturation section temperature (T_S) of 50 °C and a condensation section temperature (T_C) of 10 °C exhibits the highest relative deviation. For a particle size between 10 and 14 nm, the relative deviation in the detection efficiency for aviation soot is significant for all three temperature combinations. The combination with T_S of 50 °C and T_C of 15 °C shows the largest relative deviation. Overall, within the particle size range of 10–40 nm, the temperature combination of T_S at 45 °C and T_C at 10 °C results in the smallest deviation in detection efficiency for aviation soot. These indicate that appropriate adjustments of T_S and T_C can bring the detection efficiency for aviation soot closer to that of the calibration particles.

4. Conclusions

This study presents a quantitative analysis of the deviation in detection efficiency for aviation soot compared to calibration particles in a condensation particle counter, while also analyzing the influence of working temperature on this deviation. The results of this study reveal significant differences in surface roughness between aviation soot and calibration particles, resulting in noticeable variations in detection efficiency. Aviation soot displays agglomerated or chain-like structures, featuring a surface rougher than that of sodium chloride, which possesses a square-shaped structure. In the range of 10–40 nm, the absolute deviation in the detection efficiency of both particles can reach a maximum of 0.15, while the relative deviation can reach a maximum of 0.75.

As the saturation section temperature (T_S) decreases or the condensation section temperature (T_C) increases, the detection efficiency of aviation soot gradually declines. In the particle size of 18nm, a decrease in T_S from 50 °C to 45 °C results in a 4.3% reduction in aviation soot detection efficiency. With a further decrease in T_S from 45 °C to 40 °C, the detection efficiency of aviation soot decreases significantly by 59.5%. In comparison to the calibration temperature (T_S = 50 °C, T_C = 10 °C), the detection efficiency at a T_S of 45 °C exhibits smaller absolute and relative deviations. At this temperature, the detection efficiency of aviation soot closely aligns with the efficiency calibrated using experimental sodium chloride. It is feasible to correct for deviation in detection efficiency by establishing a functional relationship between the operating temperature and calibration temperature.

Author Contributions: Conceptualization, Q.Z. and L.C. (Liuyong Chang); methodology, Q.Z. and L.C. (Liuyong Chang); software, Q.Z. and Y.L.; validation, Q.Z. and L.C. (Liuyong Chang); formal analysis, Q.Z. and L.C. (Liuyong Chang); investigation, Q.Z. and L.C. (Liuyong Chang); resources, Q.Z. and L.C. (Liuyong Chang); data curation, Q.Z. and L.C. (Liuyong Chang); writing—original draft preparation, Q.Z.; writing—review and editing, L.C. (Liuyong Chang) and G.L.; visualization, Q.Z. and Y.L.; supervision, L.C. (Liang Chen) and L.C. (Longfei Chen); project administration, L.C. (Liang Chen), L.C. (Longfei Chen), L.C. (Liuyong Chang) and G.L.; funding acquisition, L.C. (Liang Chen), L.C. (Liuyong Chang) and G.L. All authors have read and agreed to the published version of the manuscript.

Funding: This research was funded by the National Natural Science Foundation of China (Grant Nos. 52306184, 52306181, 62075205, 61904168 and 61775203), the National Key Research and Development Program of China (Grant No. 2017YFF0210800), the Zhejiang Provincial Key Research and Development Program (Grant No. 2021C01027) and the Zhejiang Provincial Natural Science Foundation of China (Grant Nos. LZ21F050001, LZ20F050001 and LQ18F040001).

Data Availability Statement: The data used to support the findings of this study are included within this article.

Conflicts of Interest: The authors declare no conflict of interest.

References

1. Zhang, H.; Wang, J. Development and validation of a soot mechanism for RP-3 aviation fuel. *Chem. Res. Appl.* **2022**, *34*, 1872–1879.
2. You, Q.; Li, H.; Bo, X.; Zheng, Y.; Chen, S.B. Air pollution and CO_2 emission inventory of Chinese civil aviation airport. *China Environ. Sci.* **2022**, *42*, 4517–4524.
3. Forster, P.; Ramaswamy, V.; Artaxo, P.; Berntsen, T.; Betts, R.; Fahey, D.W.; Haywood, J.; Lean, J.; Lowe, D.C.; Myhre, G.; et al. Changes in Atmospheric Constituents and in Radiative Forcing. Climate Change 2007: The Physical Science Basis. Contribution of Working Group I to the 4th Assessment Report of the Intergovernmental Panel on Climate Change 2007; Cambridge University Press: Cambridge, UK, 2007.
4. Myhre, G.; Samset, B.H.; Schulz, M.; Balkanski, Y.; Bauer, S.; Berntsen, T.K.; Bian, H.; Bellouin, N.; Chin, M.; Diehl, T.; et al. Radiative forcing of the direct aerosol effect from AeroCom Phase II simulations. *Atmos. Meas. Tech. Phys.* **2013**, *13*, 1853–1877. [CrossRef]
5. Masiol, M.; Harrison, R.M. Aircraft engine exhaust emissions and other airport-related contributions to ambient air pollution: A review. *Atmos. Environ.* **2014**, *95*, 409–455. [CrossRef]
6. Lee, D.S.; Fahey, D.W.; Forster, P.M.; Newton, P.J.; Wit, R.C.N.; Lim, L.L.; Owen, B.; Sausen, R. Aviation and global climate change in the 21st century. *Atmos. Environ.* **2009**, *43*, 3520–3537. [CrossRef] [PubMed]
7. Guo, H.; Han, X.; Liu, J. Research and calibration technology of condensation particle counter. *Chin. J. Sci. Instrum.* **2021**, *42*, 1–13.
8. Sakurai, H.; Murashima, Y.; Iida, K.; Wälchli, C.; Auderset, K.; Vasilatou, K. Traceable methods for calibrating condensation particle counters at concentrations down to 1 cm^{-3}. *Metrologia* **2023**, *60*, 055012. [CrossRef]
9. Krasa, H.; Kupper, M.; Schriefl, M.A.; Bergmann, A. Toward a simplified calibration method for 23 nm automotive particle counters using atomized inorganic salt particles. *Aerosol Sci. Technol.* **2023**, *57*, 329–341. [CrossRef]

10. Hermann, M.; Wehner, B.; Bischof, O.; Han, H.-S.; Krinke, T.; Liu, W.; Zerrath, A.; Wiedensohler, A. Particle counting efficiencies of new TSI condensation particle counters. *J. Aerosol Sci.* **2007**, *38*, 674–682. [CrossRef]
11. Wang, X.; Caldow, R.; Sem, G.J.; Hama, N.; Sakurai, H. Evaluation of a condensation particle counter for vehicle emission measurement: Experimental procedure and effects of calibration aerosol material. *J. Aerosol Sci.* **2010**, *41*, 306–318. [CrossRef]
12. Liu, P.S.K.; Deshler, T. Causes of Concentration Differences Between a Scanning Mobility Particle Sizer and a Condensation Particle Counter. *Aerosol Sci. Technol.* **2003**, *37*, 916–923. [CrossRef]
13. Sem, G.J. Design and performance characteristics of three continuous-flow condensation particle counters: A summary. *Atmos. Res.* **2002**, *62*, 267–294. [CrossRef]
14. Zhou, Q.; Chen, L.; Chang, L.; Zhang, C.; Li, G.; Li, Y. Development of an optical particle counter for high particle number concentration. In Proceedings of the Third International Conference on Optics and Image Processing (ICOIP 2023), Hangzhou, China, 14–16 April 2023; pp. 143–147.
15. Fan, F.; Yang, L.; Yuan, Z. Progress and prospect in the study of heterogeneous nucleation of vapor on fine particles. *Chem. Ind. Eng. Prog.* **2009**, *28*, 1496–1500.
16. Kuang, C. A diethylene glycol condensation particle counter for rapid sizing of sub-3 nm atmospheric clusters. *Aerosol Sci. Technol.* **2018**, *52*, 1112–1119. [CrossRef]
17. Lauri, A.J. *Theoretical and Computational Approaches on Heterogeneous Nucleation*; Helsingin yliopisto: Helsinki, Finland, 2006.
18. Yoo, S.-J.; Kwon, H.-B.; Hong, U.-S.; Kang, D.-H.; Lee, S.-M.; Han, J.; Hwang, J.; Kim, Y.-J. Microelectromechanical-system-based condensation particle counter for real-time monitoring of airborne ultrafine particles. *Atmos. Meas. Tech. Phys.* **2019**, *12*, 5335–5345. [CrossRef]
19. Ambaum, M.H.P. Accurate, simple equation for saturated vapour pressure over water and ice. *Q. J. R. Meteorol. Soc.* **2020**, *146*, 4252–4258. [CrossRef]
20. Ziese, F.; Maret, G.; Gasser, U. Heterogeneous nucleation and crystal growth on curved surfaces observed by real-space imaging. *J. Phys. Condens. Matter* **2013**, *25*, 375105. [CrossRef]
21. Lv, L.; Zhang, J.; Xu, J.; Yin, J. Effects of surface topography of SiO_2 particles on the heterogeneous condensation process observed by environmental scanning electron microscopy. *Aerosol Sci. Technol.* **2021**, *55*, 920–929. [CrossRef]
22. Lu, J.; Lin, F.; Zhang, A. Discussion on calibration methods of condensation particle counters. *Shanghai Meas. Test.* **2021**, *48*, 18–20.
23. Xu, X.; Zhao, W.; Fang, B.; Gu, X.; Zhang, W. Development and performance evaluation of a standard aerosol generation system. *Acta Sci. Circumstantiae* **2016**, *36*, 2355–2361.
24. Le, K.C.; Pino, T.; Pham, V.T.; Henriksson, J.; Török, S.; Bengtsson, P.-E. Raman spectroscopy of mini-CAST soot with various fractions of organic compounds: Structural characterization during heating treatment from 25 °C to 1000 °C. *Combust. Flame* **2019**, *209*, 291–302. [CrossRef]
25. Moore, R.H.; Ziemba, L.D.; Dutcher, D.; Beyersdorf, A.J.; Chan, K.; Crumeyrolle, S.; Raymond, T.M.; Thornhill, K.L.; Winstead, D.L.; Anderson, B.E. Mapping the operation of the miniature combustion aerosol standard (Mini-CAST) soot generator. *Aerosol Sci. Technol.* **2014**, *48*, 467–479. [CrossRef]
26. Saffaripour, M.; Tay, L.-L.; Thomson, K.A.; Smallwood, G.J.; Brem, B.T.; Durdina, L.; Johnson, M. Raman spectroscopy and TEM characterization of solid particulate matter emitted from soot generators and aircraft turbine engines. *Aerosol Sci. Technol.* **2017**, *51*, 518–531. [CrossRef]
27. Marhaba, I.; Ferry, D.; Laffon, C.; Regier, T.Z.; Ouf, F.-X.; Parent, P. Aircraft and MiniCAST soot at the nanoscale. *Combust. Flame* **2019**, *204*, 278–289. [CrossRef]
28. Anderson, B.E.; Beyersdorf, A.J.; Hudgins, C.H.; Plant, J.V.; Thornhill, K.L.; Winstead, E.L.; Ziemba, L.D.; Howard, R.; Corporan, E.; Miake-Lye, R.C.; et al. *Alternative Aviation Fuel Experiment (AAFEX)*; National Aeronautics and Space Administration: Washington, DC, USA, 2011.
29. Delhaye, D.; Ouf, F.-X.; Ferry, D.; Ortega, I.K.; Penanhoat, O.; Peillon, S.; Salm, F.; Vancassel, X.; Focsa, C.; Irimiea, C.; et al. The MERMOSE project: Characterization of particulate matter emissions of a commercial aircraft engine. *J. Aerosol Sci.* **2017**, *105*, 48–63. [CrossRef]
30. Brem, B.T.; Durdina, L.; Siegerist, F.; Beyerle, P.; Bruderer, K.; Rindlisbacher, T.; Rocci-Denis, S.; Andac, M.G.; Zelina, J.; Penanhoat, O.; et al. Effects of Fuel Aromatic Content on Nonvolatile Particulate Emissions of an In-Production Aircraft Gas Turbine. *Environ. Sci. Technol.* **2015**, *49*, 13149–13157. [CrossRef]
31. Moore, R.H.; Thornhill, K.L.; Weinzierl, B.; Sauer, D.; D'Ascoli, E.; Beaton, B.; Beyersdorf, A.J.; Bulzan, D.; Corr, C.; Crosbie, E. Biofuel blending reduces aircraft engine particle emissions at cruise conditions. *Nature* **2017**, *543*, 411. [CrossRef]
32. Li, K.; Yuan, F.; Liu, Y.; Chen, C. Contrastive research on simulation of miniaturized cylindrical DMA and parallel DMA. *Transducer Microsyst. Technol.* **2022**, *41*, 14–17.
33. Rose, D.; Gunthe, S.S.; Mikhailov, E.; Frank, G.P.; Dusek, U.; Andreae, M.O.; Pöschl, U. Calibration and measurement uncertainties of a continuous-flow cloud condensation nuclei counter (DMT-CCNC): CCN activation of ammonium sulfate and sodium chloride aerosol particles in theory and experiment. *Atmos. Meas. Tech. Phys.* **2008**, *8*, 1153–1179. [CrossRef]
34. Sun, S.; Qi, T.; Xiao, J.; Hu, X.; Liu, J. Development and Calibration of High Accuracy Aerosol Electrometer. *Metrology Science and Technology.* **2021**, *2*, 54–58.
35. Fletcher, R.A.; Mulholland, G.W.; Winchester, M.R.; King, R.L.; Klinedinst, D.B. Calibration of a Condensation Particle Counter Using a NIST Traceable Method. *Aerosol Sci. Technol.* **2009**, *43*, 425–441. [CrossRef]

36. Bezantakos, S.; Biskos, G. Temperature and pressure effects on the performance of the portable TSI 3007 condensation particle counter: Implications on ground and aerial observations. *J. Aerosol Sci.* **2022**, *159*, 105877. [CrossRef]
37. Buckley, A.J.; Wright, M.D.; Henshaw, D.L. A Technique for Rapid Estimation of the Charge Distribution of Submicron Aerosols under Atmospheric Conditions. *Aerosol Sci. Technol.* **2008**, *42*, 1042–1051. [CrossRef]
38. Nicosia, A.; Manodori, L.; Trentini, A.; Ricciardelli, I.; Bacco, D.; Poluzzi, V.; Di Matteo, L.; Belosi, F. Field study of a soft X-ray aerosol neutralizer combined with electrostatic classifiers for nanoparticle size distribution measurements. *Particuology* **2018**, *37*, 99–106. [CrossRef]

Disclaimer/Publisher's Note: The statements, opinions and data contained in all publications are solely those of the individual author(s) and contributor(s) and not of MDPI and/or the editor(s). MDPI and/or the editor(s) disclaim responsibility for any injury to people or property resulting from any ideas, methods, instructions or products referred to in the content.

Article

Nonlinear Dynamic Modeling and Analysis for a Spur Gear System with Dynamic Meshing Parameters and Sliding Friction

Hao Liu [1], Dayi Zhang [1,2,*], Kaicheng Liu [3], Jianjun Wang [1], Yu Liu [4] and Yifu Long [1]

1. School of Energy and Power Engineering, Beihang University, Beijing 100191, China; lh2010@buaa.edu.cn (H.L.); wangjianjun@buaa.edu.cn (J.W.); lyfleo@buaa.edu.cn (Y.L.)
2. Beijing Key Laboratory of Aero-Engine Structure and Strength, Beijing 100191, China
3. Beijing Aerospace Propulsion Institute, Beijing 100076, China; liukc@buaa.edu.cn
4. Key Lab of Advance Measurement and Test Technology for Aviation Propulsion System, Shenyang Aerospace University, Shenyang 110136, China; 20190037@mail.sau.edu.cn
* Correspondence: dayi@buaa.edu.cn

Abstract: The performance of gear systems is closely related to the meshing parameters and sliding friction. However, the time-varying characteristics of meshing parameters caused by transverse vibration are usually not regarded and the sliding friction has always been ignored in previous studies. Therefore, the influence of the transverse vibration on meshing parameters and sliding friction have not been considered. In view of this, a nonlinear dynamic model for a spur gear system is proposed. The dynamic meshing parameters (pressure angle, backlash, etc.) and the effects of the variations of these parameters on the dynamic mesh force (DMF) and sliding friction are emphasized. The differential equations of motion are derived by the Lagrange method and solved by the Runge–Kutta method. Then, the input speed and friction coefficient are used as control parameters to compare the dynamic responses of the new and previous models. The results show that the meshing parameters and sliding friction are affected by transverse vibration, leading to distinctive nonlinear dynamic responses. This paper can provide a basis for further research and give a better understanding of system vibration control.

Keywords: spur gear system; transverse vibration; dynamic meshing parameter; sliding friction; dynamic response

1. Introduction

Gear systems have the advantages of wide transmission power range, high efficiency, and accurate transmission ratio, and are widely used in aerospace, marine, and electric power fields as typical periodic symmetrical components [1]. The performance of the gear system directly affects the overall performance of mechanical equipment. Furthermore, the noise caused by vibration and shock during the operation of the gear system is one of the important components of mechanical equipment noise, which seriously influences comfort [2], and also affects the concealment of military equipment [3]. Therefore, it is of great significance to establish a proper dynamic model of gear systems and study the dynamic characteristics for engineering design, condition monitoring, and fault diagnosis.

For simplification, the influence of transverse vibration on gear meshing parameters is not considered in most of the existing gear system models, that is, the pressure angle, contact ratio, and backlash are all assumed to be constant [4–8]. In fact, the meshing parameters change due to the transverse vibration. With the urgent demand for lightweight design of gear systems in modern industry, especially in aerospace, flexible supporting structures such as thin hollow shafts and thin wall gearboxes have been widely used [9]. As the stiffness of the supporting structure decreases, the transverse vibration of the gear becomes obvious, and the prior model could not accurately predict the dynamic behavior. Scholars have recently begun to pay attention to this problem. Chen et al. [10] deduced the relationship between the backlash and the variation of center distance, and established a dynamic model considering the sliding friction, but they did not consider the change of the other mesh parameters under the influence of transverse vibration. Focusing on the time-varying pressure angle caused by the transverse vibration, Kim et al. [11] proposed a new spur gear system model, and on this basis studied the effects of stiffness and damping on the dynamic response of the system. Then, they [12] established a 15-degrees-of-freedom (DOFs) dynamic model with a planetary gear system as an object and analyzed the influence of bearing deformation on tooth deformation, pressure angle, and contact ratio between sun and star gears and between star and ring gears, as well as vibration displacement in different directions. Afterwards, Chen [13] extended the study to a helical gear system and investigated the effects of bearing radial and axial stiffness and mass eccentricity on dynamic response. Liu et al. [14] proposed a lateral–torsional–rocking coupled model of a gear system considering time-varying center distance and backlash, and analyzed the effects of different modification methods. Yi et al. [15] established a new nonlinear dynamic model of a spur gear considering the time-varying backlash, and the results showed that the calculated dynamic response was more realistic. Wang and Zhu [16] put forward a nonlinear dynamic model of a GTF gearbox with time-varying backlash, and investigated the dynamic characteristics from the time domain and frequency domain. Jedliński [17] focused on the influence of the off-line-of-action (OLOA) direction displacement on the distance between the meshing tooth surfaces along the line of action (LOA) and established a 12-DOFs analytical model. The results show that the influence of the OLOA displacement on the distance between the meshing tooth surfaces could not be neglected when the OLOA displacement was significant. Yang et al. [18] proposed a new method for calculating the mesh stiffness of a helical gear pair, which considered the time-varying backlash. Considering the effect of bearing clearance on the radial vibration of gears, Tian et al. [19] investigated the stability of the spur gear system in depth by selecting bifurcation parameters such as rotational speed and bearing clearance.

As one of the important excitation sources, the sliding friction between the tooth surface couples the motions of LOA and OLOA. The effect of sliding friction on the vibration and noise of the gear system cannot be ignored [20]. Singh and co-authors used Floquet theory [21], harmonic balance method [22], and the numerical method [23] to comprehensively study the sliding friction of a SDOF gear pair, and later extended this to multiple DOFs systems [24,25]. Ghosh and Chakraborty [26] studied the effects of friction coefficient, damping, and modification on the system stability of a six-DOF spur gear system. Zhou et al. [27] proposed a 16-DOFs coupling dynamic model of the gear rotor system, and studied the influence of the friction coefficient on the nonlinear characteristics of the system. They found that the system entered a chaotic state with the increase in friction coefficient within a certain range. Shi et al. [28] established a SDOF gear pair considering multi-state mesh and sliding friction and analyzed the effects of load coefficient, backlash, and comprehensive error on the system dynamic response. Afterwards, they carried out a study on the nonlinear dynamic characteristics of the system under random excitation [29]. Wang [30] built a bending-torsion-shaft-coupled model of the helical gear pair and investigated the influence of sliding friction through dynamic meshing force (DMF)/speed/displacement. Aiming at the loss-of-lubrication condition of helicopter gear transmission, Hu et al. [31] predicted the friction coefficient based on the computational inverse technique, and studied the nonlinear dynamic behavior of the system considering gyroscopic effect, thermal expansion, and wear under this working condition. The results show that although friction had little effect on the natural frequency, it had significant effect on nonlinear dynamic behavior. Luo et al. [32] established a dynamic model of the planetary gear sets to research the influence of the spalling defect on the dynamic performance when the sliding friction was considered. Jiang and Liu [33] paid attention to the axial friction caused by the axial mesh force in the helical gear system and pointed out that the oscillations of the dynamic responses become more significant incorporating the effects of coupled sliding friction.

In general, although scholars have conducted many studies on gear dynamics for different factors, few studies have simultaneously considered the influences of transverse vibration on the meshing parameters and sliding friction. The direction of sliding friction changes as the pressure angle changes, and the length of friction force arm is affected by the variation of center distance. The variations of meshing parameters and sliding friction affect the dynamic characteristics of the system and then change the transverse vibration response. This interaction should be taken into account in the study of gear dynamics. For this reason, a new dynamic model for a spur gear system with dynamic meshing parameters and sliding friction is proposed, and the dynamic characteristics of the system are analyzed on this basis.

The rest of this paper is organized as follows. In the next section, the dynamic meshing parameters affected by transverse vibration and the corresponding DMF and friction force are given and a six-DOFs nonlinear dynamic model for a spur gear system is developed with time-varying pressure angle, time-varying mesh stiffness (TVMS), dynamic backlash, and sliding friction. The equations of motion are strictly derived using the Lagrange method in Section 3. In Section 4, the dynamic responses of the new and previous models are compared, with the input speed and friction coefficient as the control parameters, respectively. In the last section, some brief conclusions are presented.

The main contributions of this paper are summarized as follows:

1. We establish a new sliding friction model between the tooth surfaces considering the influence of the transverse vibration.
2. Based on the sliding friction model, a new nonlinear dynamic model with dynamic meshing parameters and sliding friction is proposed.
3. The effects of the input speed and friction coefficient on the dynamic response of the new model and the previous model are compared and analyzed.

2. Dynamic Model for Spur Gear System

Figure 1a shows the dynamic model (new model) proposed in this paper, which considers the influence of transverse vibration on meshing parameters and sliding friction. Figure 1b shows the dynamic model (previous model) commonly used in previous studies that ignores the influence of transverse vibration, which assumes that the direction of the LOA remains constant during operation. As the main difference between the two models, the effects of transverse vibration on the system can be divided into two levels. The first includes the variations of the meshing parameters such as center distance, pressure angle, and backlash caused by transverse vibration. The second is the influence of these parameters on DMF and sliding friction. This paper focuses on these two levels to model the spur gear system with the influence of transverse vibration, which will be introduced in detail.

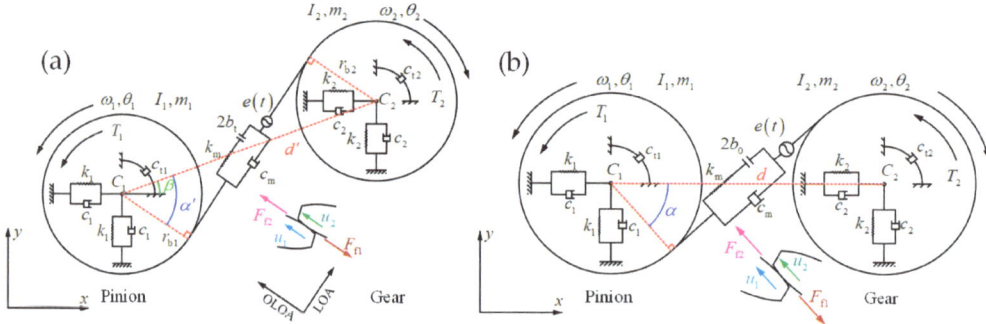

Figure 1. Dynamic model for spur gear system: (**a**) the new model, (**b**) the previous model.

2.1. Dynamic Meshing Parameters under the Influence of Transverse Vibration

In Figure 1, subscript 1 and 2 represent the pinion (driving gear) and the gear (driven gear), respectively. The supporting structure of each gear is equivalent to two sets of spring-damping units, namely k_{xi}, c_{xi}, k_{yi}, and $c_{yi}(i=1,2)$, and the torsional damping unit c_{ti}, which is used to explain the viscous loss caused by the bearings and shafts. This paper assumes that $c_{xi} = c_{yi}$ and $k_{xi} = k_{yi}$, so only c_1, c_2, k_1, and k_2 are indicated in Figure 1. r_{bi} is the radius of the base circle. m_i, I_i, and T_i are the mass, moment of inertia, and external torque, respectively. ω_i, φ_i, and θ_i are, respectively, the angular speed, angular displacement, and small angular displacement resulting from torsional vibration. Figure 2 illustrates the generalized coordinates of the meshing gears, where the black dashed and blue solid lines separately represent the meshing gears before and after motion. The initial position of the rotation center of the gear is represented by O_i and the position after the motion is denoted by C_i. G_i is the mass center of the gear. Assuming that only the translational motion is considered, the coordinates of the gear can be given by the translation coordinates x_i, y_i and the angular coordinate φ_i. The angular displacements of the pinion and gear can be written as:

$$\varphi_1 = \omega_1 t + \theta_1 + \varphi_{01}, \qquad \varphi_2 = \omega_2 t + \theta_2 + \varphi_{02} \tag{1}$$

where φ_{0i} is the initial angular displacement.

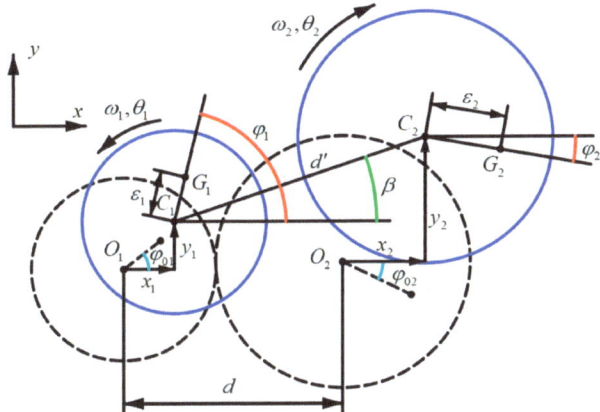

Figure 2. Generalized coordinates of gear system.

According to Figures 1a and 2, when geometric eccentricity and material defects are ignored, the displacement vectors for the mass center of the pinion and gear at an arbitrary time can be expressed as:

$$\begin{aligned}\mathbf{r}_1 &= (x_1 + \varepsilon_1 \cos \varphi_1)\mathbf{i} + (y_1 + \varepsilon_1 \sin \varphi_1)\mathbf{j} \\ \mathbf{r}_2 &= (x_2 + \varepsilon_2 \cos \varphi_2 + d)\mathbf{i} + (y_2 - \varepsilon_2 \sin \varphi_2)\mathbf{j}\end{aligned} \tag{2}$$

where ε_i is the mass eccentricity, $\mathbf{i}$ and $\mathbf{j}$ are unit vectors along the x and y axes, respectively.

The time-varying pressure angle, that is, the acute angle sandwiched between the velocity direction at the pitch point and the LOA (Figure 1a), can be expressed as:

$$\alpha' = \cos^{-1} \frac{r_{b1} + r_{b2}}{d'} \tag{3}$$

in which d' is the time-varying center distance, and the relationship between d' and the initial center distance d is $d' = \sqrt{(x_2 - x_1 + d)^2 + (y_2 - y_1)^2}$.

The contact ratio at any time is:

$$m_p = \frac{\sqrt{r_{a2}^2 - r_{b2}^2} + \sqrt{r_{a2}^2 - r_{b2}^2} - d' \sin \alpha'}{p_b} \tag{4}$$

where r_{ai} is the addendum radius of the gear, p_b is the base pitch, and $p_b = \pi m \cos \alpha$, in which m is the module and α is the pressure angle of the reference circle.

It can be seen that for Equations (3) and (4) the gear translation caused by transverse vibration directly leads to the variation of center distance, which then affects the pressure angle and contact ratio, so that they are no longer constant.

Furthermore, the center distance alone is no longer sufficient to describe the relative position of the two gears as shown in Figures 1a and 2. Therefore, it is necessary to define another position angle β, which is:

$$\beta = \tan^{-1} \frac{y_2 - y_1}{x_2 - x_1 + d} \tag{5}$$

In Figure 1a, the contact between the tooth surfaces is equivalent to a spring-damping system along the LOA. The compression direction is defined as the positive direction, and the dynamic transmission error (DTE), i.e., the gear mesh deformation along the LOA, can be given as:

$$\delta = r_{b1}\theta_1 - r_{b2}\theta_2 + (x_1 - x_2)\sin(\alpha' - \beta) + (y_1 - y_2)\cos(\alpha' - \beta) - e(t) \tag{6}$$

where the static transmission error (STE) $e(t) = e_a \sin(2\pi f_m t + \phi_0)$, in which e_a is the amplitude of the STE, $f_m (= n_1 z_1 / 60 (n_2 z_2 / 60) = \omega_1 z_1 / 2\pi (\omega_2 z_2 / 2\pi))$, is the meshing frequency and ϕ_0 is the initial phase.

k_m is the TVMS as shown in Figure 1. The analytical model proposed by Ma et al. [34] is applied to calculate the TVMS here, and the corresponding result is presented by the blue dotted line in Figure 3. The double-teeth meshing zone is from $(n + 1 - m_p)T_m$ to $nT_m (n \geq 1, n \in \mathbf{N}+)$, and the single-tooth meshing zone is from $(n - 1)T_m$ to $(n + 1 - m_p)T_m$. T_m is the meshing period. Since the periodic composition of the TVMS is related to m_p, the TVMS is also affected by transverse vibration. To simplify, the TVMS of the square wave form proposed by Kahraman and Singh [35] is adopted, as shown by the red line in Figure 3. The square wave is determined by k_h and k_l, which are, respectively, the maximum and minimum value in a single mesh period. The time-varying mesh damping c_m can be expressed as $c_m = 2\zeta_m \sqrt{k_m I_1 I_2 / (I_1 r_{b2}^2 + I_2 r_{b1}^2)}$, in which ζ_m is the damping ratio.

Dynamic backlash b_t usually consists of constant backlash and time-varying backlash [14,15], namely:

$$2b_t = 2b_0 + \Delta b \tag{7}$$

where $2b_0$ is the constant backlash, that is the initial or design backlash, which is generally guaranteed by the manufacturing tolerance or the installation center distance error. Δb is the time-varying backlash, which is generally caused by the geometric eccentricity or the transverse vibration of the gear. This paper emphasizes the influence of transverse vibration. According to the involute principle, the time-varying backlash can be expressed as:

$$\Delta b = 2(r_{b1} + r_{b2})(\text{inv}(\alpha') - \text{inv}(\alpha)) \tag{8}$$

where $\text{inv}(x)$ is the involute function and $\text{inv}(x) = \tan(x) - x$.

So far, the dynamic meshing parameters under the influence of transverse vibration have been determined.

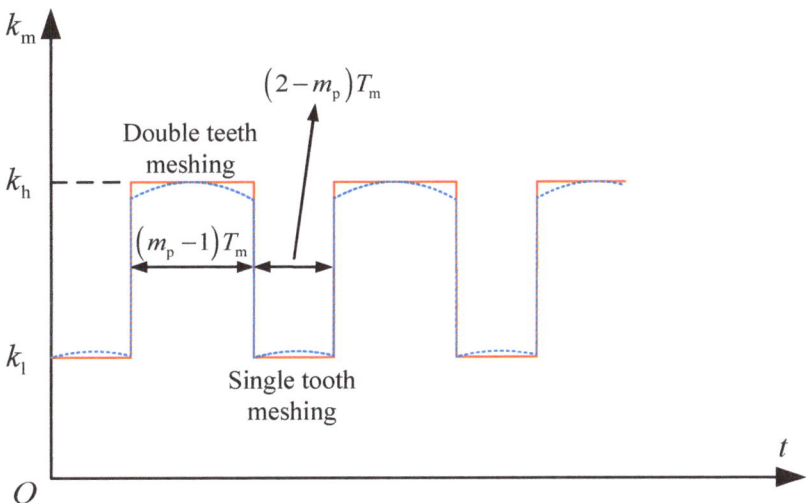

Figure 3. Time-varying mesh stiffness.

2.2. DMF and Sliding Friction Force/Torque under the Influence of Transverse Vibration

Based on the viscoelastic theory, the DMF is composed of elastic force and damping force, which can be obtained by:

$$F_m = k_m f_1(\delta) + c_m f_2(\dot{\delta}) \tag{9}$$

where $f_1(\delta)$ and $f_2(\dot{\delta})$ are the displacement and velocity functions of the backlash, respectively. $f_1(\delta)$ and $f_2(\dot{\delta})$ can be described as:

$$f_1(\delta) = \begin{cases} \delta - b_t, & \delta > b_t \\ 0, & |\delta| \leq b_t \\ \delta + b_t, & \delta < -b_t \end{cases}, \quad f_2(\dot{\delta}) = \begin{cases} \dot{\delta} - \dot{b}_t, & \delta > b_t \\ 0, & |\delta| \leq b_t \\ \dot{\delta} + \dot{b}_t, & \delta < -b_t \end{cases} \tag{10}$$

in which · represents the differential of time. The three conditions in Equation (10) correspond successively to non-impact state (drive-side tooth mesh), single-sided impact state (teeth separation), and double-sided impact state (back-side tooth mesh).

Although the sliding friction was involved in some prior studies, the influence of variations of the meshing parameters was not considered. Therefore, a new sliding friction model between contact tooth surfaces is proposed, in which the effect of variations of the meshing parameters is taken into account. Figure 4 shows the schematic diagram of sliding friction between the tooth surfaces. It should be noted that the reference coordinate system is rotated β clockwise compared with Figure 1a. The direction of sliding friction in Figure 4 is consistent with that shown in Figure 1a. $N_1 N_2$ is the theoretical LOA segment, B_1 and B_2 are the starting and ending points of the engagement, corresponding to the actual LOA segment. For the jth meshing tooth pair, the sliding velocity between the pinion tooth surface and the gear tooth surface can be expressed as:

$$v_{sj}(t) = u_{1j}(t) - u_{2j}(t) \tag{11}$$

where $j \in \mathbf{N}+$ and $j \leq \text{ceil}(m_p)$, $\text{ceil}(x)$ is the integer function, $u_{ij}(t)$ is the instantaneous tangential velocity of the tooth surface at the meshing point. If the angular velocity fluctuations of the gear are neglected, $u_{ij}(t)$ can be written as:

$$u_{ij}(t) = \omega_i R_{ij}(t) \tag{12}$$

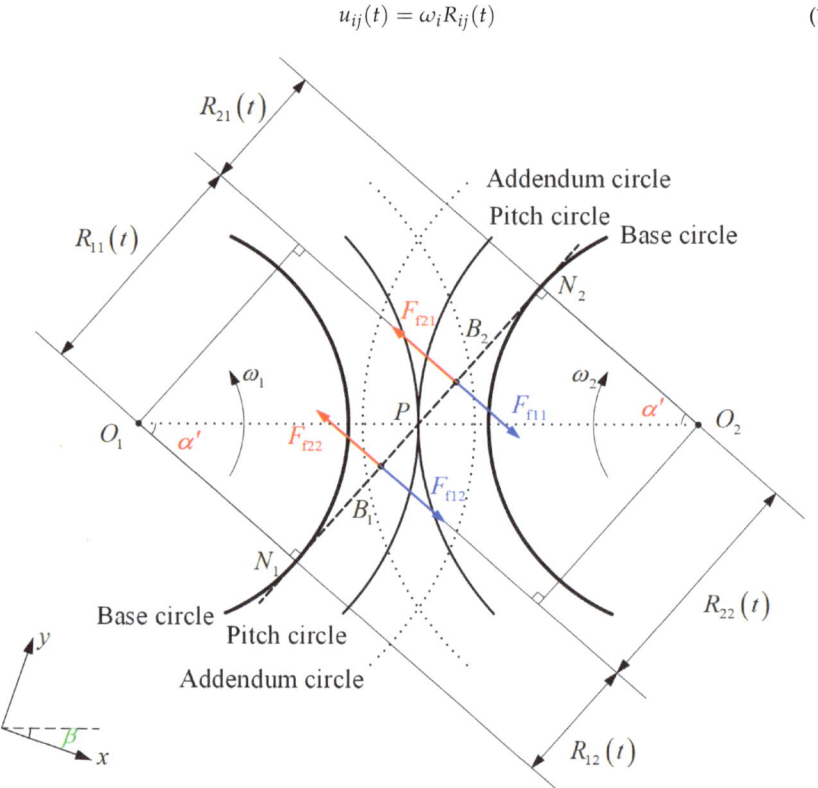

Figure 4. Schematic diagram of the sliding friction.

In Equation (12), $R_{ij}(t)$ is the contact radius, and can be obtained by:

$$\begin{aligned} R_{1j}(t) &= \sqrt{r_{a1}^2 - r_{b1}^2} + \mathrm{mod}(r_{b1}\omega_1 t, p_b) - p_b j \\ R_{2j}(t) &= (r_{b1} + r_{b2})\tan\alpha' - \sqrt{r_{a1}^2 - r_{b1}^2} + p_b j - \mathrm{mod}(r_{b1}\omega_1 t, p_b) \end{aligned} \tag{13}$$

where $\mathrm{mod}(\mathrm{num1}, \mathrm{num2})$ is the remainder function. From Figure 4 and Equation (13), it can be seen that the direction and the force arm of the sliding friction are affected by the transverse vibration.

The direction of sliding friction will change at the pitch point. In order to reconcile the difference between the actual direction and the assumed direction, a sign function is introduced here, which can be written as:

$$\lambda_j = \begin{cases} 1, & v_{sj} > 0 \\ 0, & v_{sj} = 0 \\ -1, & v_{sj} < 0 \end{cases} \tag{14}$$

In addition, the load sharing ratio model proposed by Pedrero et al. [36] is employed to determine the DMF between different meshing tooth pairs. The DMF between the jth meshing tooth pair can be determined by:

$$F_{Nij} = L_j(t)F_m \tag{15}$$

where $L_j(t)$ is the load sharing ratio of the jth meshing tooth pair.

According to the Coulomb friction law, the sliding friction force can be calculated by:

$$F_{fij} = L_j(t)F_m \tag{16}$$

Subsequently, the sliding friction torque can be expressed as:

$$T_{fij} = F_{fij}R_{ij}(t) \tag{17}$$

Thus, the DMF and sliding friction force/torque under the influence of transverse vibration are deduced.

3. Derivation of Equations of Motion

The dynamic differential equations of the spur gear system are derived in this section. As shown in Figure 1a, the gear system has six DOFs. Hence, the generalized coordinate vector of the dynamic model is $\mathbf{q} = [x_1 \ y_1 \ \theta_1 \ x_2 \ y_2 \ \theta_2]^T$. Accordingly, the system kinetic energy T, system potential energy U, and Rayleigh's dissipation function D for generalized coordinates can be expressed as:

$$
\begin{aligned}
T &= \left[m_1(\|\dot{\mathbf{r}}_1\|)^2 + m_2(\|\dot{\mathbf{r}}_2\|)^2 + I_1\dot{\varphi}_1^2 + I_2\dot{\varphi}_2^2 \right]/2 \\
U &= \left[k_1(x_1^2 + y_1^2) + k_2(x_2^2 + y_2^2) + k_m f_1^2(\delta) \right]/2 \\
D &= \left[c_1(\dot{x}_1^2 + \dot{y}_1^2) + c_2(\dot{x}_2^2 + \dot{y}_2^2) + \left(c_{t1}\dot{\theta}_1^2 + c_{t2}\dot{\theta}_2^2 \right) + c_m f_2^2(\delta) \right]/2
\end{aligned} \tag{18}
$$

The generalized force or torque subjected to the system can be given by:

$$
\mathbf{Q} = \left\{
\begin{array}{l}
\sum_{j=1}^{n=\text{ceil}(m_p)} F_{f1j} \cos(\alpha' - \beta) - \sum_{j=1}^{n=\text{ceil}(m_p)} F_{f1j} \sin(\alpha' - \beta) - m_1 g \\
T_1 - \sum_{j=1}^{n=\text{ceil}(m_p)} F_{f1j} R_{1j} - \sum_{j=1}^{n=\text{ceil}(m_p)} F_{f2j} \cos(\alpha' - \beta) \\
\sum_{j=1}^{n=\text{ceil}(m_p)} F_{f2j} \sin(\alpha' - \beta) - m_2 g - T_2 + \sum_{j=1}^{n=\text{ceil}(m_p)} F_{f2j} R_{2j}
\end{array}
\right\}^T \tag{19}
$$

The equations of motion of the spur gear system can be derived from Lagrange's equation, which can be described as:

$$\frac{d}{dt}\left(\frac{\partial T}{\partial \dot{q}_i}\right) - \frac{\partial T}{\partial q_i} + \frac{\partial D}{\partial \dot{q}_i} + \frac{\partial U}{\partial q_i} = Q_i \tag{20}$$

where q_i is the generalized coordinate of the system and $i = 1, 2, 3, \ldots, n$. n is the degree of system freedom.

Substituting Equations (18) and (19) into Equation (20), the equations of motion of the spur gear system considering the influence of transverse vibration on meshing interface can be obtained as:

$$m_1\left(\ddot{x}_1 - \varepsilon_1\ddot{\theta}_1\sin\varphi_1 - \varepsilon_1\dot{\varphi}_1^2\cos\varphi_1\right) + c_1\dot{x}_1 + c_m f_2(\delta)f_{2,\dot{x}_1} + k_1 x_1 + k_m f_1(\delta)f_{1,x_1} = \sum_{j=1}^{n=\text{ceil}(m_p)} F_{f1j}\cos(\alpha'-\beta)$$

$$m_1\left(\ddot{y}_1 + \varepsilon_1\ddot{\theta}_1\cos\varphi_1 - \varepsilon_1\dot{\varphi}_1^2\sin\varphi_1\right) + c_1\dot{y}_1 + c_m f_2(\delta)f_{2,\dot{y}_1} + k_1 y_1 + k_m f_1(\delta)f_{1,y_1} = -m_1 g - \sum_{j=1}^{n=\text{ceil}(m_p)} F_{f1j}\sin(\alpha'-\beta)$$

$$(I_1 + m_1\varepsilon_1^2)\ddot{\theta}_1 - m_1\varepsilon_1(\ddot{x}_1\sin\varphi_1 - \ddot{y}_1\cos\varphi_1) + c_{t1}\dot{\varphi}_1 + c_m f_2(\delta)f_{2,\dot{\theta}_1} + k_m f_1(\delta)f_{1,\theta_1} = T_1 - \sum_{j=1}^{n=\text{ceil}(m_p)} F_{f1j}R_{1j} \quad (21)$$

$$m_2\left(\ddot{x}_2 - \varepsilon_2\ddot{\theta}_2\sin\varphi_2 - \varepsilon_2\dot{\varphi}_2^2\cos\varphi_2\right) + c_2\dot{x}_2 + c_m f_2(\delta)f_{2,\dot{x}_2} + k_2 x_2 + k_m f_1(\delta)f_{1,x_2} = -\sum_{j=1}^{n=\text{ceil}(m_p)} F_{f2j}\cos(\alpha'-\beta)$$

$$m_2\left(\ddot{y}_2 - \varepsilon_2\ddot{\theta}_2\cos\varphi_2 + \varepsilon_2\dot{\varphi}_2^2\sin\varphi_2\right) + c_2\dot{y}_2 + c_m f_2(\delta)f_{2,\dot{y}_2} + k_2 y_2 + k_m f_1(\delta)\delta_{y_2} = -m_2 g + \sum_{j=1}^{n=\text{ceil}(m_p)} F_{f2j}\sin(\alpha'-\beta)$$

$$(I_2 + m_2\varepsilon_2^2)\ddot{\theta}_2 - m_2\varepsilon_2(\ddot{x}_2\sin\varphi_2 + \ddot{y}_2\cos\varphi_2) + c_{t2}\dot{\varphi}_2 + c_m f_2(\delta)f_{2,\dot{\theta}_2} + k_m f_1(\delta)f_{1,\theta_2} = -T_2 + \sum_{j=1}^{n=\text{ceil}(m_p)} F_{f2j}R_{2j}$$

where the comma in the subscript represents the partial differentiation, e.g., $f_{,x_1} = \partial f/\partial x_1$ and $f_{,\dot{x}_1} = \partial f/\partial \dot{x}_1$. The expressions for the remaining variables are as follows:

$$f_{1,x_i} = \begin{cases} \delta_{,x_i} - b_{t,x_i}, & \delta > b_t \\ 0, & |\delta| \leq b_t \quad (i=1,2) \\ \delta_{,x_i} + b_{t,x_i}, & \delta < -b_t \end{cases} \quad (22)$$

$$f_{1,y_i} = \begin{cases} \delta_{,y_i} - b_{t,y_i}, & \delta > b_t \\ 0, & |\delta| \leq b_t \quad (i=1,2) \\ \delta_{,y_i} + b_{t,y_i}, & \delta < -b_t \end{cases} \quad (23)$$

$$f_{1,\theta_1} = \begin{cases} r_{b1} - b_{t,\theta_1}, & \delta > b_t \\ 0, & |\delta| \leq b_t \\ r_{b1} + b_{t,\theta_1}, & \delta < -b_t \end{cases} \quad (24)$$

$$f_{1,\theta_2} = \begin{cases} -r_{b2} - b_{t,\theta_2}, & \delta > b_t \\ 0, & |\delta| \leq b_t \\ -r_{b1} + b_{t,\theta_2}, & \delta < -b_t \end{cases} \quad (25)$$

$$b_{t,x_i} = (r_{b1} + r_{b2})\alpha'_{,x_i}\tan^2\alpha' \quad (26)$$

$$b_{t,y_i} = (r_{b1} + r_{b2})\alpha'_{,y_i}\tan^2\alpha' \quad (27)$$

$$b_{t,\theta_i} = 0 \quad (28)$$

$$f_{2,\dot{x}_i} = f_{1,x_i},\ f_{2,\dot{y}_i} = f_{1,y_i},\ f_{2,\dot{\theta}_i} = f_{1,\theta_i} \quad (29)$$

$$\dot{\delta} = \dot{x}_1\delta_{,x_1} + \dot{y}_1\delta_{,y_1} + \dot{\theta}_1\delta_{,\theta_1} + \dot{x}_2\delta_{,x_2} + \dot{y}_2\delta_{,y_2} + \dot{\theta}_2\delta_{,\theta_2} - \dot{e}(t) \quad (30)$$

$$\dot{b}_t = \left[(r_{b1} + r_{b2})\tan^2\alpha'\right]\left(\dot{x}_1\alpha'_{,x_1} + \dot{y}_1\alpha'_{,y_1} + \dot{x}_2\alpha'_{,x_2} + \dot{y}_2\alpha'_{,y_2}\right) \quad (31)$$

$$\delta_{,x_1} = \sin(\alpha'-\beta) + \left[(x_1-x_2)\cos(\alpha'-\beta) - (y_1-y_2)\sin(\alpha'-\beta)\right]\left(\alpha'_{,x_1} - \beta_{,x_1}\right) \quad (32)$$

$$\delta_{,y_1} = \cos(\alpha'-\beta) + \left[(x_1-x_2)\cos(\alpha'-\beta) - (y_1-y_2)\sin(\alpha'-\beta)\right]\left(\alpha'_{,y_1} - \beta_{,y_1}\right) \quad (33)$$

$$\delta_{,\theta_1} = r_{b1} \quad (34)$$

$$\delta_{,x_2} = -\sin(\alpha' - \beta) + [(x_1 - x_2)\cos(\alpha' - \beta) - (y_1 - y_2)\sin(\alpha' - \beta)](\alpha'_{,x_2} - \beta_{,x_2}) \quad (35)$$

$$\delta_{,y_2} = -\cos(\alpha' - \beta) + [(x_1 - x_2)\cos(\alpha' - \beta) - (y_1 - y_2)\sin(\alpha' - \beta)](\alpha'_{,y_2} - \beta_{,y_2}) \quad (36)$$

$$\delta_{,\theta_2} = -r_{b2} \quad (37)$$

in which:

$$\alpha'_{,x_1} = -\alpha'_{,x_2} = -\frac{(x_2 - x_1 + d)(r_{b1} + r_{b2})}{\left[(x_2 - x_1 + d)^2 + (y_2 - y_1)^2\right]\sqrt{(x_2 - x_1 + d)^2 + (y_2 - y_1)^2 - (r_{b1} + r_{b2})^2}} \quad (38)$$

$$\alpha'_{,y_1} = -\alpha'_{,y_2} = -\frac{(y_2 - y_1)(r_{b1} + r_{b2})}{\left[(x_2 - x_1 + d)^2 + (y_2 - y_1)^2\right]\sqrt{(x_2 - x_1 + d)^2 + (y_2 - y_1)^2 - (r_{b1} + r_{b2})^2}} \quad (39)$$

$$\beta_{,x_1} = -\beta_{,x_2} = \frac{y_2 - y_1}{(x_2 - x_1 + d)^2 + (y_2 - y_1)^2} \quad (40)$$

$$\beta_{,y_1} = -\beta_{,y_2} = -\frac{x_2 - x_1 + d}{(x_2 - x_1 + d)^2 + (y_2 - y_1)^2} \quad (41)$$

It can be seen from Equation (21) that the equations of motion of the system are nonlinear and coupled with each other. This also shows that the transverse vibration will affect the meshing parameters and change the DMF and sliding friction, etc. These changes will in turn affect the dynamic characteristics of the system, and then affect the time-varying characteristics of the meshing parameters and sliding friction.

When the influence of transverse vibration is ignored, the model (Figure 1b) commonly used in previous studies can be obtained by replacing α' with α, b_t with b_0, and setting $\beta = 0$. The detailed description of the previous model refers to the work of He [20] or Chen et al. [10].

4. Comparison and Discussion of Dynamic Response

4.1. System Parameters and Model Validation

Geometrical and physical parameters of the spur gear system are listed in Table 1. Gravity and mass eccentricity are neglected in the following study. Due to the strong nonlinearity and time variation, MATLAB ode15s, which is suitable for solving stiff problems, is used to simulate [37].

Table 1. Geometrical and physical parameters of spur gear system.

Parameters	Symbols	Values
Module/(mm)	m	10
Number of teeth	z_1/z_2	20/20
Pressure angle of reference circle/(deg)	α	20
Addendum coefficient	h^*	1
Tip clearance coefficient	c^*	0.25
Face width/(mm)	L	30
Gear mass/(kg)	m_1/m_2	6.57/6.57
Moment of inertia/(kg·m^2)	I_1/I_2	0.0365/0.0365
Designed contact ratio	m_p	1.5568
Equivalent shaft-bearing stiffness/(N/m)	k_1/k_2	1×10^8
Equivalent shaft-bearing damping/(N·s/m)	c_1/c_2	512.64/512.64
Torsional damping/(N·s/m)	c_{t1}/c_{t2}	143.29/143.29

For validation purposes, the method in Ref. [15] is used here, that is, some key dynamic responses of the new and previous models are compared. The same parameters are utilized in this section. Furthermore, the input and output torque are 300 N·m. Figure 5 shows the time-domain responses of center distance, pressure angle, contact ratio, backlash, and DTE and the difference of DTE and backlash under different friction coefficients.

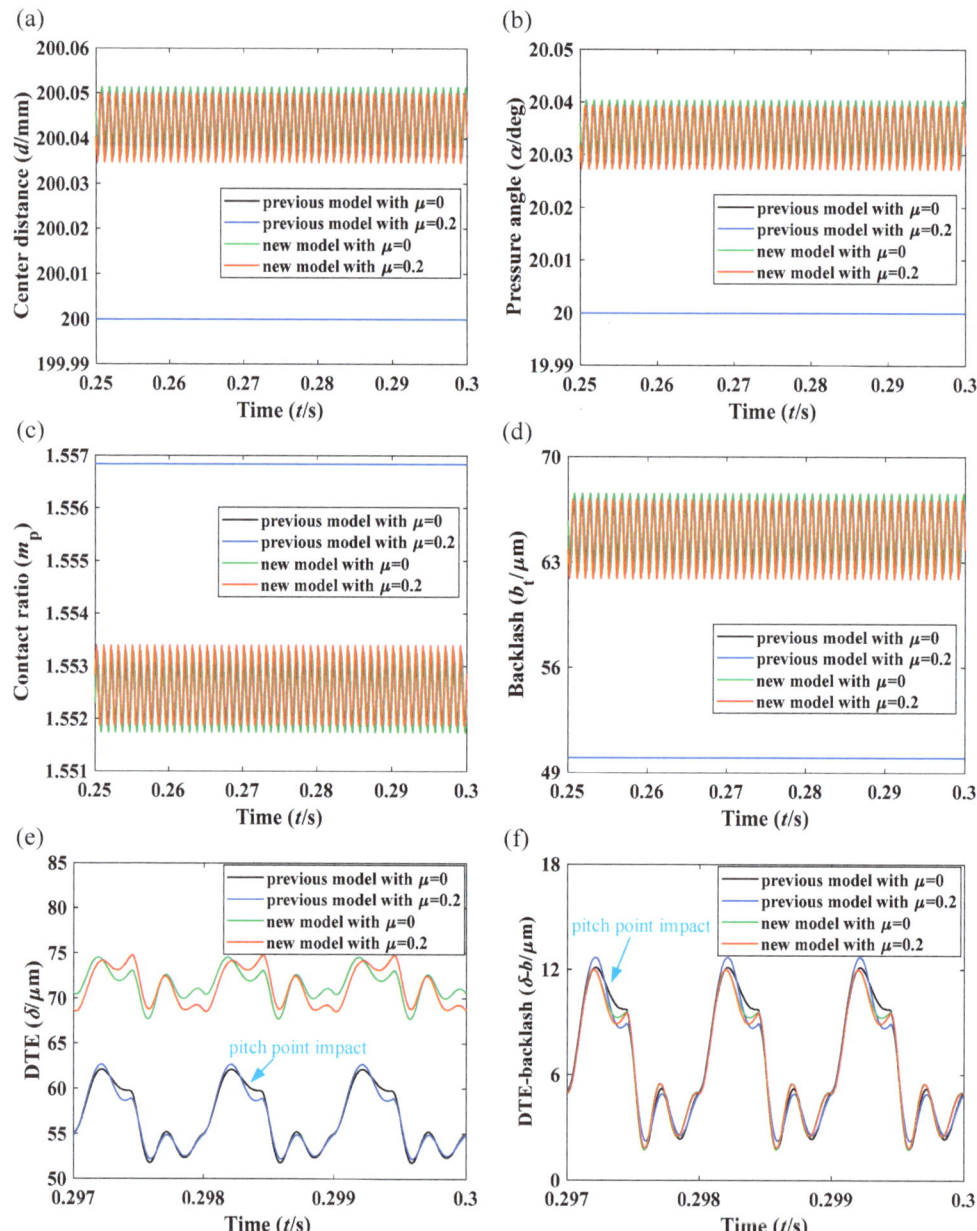

Figure 5. Time-domain curves for the various values of the friction coefficient at $n_1 = 3000$ r/min: (**a**) center distance, (**b**) pressure angle, (**c**) contact ratio, (**d**) backlash, (**e**) DTE, (**f**) difference between DTE and backlash.

As shown in Figure 5a–d, the meshing parameters (i.e., center distance, pressure angle, contact ratio, and backlash) of the previous model remain unchanged. However, these meshing parameters of the new model display time-varying characteristics due to the consideration of the influence of transverse vibration. Moreover, the change of friction coefficient has effects on the meshing parameters. The center distance, pressure angle and backlash of the new model are greater than the theoretical value or the design value, and the contact ratio is less than the theoretical value. In Figure 5e, the pitch point impact caused by the change of sliding friction direction can be clearly seen in the DTE curves of the previous model. However, the pitch point impact of the new model is not obvious. This is because the backlash of the previous model is constant, and the influence of sliding friction is directly reflected in the DTE response, thus affecting the DMF. For the new model, sliding friction not only affects the DTE but also affects the backlash. With reference to Equations (7), (8) and (21), the influence of sliding friction on DMF depends on the difference between the DTE and the backlash. For better illustration, Figure 5f illustrates the response curve of the difference between DTE and backlash. Compared with Figure 5e, the shape of the difference curve of the previous model is the same as the DTE due to the constant backlash, with translation. The difference curve of the new model changes obviously, similar to the difference curve of the previous model, and the pitch point impact can also be observed clearly.

It can be seen that the transverse vibration has an obvious influence on the meshing parameters and sliding friction, which affects the dynamic characteristics of the system. Therefore, compared with the previous model, the new model can to some extent provide a more realistic response prediction. Next, this paper will further study the influence of input speed and sliding friction coefficient on the system dynamic response.

4.2. Effect of the Input Speed on System Dynamic Response

The gear system exhibits different dynamic behavior under different input speed. In this section, the input speed is used as the control parameter to study the dynamic behaviors of the system.

4.2.1. Chaotic Response of the System with $\mu = 0$ and $\mu = 0.2$

The global characteristics of the new and previous models are compared via bifurcation diagrams with respect to the input speed n_1, as shown in Figure 6. In order to further illustrate the necessity of considering the sliding friction, two cases $\mu = 0$ and $\mu = 0.2$ are considered. Figure 6 indicates that the steady-state motion of the system presents complex bifurcation characteristics with the variation of the input speed in both new and previous models. It can be observed that the two models have obvious differences in predicting the global characteristics.

When $\mu = 0$, both two models perform periodic motion in the range of $[500, 8200]$ r/min as shown in Figure 6a,c. With the increase in input speed, the new model undergoes chaotic motion in the range of $(8200, 8720]$ r/min and periodic motion in the range of $(8720, 9620]$ r/min. In the range of $(9620, 10,500]$ r/min, the motion of the new model is mainly chaotic, but there also appears quasi-periodic motion. The previous model mainly experiences chaotic motion in the range of $(8220, 9960]$ r/min, turning into periodic motion in the regions of $(9960, 10,500]$ r/min. The new model resonates at 1200 r/min, earlier than 1300 r/min in the previous model. Comparing Figure 6b,d, the new model and the previous model perform periodic motion in the regions of $[500, 8240]$ r/min and $[500, 8140]$ r/min, respectively. In the ranges of $(8240, 8820]$ r/min and $(8820, 9860]$ r/min, the new model undergoes chaotic and periodic motion, respectively. Then, the new model carries out several bifurcations of quasi-periodic motion and chaotic motion at 9880 r/min, 10,040 r/min, and 10,060 r/min, and finally enters chaos. The previous model performs chaotic motion in the regions of $(8140, 8640]$ r/min and $(9140, 9900]$ r/min, and mainly undergoes periodic and quasi-periodic motions in the range of $(8640, 9140]$ r/min. The previous model returns to periodic motion at $n_1 > 9900$ r/min. It can be seen in Figure 6b,d that both models have three resonance peaks at 620 r/min, 940 r/min, and 1880 r/min, as indicated by the red circle numbers. Compared with the case of $\mu = 0$, the occurrence of these resonance points is caused by sliding friction. The region 1880 r/min corresponds to the torsional natural frequency of the system (620 Hz), 940 r/min and 620 r/min correspond to the first- and second-order super-harmonic resonances, respectively. These resonance peaks greatly affect the observation of the resonance peaks at 1200 r/min and 1300 r/min. Furthermore, the sliding friction has a significant effect on the hopping and bifurcation points of the two models. Considering the complex operation conditions of the gear system, the rotational speed may fluctuate slightly. In order to avoid the instability caused by the system rotational speed falling near the corresponding speed of the bifurcation point, it is necessary to consider the sliding friction in order to predict more accurately. Consequently, the subsequent part of this section gives only the result comparisons for $\mu = 0.2$.

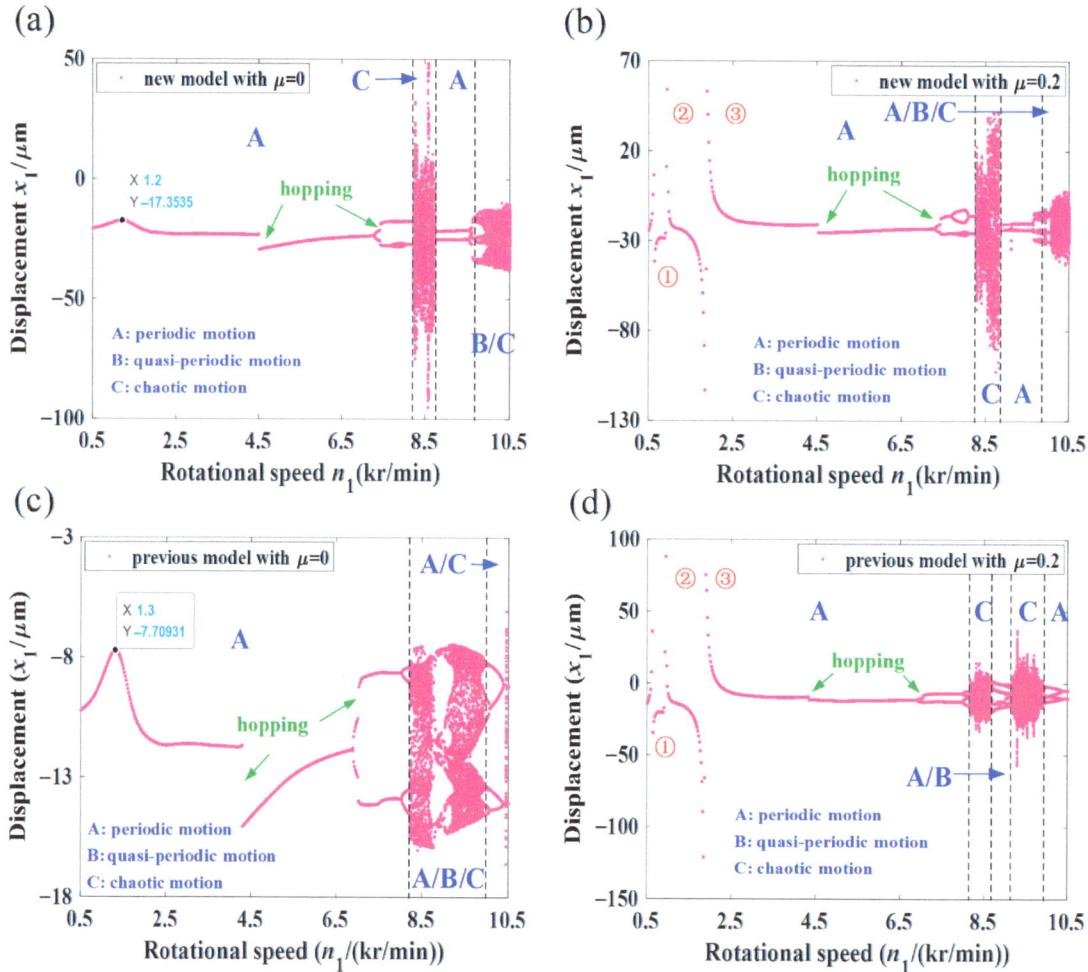

Figure 6. Bifurcation diagrams of x_1 vs. n_1: (**a**) new model with $\mu = 0$, (**b**) new model with $\mu = 0.2$. (**c**) previous model with $\mu = 0$, (**d**) previous model with $\mu = 0.2$.

4.2.2. Comparison of Dynamic Responses at Different Input Speed with $\mu = 0.2$

Based on the bifurcation diagram, $n_1 = 3000\,\text{r/min}$, $7000\,\text{r/min}$, $9500\,\text{r/min}$, and $10,400\,\text{r/min}$ are selected as examples to analyze the dynamic response of the system.

Figure 7 shows the comparison of dynamic responses of the new and previous models at $n_1 = 3000\,\text{r/min}$. From Figure 7a,b, the DTE amplitude of the new model is larger than that of the previous model, and the amplitude difference of DTE between the two models is mainly concentrated in f_m, $2f_m$ and $4f_m$. In Figure 7c, the mean value of vibration displacement x_1 of the new model is larger and the fluctuation of the new model is enhanced. The frequency-domain responses of x_1 in both models consist of the meshing frequency multiplication components (nf_m), and the amplitude difference of f_m is the most obvious, as shown in Figure 7d. As shown in Figure 7e, the backlash of the previous model is constant and the relative relationship between the DTE and backlash is intuitive. Therefore, it is generally not necessary to draw the backlash curve along with the DTE curve when judging the meshing state. However, the backlash of the new model is time-variable. It may not be so intuitive to determine the relative relationship between the two time variables (δ and b_t), and the response curve of the backlash is required to assist, as indicated by the black dashed line in Figure 7a (the backlash of the previous model is also drawn). It can be seen that the DTEs of the two models are always greater than the backlash, so they are each in a non-impact state, and F_m (Figure 7f) is always larger than 0. Although the DTE of the new model is much higher than that of the previous model, the difference between the DMF in the two models is not evident. This is because the backlash also becomes larger after considering the influence of transverse vibration. Figure 5f can also give a certain explanation, that is, the difference curves of the DTE and backlash of the two models are similar, and similar DMF results are obtained. As shown in Figure 7g, the mean value of the OLOA displacement of the previous model ($\delta_{oloa1} = x_1 \cos\alpha - y_1 \sin\alpha$) is far less than that of the new model ($\delta_{oloa1} = x_1 \cos(\alpha' - \beta) - y_1 \sin(\alpha' - \beta)$). The difference in DMF between the two models is not obvious, which means that the difference in sliding friction force is also not obvious. Therefore, the difference between δ_{oloa1} in the two models is obviously due to the consideration of the influence of the transverse vibration in the new model. The Poincaré map for the two models, which reveals that both models experience period-1 motion, is shown in Figure 7h.

Figure 8 performs the same comparison as Figure 7 at $n_1 = 7000\,\text{r/min}$. From Figure 8a,c, the DTE amplitude, the mean value, and the fluctuation of x_1 of the new model are still larger than those of the previous model. Compared with the new model, in addition to the amplitude difference at f_m, $2f_m$, and $3f_m$, the frequency components of $nf_m/2$ (n is an odd number) also appear in the DTE and x_1 frequency-domain responses of the previous model, as shown in Figure 8b,d. It can be seen from Figure 8e that the backlash of the new model is still larger than the initial one. Compared with Figure 7e, the amplitude change of the dynamic backlash is not obvious with the increase in input speed. Comparing the real and dashed lines in Figure 8a, it can be found that $\delta < b_t(b_0)$ occurs in both models, but δ is always greater than $-b_t(b_0)$, so both models are in a single-sided impact state. The situation of $F_m = 0$ appears as illustrated in Figure 8f, and the phenomenon of teeth separation occurs periodically. When the meshing state is switched from drive-side tooth mesh to teeth separation, the DMF first decreases to 0 and then increases in the reverse direction, resulting in tooth surface impact as shown in the magnified figure in Figure 8f. The transition process of the DMF can be described as $F_m > 0 \to F_m = 0 \to F_m < 0 \to F_m = 0$. Afterwards, the gear system enters teeth separation and the DMF is equal to 0 until the gear tooth meshes again. Both the mean value and amplitude of δ_{oloa1} in the new model are greater than those of the previous model, as shown in Figure 8g. It can be seen from the time-domain response and the Poincaré map (Figure 8h) that the previous model undergoes period-2 motion, while the new model still experiences period-1 motion.

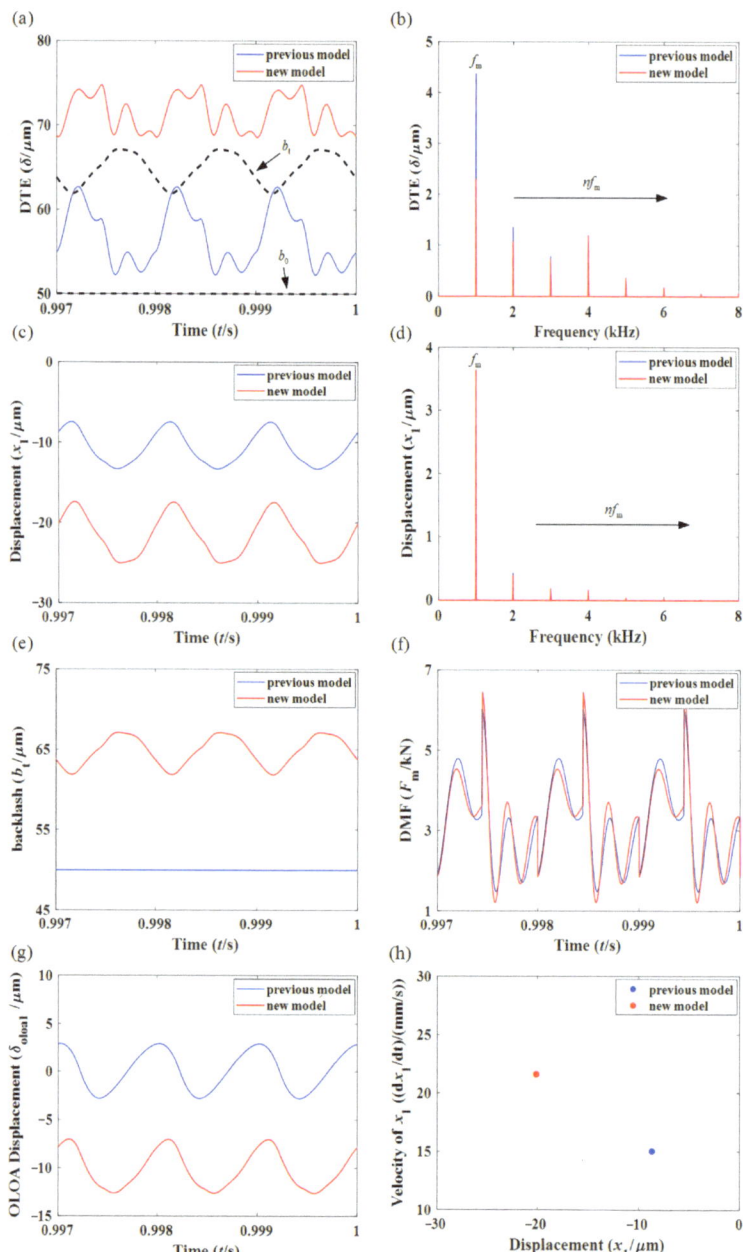

Figure 7. Comparison of dynamic responses at $n_1 = 3000$ r/min: (**a**,**b**) DTE, (**c**,**d**) displacement x_1, (**e**) backlash, (**f**) DMF, (**g**) OLOA displacement δ_{oloa1}, (**h**) Poincaré map.

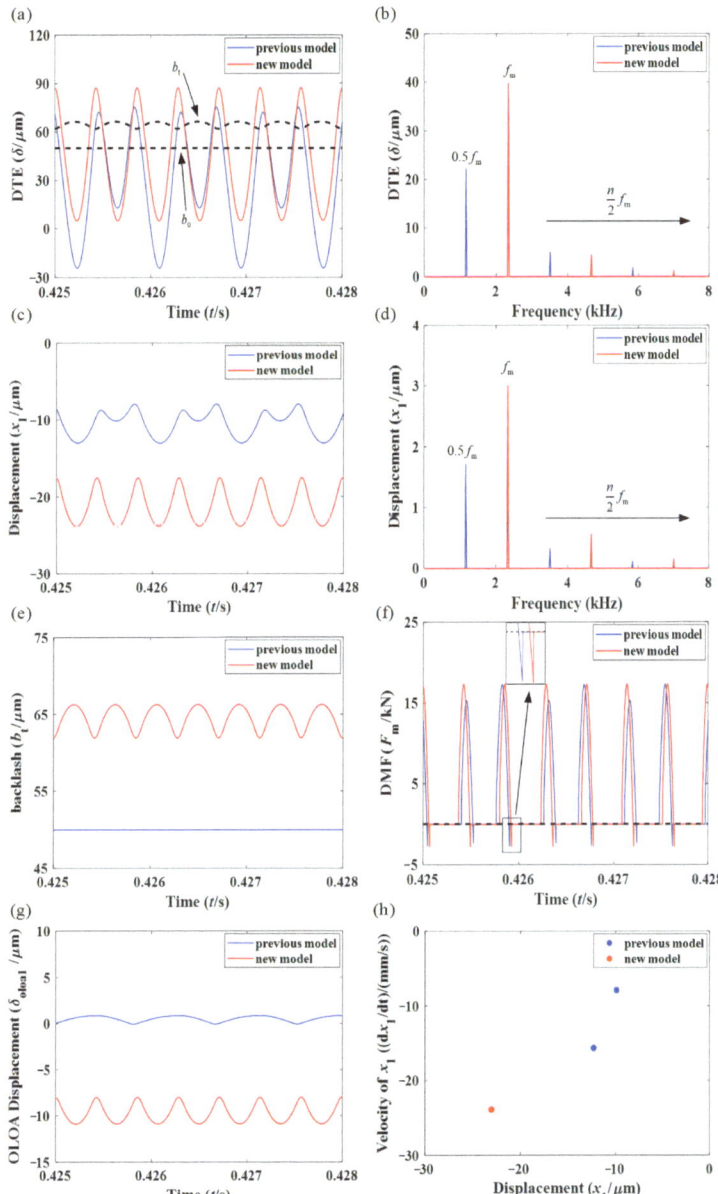

Figure 8. Comparison of dynamic responses at $n_1 = 7000$ r/min: (**a**,**b**) DTE, (**c**,**d**) displacement x_1, (**e**) backlash, (**f**) DMF, (**g**) OLOA displacement δ_{oloa1}, (**h**) Poincaré map.

Figure 9 illustrates the same comparison at $n_1 = 9500$ r/min. The complex time-domain response, the continuous spectrum in the frequency-domain response, and the Poincaré map show that the previous model has entered chaos. The period of the DTE and x_1 responses, the presence of $nf_m/2$ (n is an odd number) components, and the Poincaré map indicate that the new model experiences period-2 motion. The new model is still in a single-sided impact state. As for the previous model, $\delta < -b_0$ occurs and it is in a double-sided impact state. As shown in Figure 9e, the dynamic backlash is still higher than the initial backlash. Compared with Figure 8f, the previous model in Figure 9f not only converts the drive-side tooth mesh into teeth separation as shown in magnified figure A (in Figure 9f), but also changes from teeth separation to back-side tooth mesh as shown in magnified figure B (in Figure 9f). As the system moves from teeth separation to back-side tooth mesh, the DMF changes from $F_m = 0 \to F_m < 0 \to F_m = 0 \to F_m > 0 \to F_m = 0$, resulting in tooth surface impact. The DMF transition process of the new model is similar to that shown in Figure 8f.

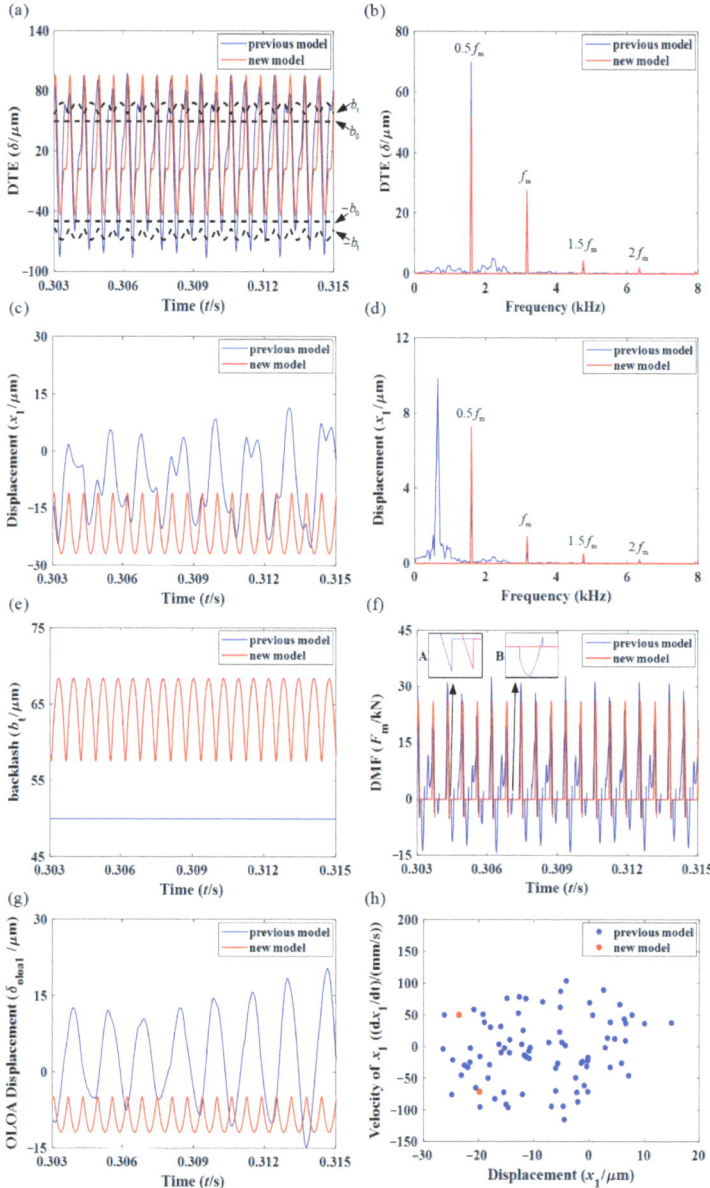

Figure 9. Comparison of dynamic responses at $n_1 = 9500$ r/min: (**a**,**b**) DTE, (**c**,**d**) displacement x_1, (**e**) backlash, (**f**) DMF, (**g**) OLOA displacement δ_{oloa1}, (**h**) Poincaré map.

When $n_1 = 10{,}400$ r/min, the comparison of dynamic responses of the two models is shown in Figure 10. It can be seen that the previous model undergoes period-2 motion, and obvious $0.5f_m$ and f_m components appear in the frequency-domain responses of DTE and x_1. The new model undergoes chaotic motion, continuous spectra appear in the frequency-domain responses of DTE and x_1, but $0.5f_m$ and f_m components are still observed. From Figure 10a, the previous model is in a double-sided impact state. Therefore, the DMF response curve contains the transition process from drive-side tooth mesh to teeth separation ($F_m > 0 \to F_m = 0 \to F_m < 0 \to F_m = 0$) and the transition process from teeth separation to back-side tooth mesh ($F_m = 0 \to F_m < 0 \to F_m = 0 \to F_m > 0 \to F_m = 0$) as shown in Figure 10f. The new model is in a state of constant transition between single-sided impact state and double-sided impact state. From the above analysis, it can be found that the meshing state can be judged not only by comparing DTE and backlash, but also by the transition process of DMF. The Poincaré map shown in Figure 10h confirms the previous analysis of the motion of the system.

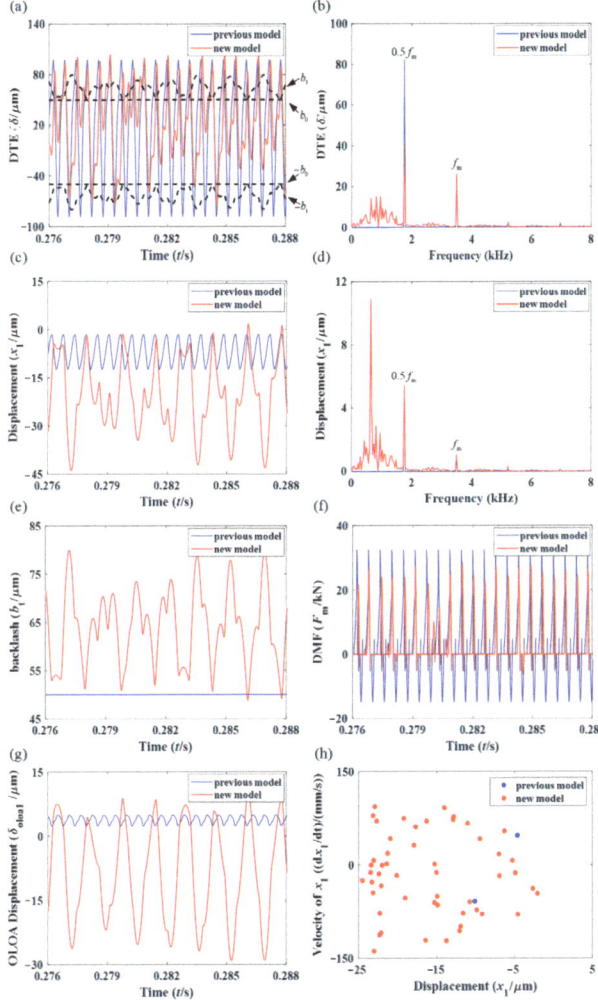

Figure 10. Comparison of dynamic responses at $n_1 = 10{,}400$ r/min: (**a**,**b**) DTE, (**c**,**d**) displacement x_1, (**e**) backlash, (**f**) DMF, (**g**) OLOA displacement δ_{oloa1}, (**h**) Poincaré map.

4.3. Effect of the Friction Coefficient on System Dynamic Response

Based on the previous analysis, it is obvious that the friction coefficient is one of the key parameters affecting the dynamic characteristics of the gear system. In this section, the friction coefficient is used as the control parameter to study the dynamic behaviors of the system.

4.3.1. Chaotic Response of the System at $n_1 = 3000$ r/min and $n_1 = 9000$ r/min

In this section, a low speed $n_1 = 3000$ r/min and a high speed 9000 r/min are chosen according to Figure 6 and other parameters remain the same. When $n_1 = 3000$ r/min, both two models undergo period-1 motion and are far away from the resonance, bifurcation, and hopping points; when $n_1 = 9000$ r/min, the motions of both models are relatively complicated. Referring to the literature [27,38], [0, 0.5] is selected as the variation range of the friction coefficient. Figure 11 shows the bifurcation diagrams with respect to the friction coefficient μ.

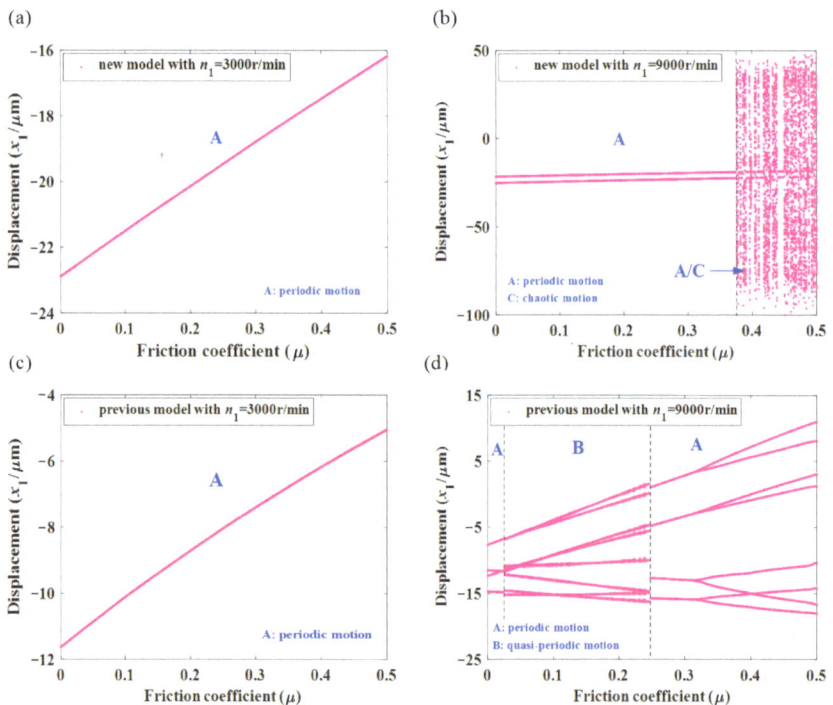

Figure 11. Bifurcation diagrams of x_1 vs. μ for the new and previous models.

As shown in Figure 11a,c, when $n_1 = 3000$ r/min, the two models maintain period-1 motion as the friction coefficient increases from 0 to 0.5. The motion form of the system is not affected by the variation of friction coefficient, but the response of the system is affected by the change of friction coefficient. When $n_1 = 9000$ r/min, the new model undergoes period-1 motion in the range of $\mu \in [0, 0.374]$. In the range of $\mu \in (0.374, 0.5]$, the new model mainly performs chaotic motion. The previous model experiences period-4 motion or period-8 motion in the regions of $\mu \in [0, 0.024] \cup [0.248, 0.5]$ and period-8 motion or quasi-periodic motion in the range of $\mu \in (0.024, 0.248)$. The different nonlinear dynamic characteristics of the new and previous models are evidently caused by the variation of meshing parameters and sliding friction induced by transverse vibration.

4.3.2. Comparison of Dynamic Responses under Different Friction Coefficients at $n_1 = 3000$ r/min

As the friction coefficient increases, the motion forms of the two models remain unchanged at $n_1 = 3000$ r/min within the parameters covered in this paper. Thus, the influence of friction coefficient variation on the amplitudes of system dynamic responses is emphasized. The dynamic responses of the new and previous models under different friction coefficients are shown in Figures 12 and 13, respectively. Compared with Figures 12a and 13a, as the friction coefficient increases, the DTE fluctuation of the new model is enhanced, while that of the previous model is not significantly changed. As reflected in the frequency domain, in addition to the $2f_m$ amplitude of DTE increasing with the increase in friction coefficient as in the previous model, the f_m amplitude of the new model also increases significantly, as shown in Figures 12b and 13b. As indicated in Figures 12c and 13c, the vibration displacement x_1 of the new model is larger than that of the previous model. With the increase in friction coefficient, the x_1 fluctuation of the previous model increases, but that of the new model decreases first and then increases. This may be due to the change of meshing interface. Accordingly, the f_m amplitude of x_1 in the new model first decreases and then increases, while that of the previous model gradually increases. However, the $2f_m$ amplitude of x_1 in both models increases with the increase in friction coefficient, as shown in Figures 12d and 13d. In Figures 12e and 13e, the variation trend of the backlash in the new model is similar to that of x_1, while the backlash in the previous model is constant. As mentioned in Section 4.1, for the previous model, the pitch point impact can be directly observed in the DTE time-domain response (Figure 13a), while the pitch point impact cannot be observed in the DTE time-domain response (Figure 12a) of the new model due to the dynamic backlash. In consequence, the response curve of the difference between DTE and backlash is given, i.e., Figure 12f. The pitch point impact becomes more and more obvious with the increase in friction coefficient as shown in Figures 12f and 13a,f. Compared with Figures 12g and 13g, the DMF of the new model is less affected by the change of friction coefficient than that of the previous model. Figures 12h and 13h show that the OLOA displacements of both models intensify due to the increase in friction coefficient. Nevertheless, the mean value and amplitude of δ_{oloa1} of the new model are greater than those of the previous model. When $\mu = 0$, the OLOA displacement of the previous model is 0, while that of the new model is not 0. This is because the direction of the LOA of the new model changes constantly under the influence of transverse vibration.

4.3.3. Comparison of Dynamic Responses under Different Friction Coefficients at $n_1 = 9000$ r/min

Based on Figure 11b,d, $\mu = 0.24$, 0.3, and 0.42 are taken as examples to carry out dynamic response analysis of the system at $n_1 = 9000$ r/min.

Figure 14 illustrates the comparison of dynamic responses of the two models with $\mu = 0.24$. Figure 14a shows that there is little difference in the amplitudes of DTE between the two models. From Figure 14b, the DTE frequency-domain response of the new model is mainly located at $0.5f_m$ and f_m. The DTE frequency-domain response of the previous model also contains $0.25f_m$ and $0.75f_m$ components in addition to the components of the new model. As shown in Figure 14c, the x_1 amplitude of the new model is still greater than that of the previous model. Compared with the new model (Figure 14d), the x_1 frequency-domain response of the previous model consists of more components, such as $0.125f_m$ and $0.25f_m$. As shown in Figure 14e, the backlash of the new model is obviously greater than that of the previous model. According to the transition process of DMF shown in Figure 14f, it can be concluded that the new model is in a single-sided impact state while the previous model is in a double-sided impact state. Both the mean value and amplitude of δ_{oloa1} in the new model are still greater than those in the previous model, as shown in Figure 14g. It can be seen that the new model experiences period-2 motion and the previous model undergoes quasi-periodic motion.

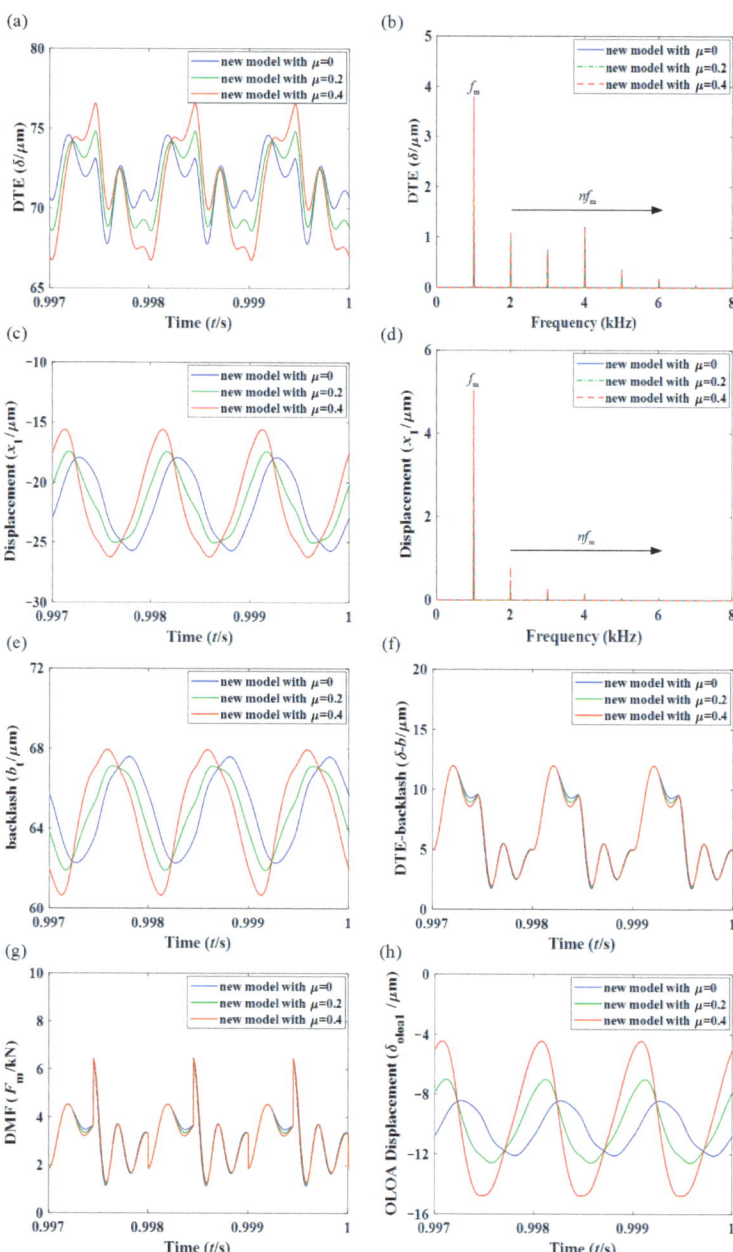

Figure 12. Dynamic responses of the new model with different friction coefficients at $n_1 = 3000$ r/min: (**a**,**b**) DTE, (**c**,**d**) displacement x_1, (**e**) backlash, (**f**) difference between DTE and backlash, (**g**) DMF, (**h**) OLOA displacement δ_{oloa1}.

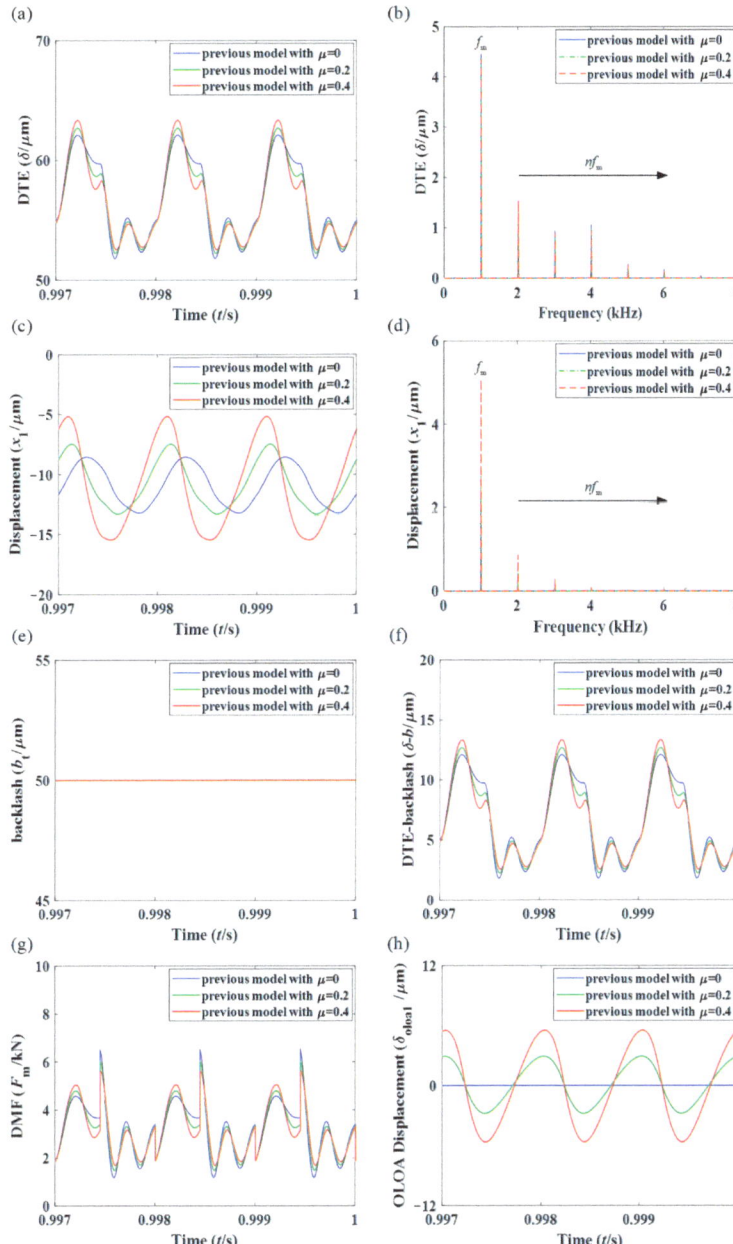

Figure 13. Dynamic responses of the previous model with different friction coefficients at $n_1 = 3000$ r/min: (**a**,**b**) DTE, (**c**,**d**) displacement x_1, (**e**) backlash, (**f**) difference between DTE and backlash, (**g**) DMF, (**h**) OLOA displacement δ_{oloa1}.

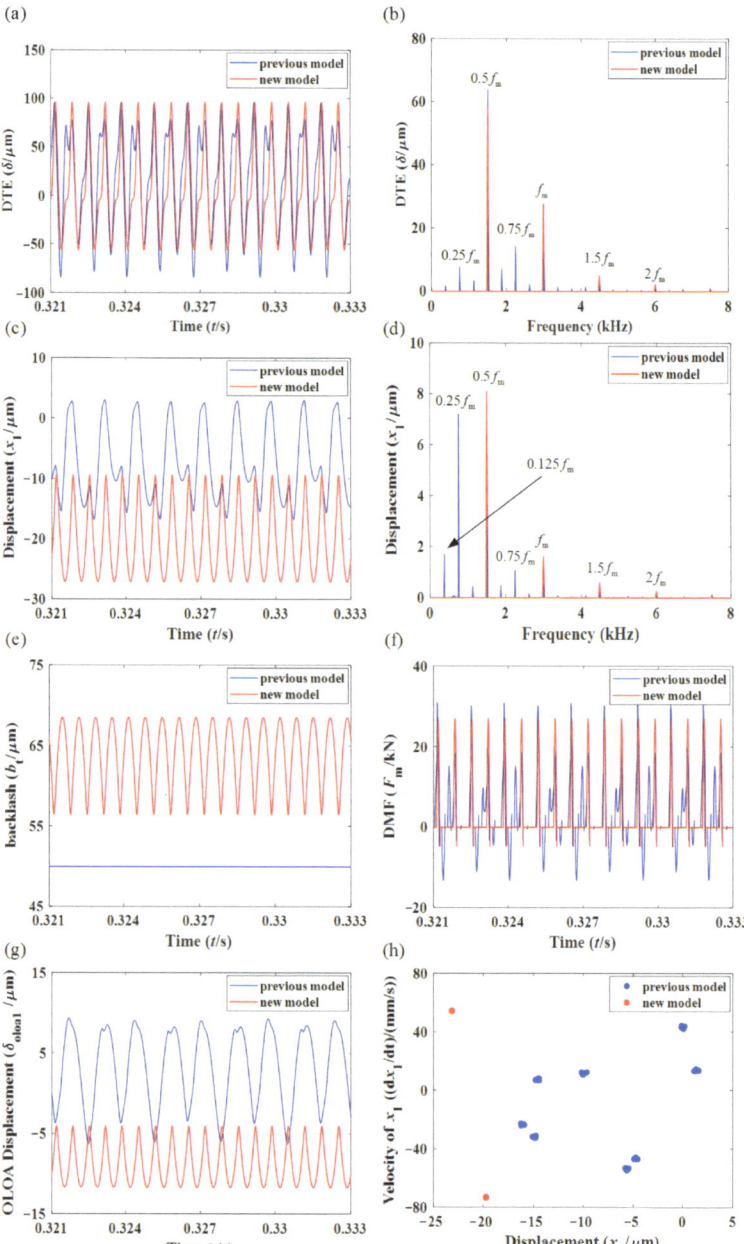

Figure 14. Comparison of dynamic responses with $\mu = 0.24$: (**a**,**b**) DTE, (**c**,**d**) displacement x_1, (**e**) backlash, (**f**) DMF, (**g**) OLOA displacement δ_{oloa1}, (**h**) Poincaré map.

Considering $\mu = 0.3$, the comparison of dynamic responses of the two models is presented in Figure 15. Figure 15a illustrates that there is still little difference in the amplitudes of DTE between the two models. The x_1 amplitude of the new model is still greater than that of the previous model, as shown in Figure 15c. As indicated in Figure 15e, the backlash of the new model is still larger than that of the previous model. From the

DMF response curve plotted in Figure 15f, it can be seen that the previous model is still in a double-sided impact state and the new model is still in a single-sided impact state. The OLOA displacements of the two models in Figure 15g are still significantly different. According to Figure 15b,d,h, the previous model and the new model undergo period-4 motion and period-2 motion, respectively.

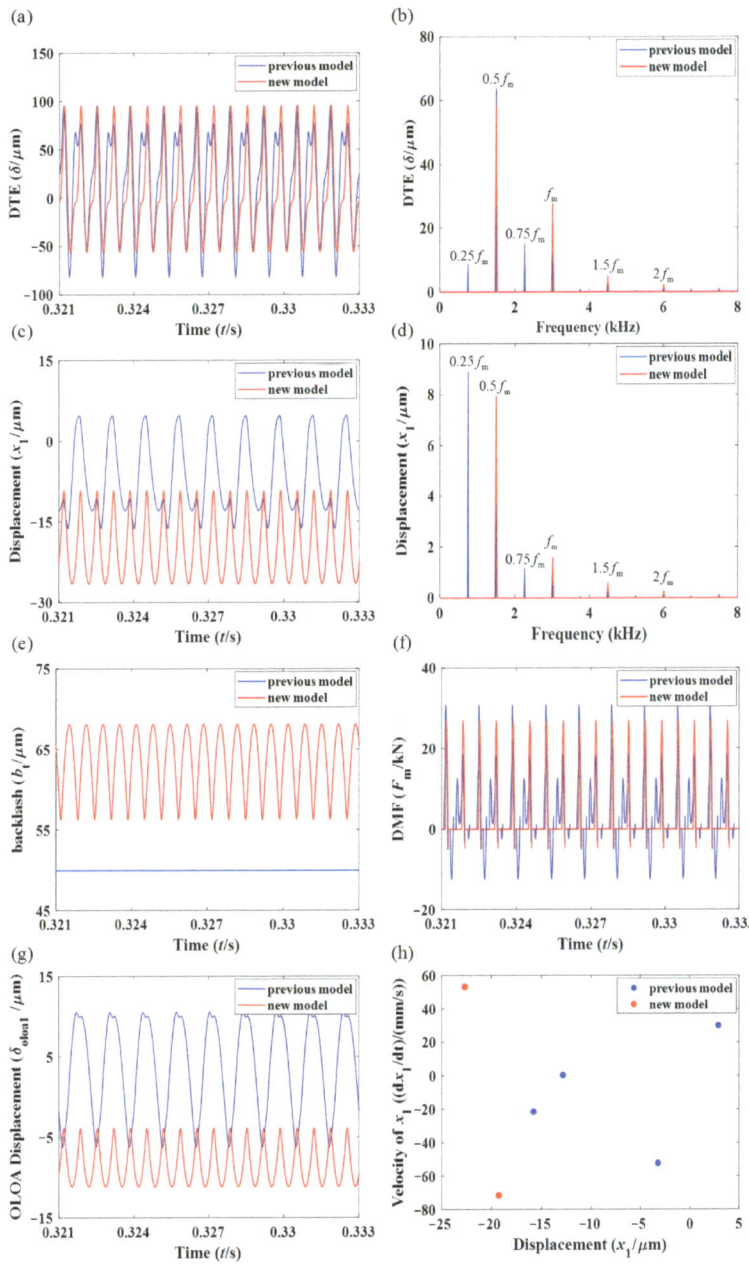

Figure 15. Comparison of dynamic responses with $\mu = 0.3$: (**a**,**b**) DTE, (**c**,**d**) displacement x_1, (**e**) backlash, (**f**) DMF, (**g**) OLOA displacement δ_{oloa1}, (**h**) Poincaré map.

Figure 16 shows the comparison of dynamic responses of the two models with $\mu = 0.42$. The black circles and boxes in Figure 16f indicate local magnification. It can be seen that the new model has entered chaos, and the previous model transfers from period-4 motion to period-8 motion by period-doubling bifurcation. S shown in Figure 16c, the fluctuation of x_1 of the new model clearly intensifies. The fluctuation of the backlash becomes more obvious, and even appears to be smaller than the initial backlash, as shown in Figure 16e. Both models are in a double-sided impact state, as shown in Figure 16f.

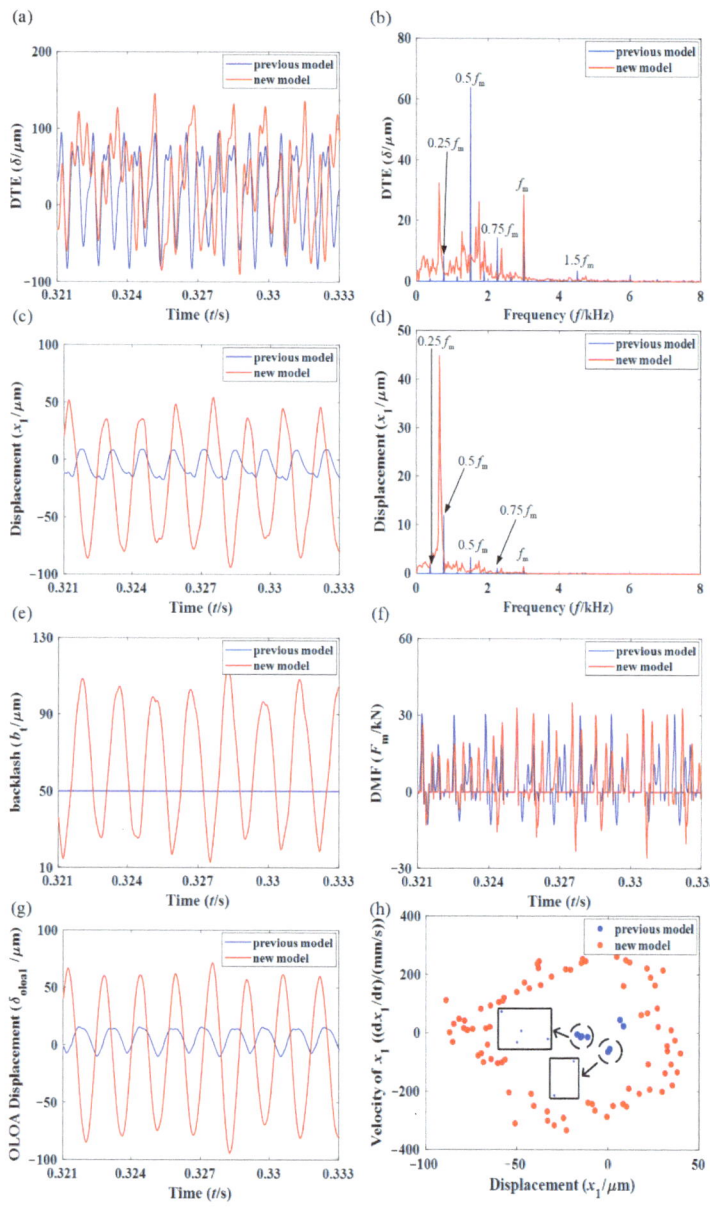

Figure 16. Comparison of dynamic responses with $\mu = 0.42$: (**a**,**b**) DTE, (**c**,**d**) displacement x_1, (**e**) backlash, (**f**) DMF, (**g**) OLOA displacement δ_{oloa1}, (**h**) Poincaré map.

5. Conclusions

In this paper, a new six-DOFs dynamic model for a spur gear system with dynamic meshing parameters and sliding friction is presented, in which the sliding friction modeling under the influence of the transverse vibration is emphasized. Compared with the previous model, the new model can to some extent provide more realistic system-response prediction.

The effects of the input speed and friction coefficient on the dynamic response were investigated. Both the input speed and the friction coefficient have a significant influence on the dynamic characteristics of the spur gear system. Under a certain friction coefficient, with the increase in the input speed, the mean value and amplitude of vibration displacement x_1, backlash, and OLOA displacement of the new model are greater than those of the previous model in the periodic and quasi-periodic regions. The DTE and DMF of the new model are not always greater than those of the previous model. Considering a low speed, with the increasing friction coefficient, the x_1 fluctuation of the new model decreases first and then increases, but that of the previous model keeps increasing. In addition, the transition process may be a better choice to judge the meshing state of the new model due to the variation of the backlash.

The new model could be used in the dynamic analysis of spur gear systems with highly flexible support structures or under extreme vibration conditions. Certainly, experimental work relevant to this study is the first task to be carried out next. In the future, research can also be extended to different kinds of gear systems such as helical gears, bevel gears, and planetary gears. Furthermore, the flexible rotor and box should be considered and research should be extended to gear-rotor systems and gearbox systems in the future. This paper provides a basis for the study of dynamic characteristics, vibration, and noise control of gear systems.

Author Contributions: Conceptualization, H.L., D.Z. and J.W.; methodology, H.L., J.W. and Y.L. (Yu Liu); software, H.L., K.L. and Y.L. (Yifu Long); validation, D.Z. and J.W.; investigation, D.Z., J.W. and Y.L. (Yu Liu); resources, H.L., D.Z. and Y.L. (Yu Liu); data curation, H.L., K.L. and Y.L. (Yifu Long); writing—original draft preparation, H.L., D.Z. and J.W.; writing—review and editing, H.L., K.L., J.W. and Y.L. (Yifu Long); visualization, H.L. and Y.L. (Yifu Long); supervision, D.Z., J.W. and Y.L. (Yu Liu); project administration, D.Z., J.W. and K.L.; funding acquisition, D.Z. and Y.L. (Yu Liu). All authors have read and agreed to the published version of the manuscript.

Funding: This work was financially supported by the National Natural Science Foundation of China (Grant nos. are 52105083, 52175071 and 52205115), the major projects of aero-engines and gas turbines (J2019-I-0008-0008), and the Innovation Centre for Advanced Aviation Power (HKCX2020-02-016).

Institutional Review Board Statement: Not applicable.

Informed Consent Statement: Not applicable.

Data Availability Statement: Not applicable.

Conflicts of Interest: The authors declare no conflict of interest.

References

1. Wang, Y.; Ye, H.; Yang, L.; Tian, A. On the Existence of Self-Excited Vibration in Thin Spur Gears: A Theoretical Model for the Estimation of Damping by the Energy Method. *Symmetry* **2018**, *10*, 664. [CrossRef]
2. Chen, Z.G.; Shao, Y.M.; Lim, T.C. Non-Linear Dynamic Simulation of Gear Response under the Idling Condition. *Int. J. Automot. Technol.* **2012**, *13*, 541–552. [CrossRef]
3. Yin, M. Study on Dynamics of Herringbone Gear-Rotor-Journal Bearing System with Lubrication Effects. Ph.D. Thesis, Northwestern Polytechnical University, Xi'an, China, 2017.
4. Kahraman, A.; Singh, R. Non-Linear Dynamics of a Geared Rotor-Bearing System with Multiple Clearances. *J. Sound Vib.* **1991**, *144*, 469–506. [CrossRef]
5. Li, S.; Kahraman, A. A Tribo-Dynamic Model of a Spur Gear Pair. *J. Sound Vib.* **2013**, *332*, 4963–4978. [CrossRef]
6. Zhao, B.; Huangfu, Y.; Ma, H.; Zhao, Z.; Wang, K. The Influence of the Geometric Eccentricity on the Dynamic Behaviors of Helical Gear Systems. *Eng. Fail. Anal.* **2020**, *118*, 104907. [CrossRef]

7. Cao, Z.; Chen, Z.; Jiang, H. Nonlinear Dynamics of a Spur Gear Pair with Force-Dependent Mesh Stiffness. *Nonlinear Dyn.* **2020**, *99*, 1227–1241. [CrossRef]
8. Geng, Z.; Li, J.; Xiao, K.; Wang, J. Analysis on the Vibration Reduction for a New Rigid–Flexible Gear Transmission System. *J. Vib. Control* **2022**, *28*, 2212–2225. [CrossRef]
9. Kong, X.; Hu, Z.; Tang, J.; Chen, S.; Wang, Z. Effects of Gear Flexibility on the Dynamic Characteristics of Spur and Helical Gear System. *Mech. Syst. Signal Process.* **2023**, *184*, 109691. [CrossRef]
10. Siyu, C.; Jinyuan, T.; Caiwang, L.; Qibo, W. Nonlinear Dynamic Characteristics of Geared Rotor Bearing Systems with Dynamic Backlash and Friction. *Mech. Mach. Theory* **2011**, *46*, 466–478. [CrossRef]
11. Kim, W.; Yoo, H.H.; Chung, J. Dynamic Analysis for a Pair of Spur Gears with Translational Motion Due to Bearing Deformation. *J. Sound Vib.* **2010**, *329*, 4409–4421. [CrossRef]
12. Kim, W.; Lee, J.Y.; Chung, J. Dynamic Analysis for a Planetary Gear with Time-Varying Pressure Angles and Contact Ratios. *J. Sound Vib.* **2012**, *331*, 883–901. [CrossRef]
13. Chen, Y.-C. Time-Varying Dynamic Analysis for a Helical Gear Pair System with Three-Dimensional Motion Due to Bearing Deformation. *Adv. Mech. Eng.* **2020**, *12*, 168781402091812. [CrossRef]
14. Liu, H.; Zhang, C.; Xiang, C.L.; Wang, C. Tooth Profile Modification Based on Lateral-Torsional-Rocking Coupled Nonlinear Dynamic Model of Gear System. *Mech. Mach. Theory* **2016**, *105*, 606–619. [CrossRef]
15. Yi, Y.; Huang, K.; Xiong, Y.; Sang, M. Nonlinear Dynamic Modelling and Analysis for a Spur Gear System with Time-Varying Pressure Angle and Gear Backlash. *Mech. Syst. Signal Process.* **2019**, *132*, 18–34. [CrossRef]
16. Wang, S.; Zhu, R. Theoretical Investigation of the Improved Nonlinear Dynamic Model for Star Gearing System in GTF Gearbox Based on Dynamic Meshing Parameters. *Mech. Mach. Theory* **2021**, *156*, 104108. [CrossRef]
17. Jedliński, Ł. Influence of the Movement of Involute Profile Gears along the Off-Line of Action on the Gear Tooth Position along the Line of Action Direction. *Eksploat. Niezawodn.–Maint. Reliab.* **2021**, *23*, 736–744. [CrossRef]
18. Yang, H.; Shi, W.; Chen, Z.; Guo, N. An Improved Analytical Method for Mesh Stiffness Calculation of Helical Gear Pair Considering Time-Varying Backlash. *Mech. Syst. Signal Process.* **2022**, *170*, 108882. [CrossRef]
19. Tian, G.; Gao, Z.; Liu, P.; Bian, Y. Dynamic Modeling and Stability Analysis for a Spur Gear System Considering Gear Backlash and Bearing Clearance. *Machines* **2022**, *10*, 439. [CrossRef]
20. He, S. Effect of Sliding Friction on Spur and Helical Gear Dynamics and Vibro-Acoustics. Ph.D. Thesis, The Ohio State University, Columbus, OH, USA, 2008.
21. Vaishya, M.; Singh, R. Analysis of periodically varying gear mesh systems with coulomb friction using floquet theory. *J. Sound Vib.* **2001**, *243*, 525–545. [CrossRef]
22. Vaishya, M.; Singh, R. Sliding friction-induced non-linearity and parametric effects in gear dynamics. *J. Sound Vib.* **2001**, *248*, 671–694. [CrossRef]
23. Gunda, R.; Singh, R. Dynamic Analysis of Sliding Friction in a Gear Pair. In Proceedings of the 9th International Power Transmission and Gearing Conference, Parts A and B, ASMEDC, Chicago, IL, USA, 1 January 2003; Volume 4, pp. 441–448.
24. He, S.; Gunda, R.; Singh, R. Effect of Sliding Friction on the Dynamics of Spur Gear Pair with Realistic Time-Varying Stiffness. *J. Sound Vib.* **2007**, *301*, 927–949. [CrossRef]
25. He, S.; Rook, T.; Singh, R. Construction of Semianalytical Solutions to Spur Gear Dynamics Given Periodic Mesh Stiffness and Sliding Friction Functions. *J. Mech. Des.* **2008**, *130*, 122601. [CrossRef]
26. Ghosh, S.S.; Chakraborty, G. Parametric Instability of a Multi-Degree-of-Freedom Spur Gear System with Friction. *J. Sound Vib.* **2015**, *354*, 236–253. [CrossRef]
27. Zhou, S.; Song, G.; Sun, M.; Ren, Z. Nonlinear Dynamic Response Analysis on Gear-Rotor-Bearing Transmission System. *J. Vib. Control* **2018**, *24*, 1632–1651. [CrossRef]
28. Shi, J.; Gou, X.; Zhu, L. Modeling and Analysis of a Spur Gear Pair Considering Multi-State Mesh with Time-Varying Parameters and Backlash. *Mech. Mach. Theory* **2019**, *134*, 582–603. [CrossRef]
29. Shi, J.; Gou, X.; Jin, W.; Feng, R. Multi-Meshing-State and Disengaging-Proportion Analyses of a Gear-Bearing System Considering Deterministic-Random Excitation Based on Nonlinear Dynamics. *J. Sound Vib.* **2023**, *544*, 117360. [CrossRef]
30. Wang, C. Dynamic Model of a Helical Gear Pair Considering Tooth Surface Friction. *J. Vib. Control* **2020**, *26*, 1356–1366. [CrossRef]
31. Hu, B.; Zhou, C.; Wang, H.; Chen, S. Nonlinear Tribo-Dynamic Model and Experimental Verification of a Spur Gear Drive under Loss-of-Lubrication Condition. *Mech. Syst. Signal Process.* **2021**, *153*, 107509. [CrossRef]
32. Luo, W.; Qiao, B.; Shen, Z.; Yang, Z.; Cao, H.; Chen, X. Investigation on the Influence of Spalling Defects on the Dynamic Performance of Planetary Gear Sets with Sliding Friction. *Tribol. Int.* **2021**, *154*, 106639. [CrossRef]
33. Jiang, H.; Liu, F. Dynamic Characteristics of Helical Gears Incorporating the Effects of Coupled Sliding Friction. *Meccanica* **2022**, *57*, 523–539. [CrossRef]
34. Ma, H.; Song, R.; Pang, X.; Wen, B. Time-Varying Mesh Stiffness Calculation of Cracked Spur Gears. *Eng. Fail. Anal.* **2014**, *44*, 179–194. [CrossRef]
35. Kahraman, A.; Singh, R. Non-Linear Dynamics of a Spur Gear Pair. *J. Sound Vib.* **1990**, *142*, 49–75. [CrossRef]
36. Pedrero, J.I.; Pleguezuelos, M.; Artés, M.; Antona, J.A. Load Distribution Model along the Line of Contact for Involute External Gears. *Mech. Mach. Theory* **2010**, *45*, 780–794. [CrossRef]

37. Shampine, L.F.; Reichelt, M.W. The MATLAB ODE Suite. *SIAM J. Sci. Comput.* **1997**, *18*, 1–22. [CrossRef]
38. Wen, S.; Huang, P. *Principles of Tribology*, 2nd ed.; John Wiley & Sons, Ltd.: Hoboken, NJ, USA, 2017; ISBN 978-1-119-21490-8.

Disclaimer/Publisher's Note: The statements, opinions and data contained in all publications are solely those of the individual author(s) and contributor(s) and not of MDPI and/or the editor(s). MDPI and/or the editor(s) disclaim responsibility for any injury to people or property resulting from any ideas, methods, instructions or products referred to in the content.

Article

Stiffness Characteristics and Analytical Model of a Flange Joint with a Spigot

Hao Liu [1], Jianjun Wang [1], Yu Liu [2,*], Zhi Wang [2] and Yifu Long [1]

[1] School of Energy and Power Engineering, Beihang University, Beijing 100191, China; lh2010@buaa.edu.cn (H.L.); wangjianjun@buaa.edu.cn (J.W.); lyfleo@buaa.edu.cn (Y.L.)
[2] Key Lab of Advance Measurement and Test Technology for Aviation Propulsion System, Shenyang Aerospace University, Shenyang 110136, China; wangzi629@sau.edu.cn
* Correspondence: 20190037@mail.sau.edu.cn

Abstract: Flange joints with spigots are widely used in aero-engines. The spigot will restrict the shear slipping between flanges, which, in turn, affects the stiffness characteristics of the joint. The current model and research on flange joints without spigots may not be suitable for the dynamic characteristics of aero-engines. Moreover, the complexity of contact pairs limits the application of the flange joint finite element (FE) model in aero-engine dynamics analysis. Therefore, a simplified analytical model of a flange joint with a spigot is proposed in this paper. First, the stiffness characteristic of the flange joint with a spigot is studied using the FE method. Second, a corresponding experiment is executed to verify the result of the FE analysis. Furthermore, based on the former FE and experimental analysis, one section of a flange joint is simulated by the Jenkins friction model and a spring. Then, a simplified analytical model of the entire flange joint is built according to the different statuses of each section. Finally, a simulation analysis of the stiffness characteristic is performed. The result shows that the simplified analytical model can be utilized to describe the bending stiffness characteristic of the flange joint with a spigot.

Keywords: flange joint; spigot; finite element analysis and experiment; hysteresis; analytical model

Citation: Liu, H.; Wang, J.; Liu, Y.; Wang, Z.; Long, Y. Stiffness Characteristics and Analytical Model of a Flange Joint with a Spigot. Symmetry 2023, 15, 1221. https://doi.org/10.3390/sym15061221

Academic Editor: Raffaele Barretta

Received: 21 April 2023
Revised: 30 May 2023
Accepted: 6 June 2023
Published: 7 June 2023

Copyright: © 2023 by the authors. Licensee MDPI, Basel, Switzerland. This article is an open access article distributed under the terms and conditions of the Creative Commons Attribution (CC BY) license (https:// creativecommons.org/licenses/by/ 4.0/).

1. Introduction

As an important symmetrical joint structure, the flange joint has several advantages: easy installation, stable performance, and good centering. Thus, the flange joint is widely used in the rotor and stator of aero-engines [1–3]. Due to the non-negligible stiffness loss, the stiffness characteristic of the flange joint has a great influence on the dynamic characteristics of aero-engines.

The stiffness characteristic of a flange joint depends on many parameters, making it rather expensive to study experimentally [4]. FE analysis could provide good results [5–7] and is not as expensive as experiments, but it is still highly time-consuming. Furthermore, if the actual structure of a flange joint is taken into consideration, the numerical simulation would face serious difficulties due to the complex structures and the large number of contact surfaces. Therefore, scholars have begun to pay attention to the simplified modeling of flange joints and have carried out in-depth research on the stiffness characteristics of flange joints [8,9]. Some linear, simplified models were developed in order to be utilized in modal analysis [10–12]. Besides, a variety of nonlinear simplified models were also established to simulate the nonlinear characteristics of flange joints. Luan et al. [13] developed a simplified nonlinear model by using bilinear springs to simulate the axial stiffness of a flange joint. Based on the stiffness model of Luan, Wang et al. [14] introduced a bending stiffness model of a flange joint to simulate the influence of the stiffness loss on the rotor system. Bouzid et al. [15] studied fiber-reinforced plastic bolted flange joints integrity and bolt tightness. However, all of the above studies neglect friction behavior, which may lead to significant damping effects.

Because of the friction between the mating surfaces, micro/macro slip occurs in the tangential direction of the contact surface when a cyclic load is applied, resulting in hysteresis behavior [8]. Hysteresis has a significant impact on dynamic response, since the area of the hysteresis loop presents energy dissipation in one cycle. Thus, the friction behavior of the joint should be considered in the analytical model. Ahmadian and co-authors [16,17] considered the damping effects of friction, but their studies focused on a simple bolt joint rather than a flange joint. Bograd et al. [18] studied hysteresis by using the Jenkins friction model when the structure is under transverse load. Oldfield et al. [19,20] used the Jenkins friction model to simulate the hysteresis of the joint when the structure is under torsional load. These studies aimed at the simulation of the basic structure of a bolted joint. Nevertheless, the complex structure of the flange joint may have a direct impact on the stiffness characteristics. Hence, scholars further studied the hysteresis characteristics and the simulation of the flange joint. Van-Long et al. [21] experimentally attained the hysteresis of the flange joint without a spigot. Firrone et al. [22] studied the microslip, which leads to hysteresis occurring at the contact interface between two turbine disks in aero-engines. Shi and Zhang [23] proposed an improved contact parameter model for bolted joint interfaces and analyzed the flanged-bolted joints incorporating the proposed model. He and Li [24] established an axial double spring-bending beam model to simulate the bolted flange joint.

However, the spigot structure was not considered in the above studies. In fact, the shear slipping between the two pieces of flange is restricted due to the presence of the spigot. When the flange joint is under transverse load, one part of the flange plate is subject to tension and tends to be separated, and the other part is subject to compression [25]. Therefore, there is a noteworthy difference between the simulation of the flange joint with a spigot and the one without a spigot. Shuguo et al. [26] built an FE model of a flange joint with a spigot to study the stiffness characteristics under transverse load. It was found that the bending stiffness decreases suddenly once the load reaches a certain value in the process of loading, and the values of the bending stiffness before and after the decrease are both constant. Liu et al. [25] further studied the sudden decrease in the bending stiffness, and pointed out that the angle of the rotation–loading curve presents hysteresis characteristics when the joint with a spigot is under a certain transverse harmony load. However, the above studies are based on the FE method. Yu et al. [27] developed a simplified analytical model of the flange joint with a spigot; nevertheless, the simulation of the single sector model did not consider the deformation characteristic of flange closure. Due to the closure of the flange, the symmetrical deformation and sliding of the Jenkins friction model are limited, and the spigot can only undergo unidirectional deformation in the direction of opening. The influence of this feature on overall stiffness characteristics has not been studied. Moreover, these studies lack experimental validation.

The review of the literature above indicates that it is necessary to study the stiffness characteristic of the flange joint with a spigot using the FE method and experiments, and establish the corresponding simplified analytical model for convenience of use. For this reason, based on the authors' previous finite element studies [25,28], an experimental instrument is built to study and verify the stiffness characteristics of the flange joint, and a simplified analytical model is proposed on this basis.

The remainder of this paper is as follows: Part 2 briefly introduces the FE model of a flange joint and the analysis of angular bending stiffness characteristics completed in reference [25]. In Part 3, the experimental instrument used in this paper is illustrated, and the stiffness characteristic mentioned in Part 2 is verified. In Part 4, a simplified analytical model of the flange joint is proposed based on the former analysis; numerical simulations are carried out to verify accuracy and suitability. Finally, Part 5 gives some brief conclusions.

The main contributions of this paper are summarized as follows:

1. We designed and built an experimental instrument to research the angular bending stiffness characteristics of a flange joint with a spigot.

2. Based on the experimental method, our previous FE model and study is validated.
3. From the viewpoint of convenient engineering applications, we proposed a simplified analytical model for the deformation and sliding of the flange contact interface. The accuracy and applicability are verified by comparison with the FE model and experimental test results. It can simulate the stiffness characteristics of a flange joint with a spigot well.

2. FE Modeling and Analysis of Flange Joint with a Spigot

The analysis method in this section is the same as in our previous work [25]. The main structure parameters of the flange joint, including the cylinder, nut, head of bolt, contact interface, and spigot, are exactly the same as those in reference [25], as shown in Figure 1 and listed in Table 1. The corresponding FE model, composed of SOLID185 elements, is shown in Figure 2. The contact surfaces, which are signified by a yellow line in Figure 1, are modeled by CONTAC174 and TARGE170 in ANSYS. The contact interface of the spigot is set as interference fit for strict centering by setting KEYOPT CNOF to a positive value. PRETS179 is built into the bolt to simulate the preload. The FE model has 121,344 elements and 131,527 nodes. To guarantee the mesh quality, the flange, bolts, and accessories were divided into a large number of elements, and a coarse mesh was used far from the contact interface. The partial magnification in Figure 2 shows the details of the finite element mesh of the bolt and spigot. As shown in Figure 2, all nodes on the left edge of the cylinder are fixed constraints, and all nodes on the right edge of the cylinder are rigidly bound to a node in the axis of the cylinder. This node is built to apply transverse loads. This paper focuses on the angle-bending stiffness of the flange joint, which has an obvious influence on the transverse vibration of the structure. For simplicity, "stiffness" is used to represent "angle-bending stiffness" in the following text.

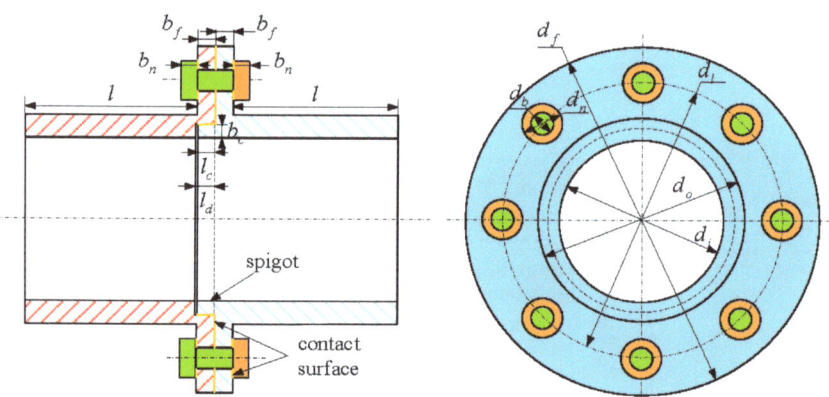

Figure 1. Sketch of the model for the flange joint.

Table 1. Main parameters of the flange joint.

Parameters	Values	Parameters	Values
b_f	2.5 mm	d_f	58 mm
b_n	3 mm	d_o	40 mm
l	27.5 mm	d_i	30 mm
b_c	2.5 mm	d_l	49 mm
l_c	1.7 mm	d_n	7 mm
l_d	2 mm	d_b	5 mm

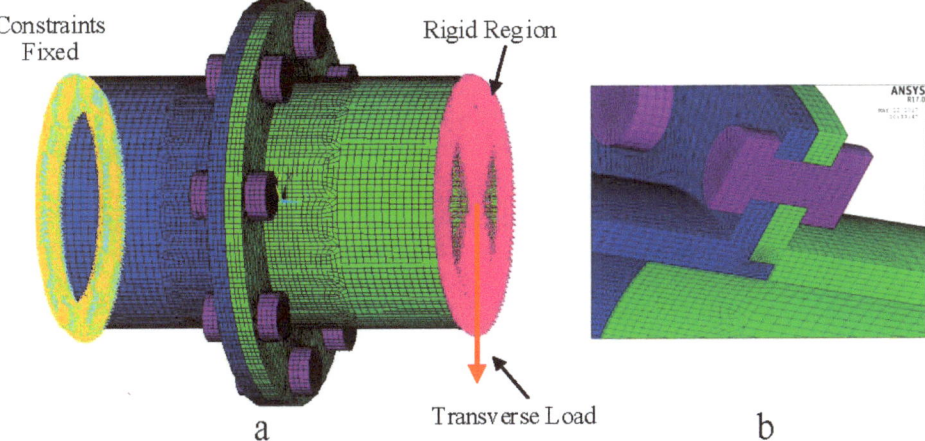

Figure 2. FE model of the flange joint: (**a**) the structural integral FE model, (**b**) the partial magnification of the bolt and spigot.

In order to study the spigot's influence on the stiffness of the flange joint, a full transient dynamic analysis, which can take the nonlinear factors of the contact surfaces into consideration, is carried out. The transverse load applied to the FE model is shown in Figure 3. The entire loading procedure can be divided into two stages. In the first stage, which begins at 0 s and ends at 1 s, the load remains zero, and time integration of ANSYS is turned off so that the preload can be applied as prestressing of the structure. The second stage begins at 1 s and ends at 3 s, in which two cycles of harmonic transverse load are applied to the model. The frequency of the harmonic load is set at 1 Hz to avoid the influence of the inertial force of the model.

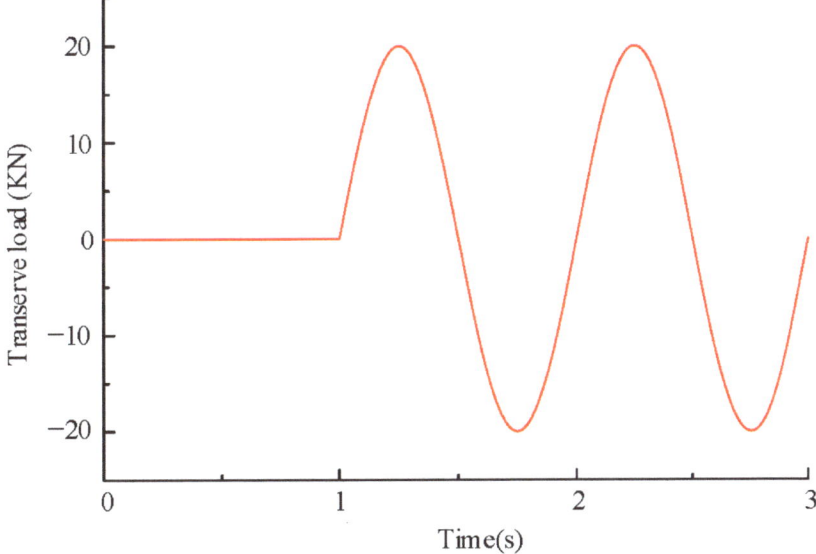

Figure 3. Time–domain curve of the transverse harmonic load.

Considering the rigid region, the angle of rotation of the node on which the transverse load is applied can represent the rotation of the right edge of the cylinder, and it can be an object of the angle-bending stiffness analysis. The curve of angle of rotation-load under the transverse load shown in Figure 3 is performed in Figure 4. The stiffness is indicated by the slope of the curve. Initially, the angle of rotation increases linearly with the increase in load. This process is named "initial loading". At the end of initial loading, the slope of the curve decreases suddenly, which means the bending stiffness decreases. k_1 indicates the stiffness before change, and k_2 presents the stiffness after change. When the load reaches its peak, the process of loading transforms into unloading. In the initial stage of the unloading process, the stiffness is equal to k_1 (the value of initial loading). The longitude of the curve of the initial stage of unloading is almost double that of the initial loading stage. When the initial stage of unloading process ends, the stiffness changes to k_2 again, and almost remains at this value until the end of the process of reverse loading. The curve of the process of reverse unloading is nearly symmetric to the one of unloading about the origin. After the process of reverse unloading comes the second cycle, which starts with the process of loading but not the process of initial loading. It can be distinguished that the processes of loading and unloading present different routes that form a cycle, which is usually called the hysteresis loop.

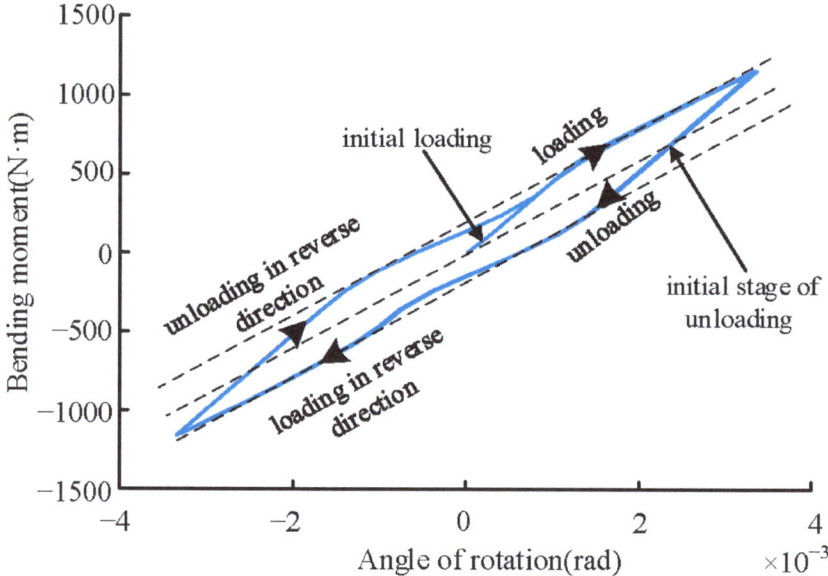

Figure 4. Angle of rotation–load curve under transverse harmonic load.

Figure 5 presents the angle of rotation–load curves under different amplitudes of harmonic load. The loading processes of the three curves have nearly the same route and are only different in longitude. The hysteresis loop can be clearly observed when the load amplitude is 900 N · m or 1200 N · m. The curve under a load of 600 N · m amplitude is nearly a straight line. This is because the load is too small to reach the point where the stiffness changes, thus the routes of loading and unloading coincide. Up to now, research on the stiffness characteristics of the flange joint with a spigot based on the FE method has been completed.

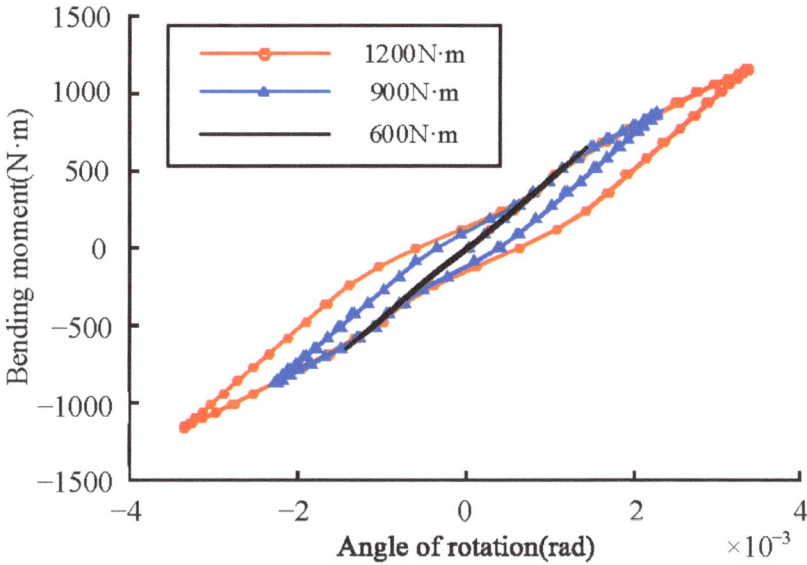

Figure 5. Angle of rotation–load curves under different amplitudes of harmonic load.

3. Experimental Verification

In order to validate the result of the FE model, the stiffness of the flange joint is measured in this paper. The instrument setup of the experiment is shown in Figure 6. The constraint of the experimental sample is set the same as the FE model. The shape of the spigot is illustrated in Figure 6b. In order to impose the force both upward and downward and make the right end rotate freely, the force load is applied by a pin that is plugged through the cylinder, as shown in Figure 6c. Two dial indicators are utilized to measure the deformations on the upside and downside of the right end, through which the angle of rotation can be calculated.

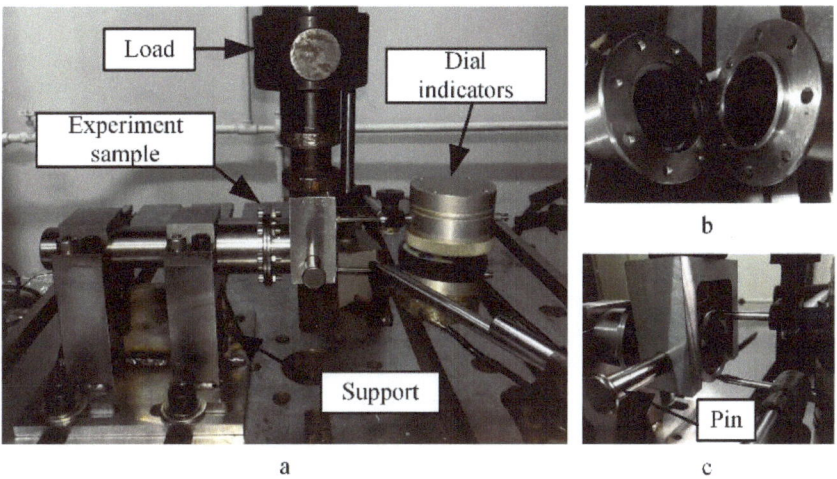

Figure 6. Stiffness measurement experiment: (**a**) the instrument setup of the experiment, (**b**) the spigot, (**c**) the application of load and points of measure.

Considering the cyclic load of 1200 N · m amplitude, the measurement result of the angle of rotation–load is presented in Figure 7. Compared with the results of the FE method, as shown in Figures 4 and 5, the processes of loading and unloading and the reverse processes can be clearly observed. The difference is that the stage of initial loading is not obvious. This is because the stage of initial loading in the FE method is based on the condition that there is no prestress or pre-deformation on the contact surface of the spigot. However, this condition is hard to attain in the actual assembly of the flange joint. Besides, the hysteresis loop measured by experiment is thinner and longer than that of the FE method, as shown in Figures 5 and 7. This means that the stiffness of the experiment example is smaller than that of the FE model. There are two main reasons for this difference. One is that the spigot interface of the experiment example is smaller than that of the FE model for reasons of machining due to interference control, and the other is that the support structure is not entirely rigid. By comparison, the validity of the FE analysis on the stiffness characteristic of the flange joint with a spigot is verified. The relationship between the angle of rotation and load provides the direction and foundation for the creation of the simplified analytical model.

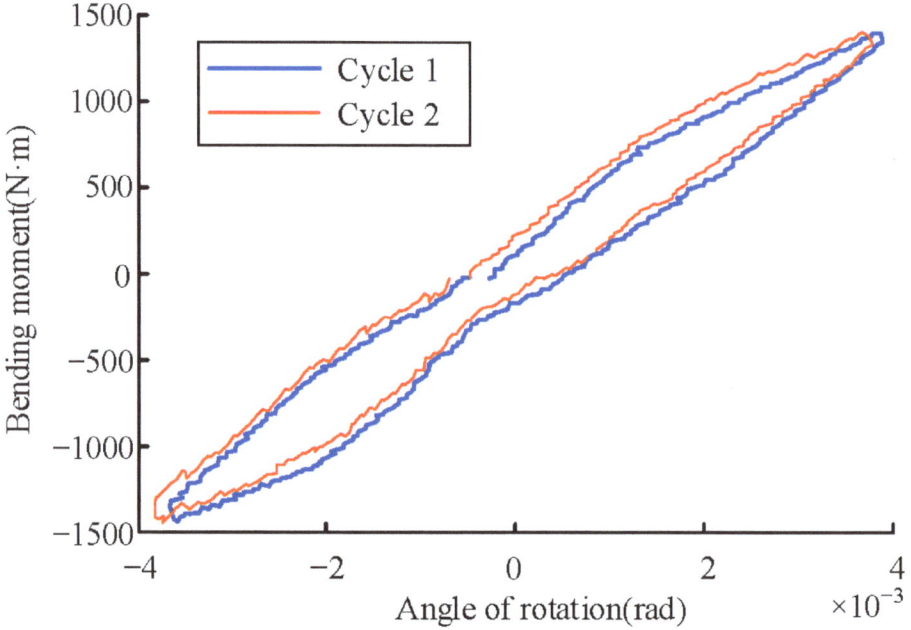

Figure 7. Curves of the experimental result of the angle of rotation–load.

4. Simplified Analytical Modeling and Analysis

Since the stiffness characteristic of the flange joint with a spigot is analyzed through the FE and experimental methods, simplified analytical modeling can be constructed. The processes for simplified analytical modeling and analysis are given below.

4.1. Division of the Sections of a Flange Joint

As a symmetrical structure, the entire flange joint can be divided into eight sections based on the number of bolts, as shown in Figure 8.

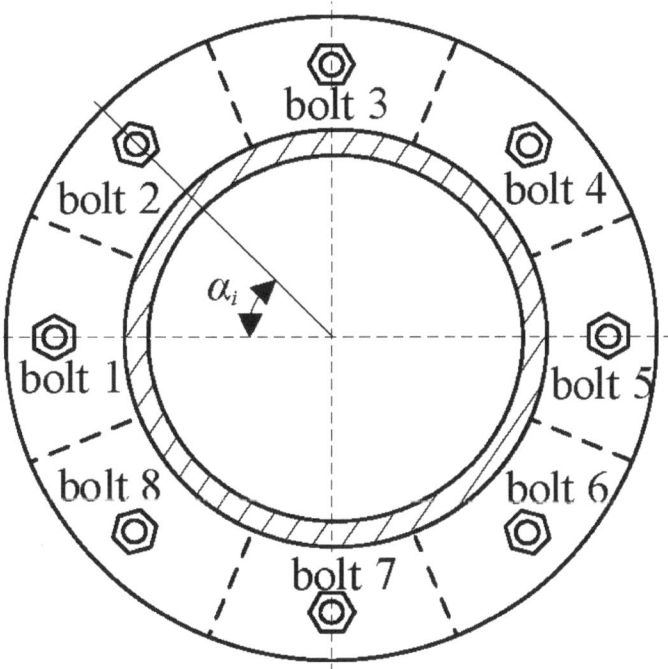

Figure 8. Division of the flange joint.

Some assumptions are made, as given below:
1. The flange, bolts, and spigot conform to the small deformation assumption;
2. The cylinder is assumed to be rigid;
3. The flange is assumed to be a plane when the joint is subject to a bending load;
4. Every section is independent, and there is no interaction between the two adjacent sections;
5. As the compression stiffness is much larger than the tension stiffness, the section's deformation under compression force is ignored [6].

When the bending load is applied to the flange joint, a part of the section is under compression, and the other part is subjected to tension. The bending load can be converted to a tension force or compression force, which is applied to the section.

The section's displacement δ under tension force F is illustrated in Figure 9. In the presence of a spigot, the tension force gives rise to displacements of both the flange and the spigot. The displacement of the spigot may be due to deformation or the sliding of the contact surface. Figure 10 presents the different conditions of the spigot's contact surface. The tangential force F_s is a component of the tension force F, which is acting on the spigot; P is the contact pressure caused by the interference of the spigot. If $F_s \leq \mu P$ (μ is the sliding friction coefficient of the contact surface), the contact surface of the spigot will be in the sticking condition. In this condition, the section's displacement δ is equal to the deformation of the spigot. According to the Coulomb friction law, the contact surface begins to slide if $F_s > \mu P$. Based on the analysis above, the contact surface of the spigot can be simplified into a Jenkins friction model [18], as shown in Figure 10b.

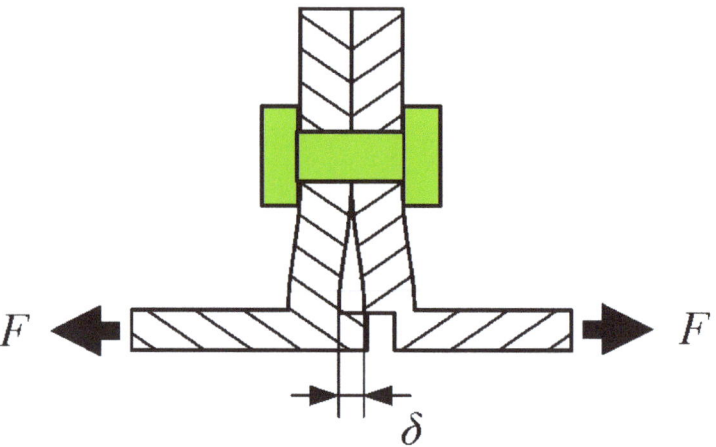

Figure 9. Deformation of one section under tension force.

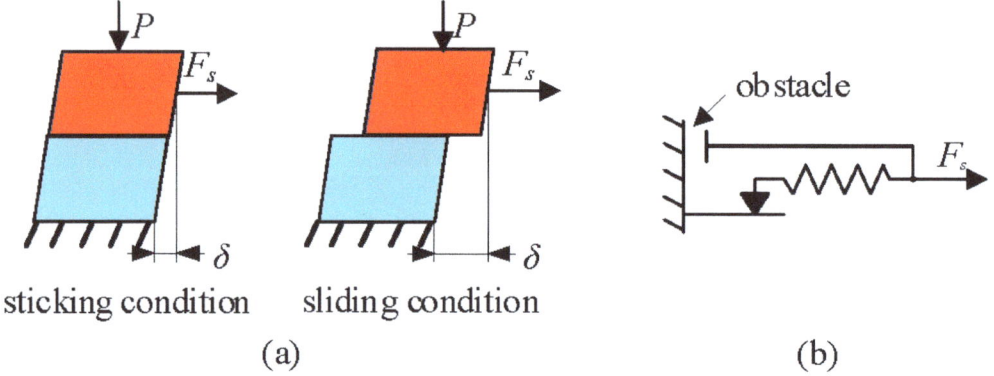

Figure 10. Displacement of the spigot: (**a**) the sticking and sliding conditions of the contact surface, (**b**) the simplified model of the contact surface.

4.2. Force and Displacement of the Spigot

Considering the loading and unloading processes of the load, the contact status of the spigot can be divided into four stages according to the different statuses of deformation and sliding, as shown in Figure 11. Stage *a* is the elastic deformation, and stage *b* is the sliding. Stage *c* is the beginning of unloading. With the decreasing of the load, the elastic deformation that occurred in stage *a* is restored first, and the elastic deformation occurs subsequently in the opposite direction. In stage *d*, the spigot slips again until the two pieces of the flange plate are closed up.

Take the case of a load of 600 N · m amplitude as an example (as seen in Figure 5). The four stages will not all appear in the loading and unloading process because the load is not large enough. The load of displacement is regarded as the criterion to distinguish the contact status, as shown in Figure 12. From Figure 12, δ_{max} is the peak value of displacement load; δ_c is the critical value at which the spigot begins to slip. Moreover, the value of δ cannot be negative, based on assumption 5.

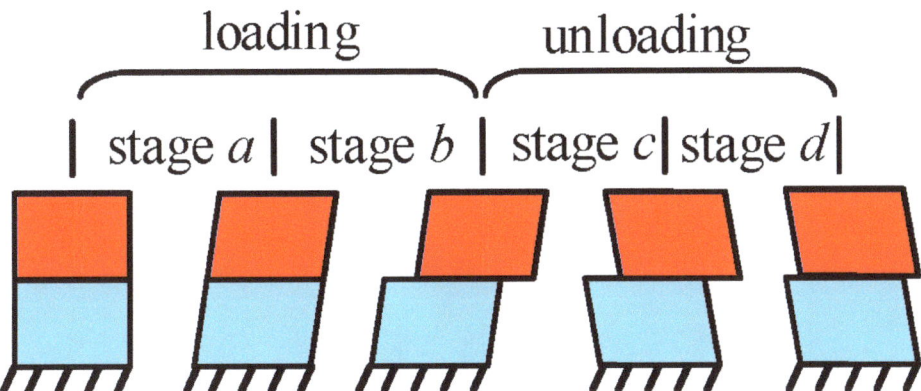

Figure 11. Different stages in the processes of loading and unloading.

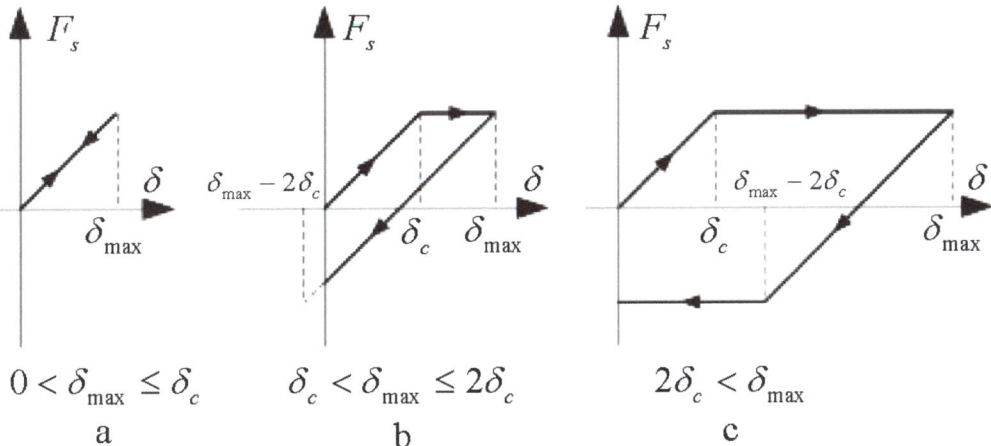

Figure 12. Relationship between displacement load and tension force: (**a**) the stage of elastic deformation, (**b**) the stage of partial slip, (**c**) the stage of complete slip.

In Figure 12a, the spigot only experiences the stage of elastic deformation (stage *a*) because of $0 < \delta_{max} \leq \delta_c$. The routes of loading and unloading coincide in the case.

In Figure 12b, the first stage is elastic deformation (stage *a*), and the spigot begins to slip when the displacement load exceeds δ_c (stage *b*). In stage *b*, the tension force remains unchanged with the increase in δ. In the process of unloading, the elastic deformation is restored first, and elastic deformation occurs subsequently in the opposite direction (stage *c*). Thus, the absolute value of displacement in stage *c* will be twice that of δ_c. However, because of $\delta_c < \delta_{max} \leq 2\delta_c$, the absolute value of elastic deformation ($\delta_{max} - \delta_{min}$, where $\delta_{min} = 0$) will not reach $2\delta_c$ only when $\delta_{max} = 2\delta_c$.

In Figure 12c, δ_{max} is large enough to make the absolute value of elastic deformation reach $2\delta_c$. After the end of stage *c*, the spigot begins to slip again (stage *d*) until δ reduces to zero.

Based on the above analysis, different relationships between force and displacement in the three kinds of contact status are deduced.

If δ_{max} matches $0 < \delta_{max} \leq \delta_c$, the relationship between the force and the displacement can be expressed as

$$F_s = k_s \delta \qquad (1)$$

where k_s is the tangential contact stiffness of the spigot.

If δ_{max} matches $\delta_c < \delta_{max} \leq 2\delta_c$, based on the different processes of loading and unloading, the relationships between the force and the displacement can be written as

$$F_s = \begin{cases} k_s \delta & \dot{\delta} > 0 \ \& \ 0 \leq \delta < \delta_c \\ k_s \delta_c & \dot{\delta} > 0 \ \& \ \delta_c \leq \delta < \delta_{max} \\ k_s \delta + k_s(\delta_c - \delta_{max}) & \dot{\delta} \leq 0 \ \& \ 0 \leq \delta \leq \delta_{max} \end{cases} \qquad (2)$$

where $\dot{\delta} > 0$ is the process of loading, and $\dot{\delta} \leq 0$ is the process of unloading.

If δ_{max} matches $\delta_{max} > 2\delta_c$, the relationships between the force and the displacement can be described as

$$F_s = \begin{cases} k_s \delta & \dot{\delta} > 0 \ \& \ 0 \leq \delta < \delta_c \\ k_s \delta_c & \dot{\delta} > 0 \ \& \ \delta_c \leq \delta < \delta_{max} \\ k_s \delta + k_s(\delta_c - \delta_{max}) & \dot{\delta} \leq 0 \ \& \ \delta_{max} - 2\delta_c < \delta \leq \delta_{max} \\ -k_s \delta_c & \dot{\delta} \leq 0 \ \& \ 0 \leq \delta \leq \delta_{max} - 2\delta_c \end{cases} \qquad (3)$$

Taking the stiffness of the flange plate into consideration, the section in Figure 9 can be simplified into the model in Figure 13. The Jenkins friction model is parallel with a spring, which reflects the stiffness of the flange plate k_t, which can be obtained using the method in reference [13]. The stiffness of the spring in the Jenkins friction model k_s, which is relevant to the contact status, the qualities of the contact surface, the shape of the spigot, and so on, is difficult to acquire with the analytical method. The preferable way to obtain k_s may using the FE or experiment methods clarified above.

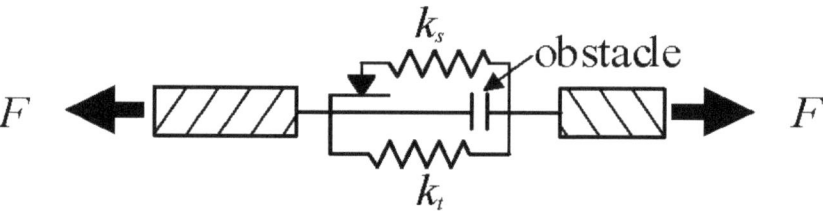

Figure 13. Simplified model of a section of the flange joint.

Thus, the total tension force F can be written as

$$F = F_s + F_t = F_s + k_t \delta \qquad (4)$$

4.3. Building of the Analytical Model

Figure 14 shows the sketch map of the deformation when the flange joint is under the load of a bending moment. Due to the turning of the flange plate, some sections are subject to tension, and others to compression. Based on the previous assumptions, only one section is subject to compression. The deformation of each section, which is simplified into a Jenkins friction model, is shown in Figure 14b (the obstacle is not shown in the figure). As the deformation under compression force is ignored (assumption 5), the deformation of section 5 is 0. θ is the angle of rotation, and clockwise is positive. The deformations of section 2 and that of section 8 are coincident, as are sections 3 and 7 (sections 4 and 6).

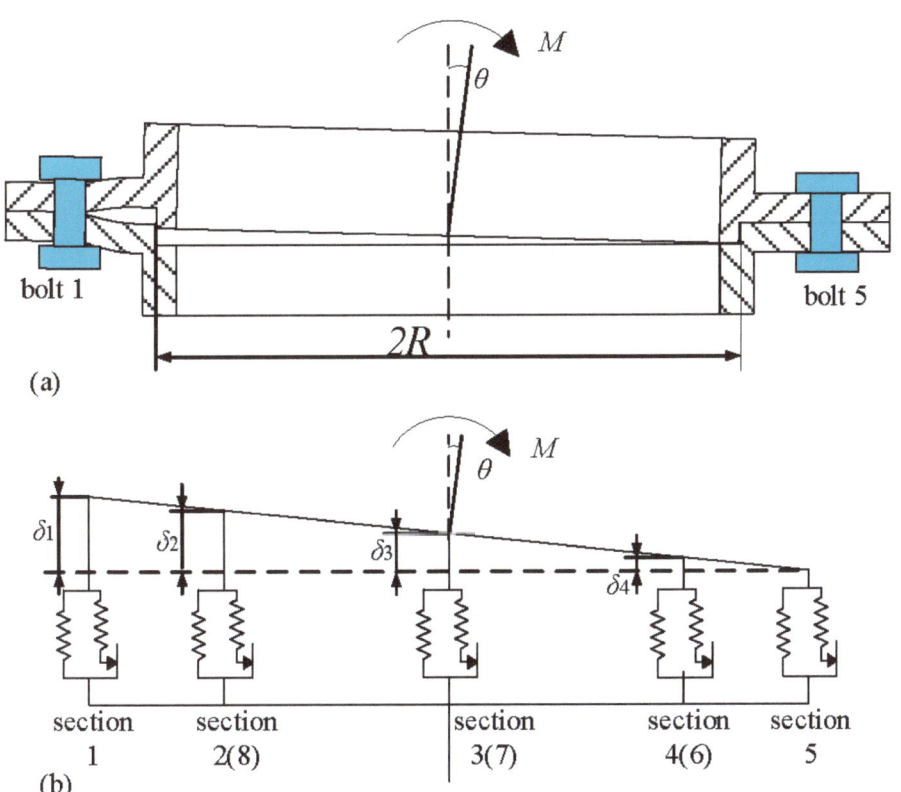

Figure 14. Deformation of the flange joint: (**a**) sketch of the deformation; (**b**) simplified model.

Considering the positive angle of rotation θ, the deformation δ_i of the ith section or bolt can be calculated by

$$\delta_i = R \sin\theta (1 + \cos\alpha_i) \approx R\theta(1 + \cos\alpha_i) \tag{5}$$

where R is the radius of the spigot. α_i is the position angle of each bolt, as shown in Figure 8, which can be described as

$$\alpha_i = 2\pi(i-1)/n \tag{6}$$

where n is the number of bolts.

If the angle of rotation θ is negative, the corresponding deformation δ_i can be expressed as

$$\delta_i = R \sin\theta (\cos\alpha_i - 1) \approx R\theta(\cos\alpha_i - 1) \tag{7}$$

As shown by FE and experimental analysis, the angle of rotation in one cycle can be divided into four processes based on the different conditions of loading and unloading in Figure 15. The value of $\dot{\delta}_i$ in any process can be obtained by calculating the derivative of Equation (5) or Equation (7).

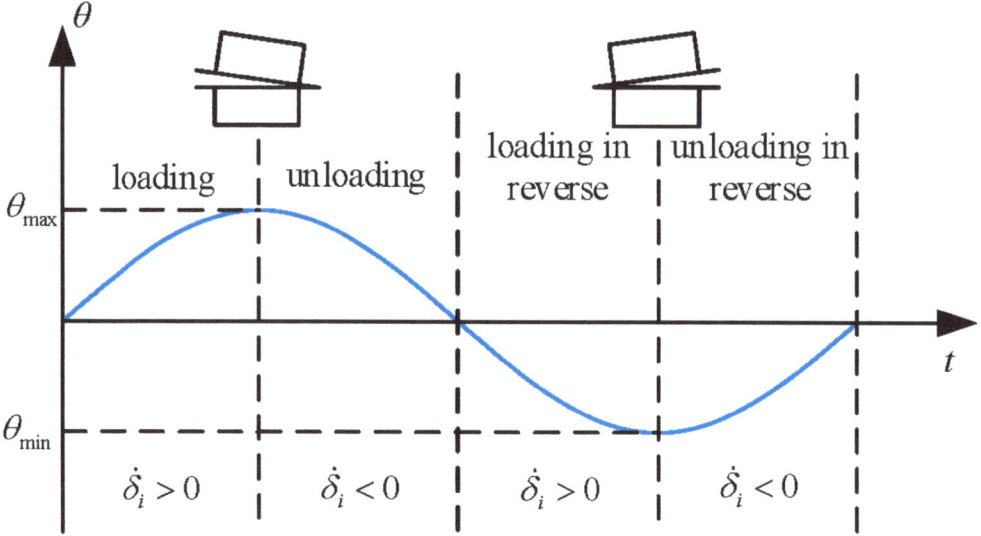

Figure 15. Diagram of the angle of rotation variation in one cycle.

According to the different values of δ_i and $\dot{\delta}_i$ in different process, the tension force of the ith section can be calculated by Equation (4). It is noteworthy that when θ is positive, the deformation of section 5 is ignored because it is subject to compression. Therefore, F_5 could not be obtained by Equation (4). For the reason of $\sum_{i=1}^{n} F_i = 0$, F_5 can be written as

$$F_5 = -2F_4 - 2F_3 - 2F_2 - F_1 \tag{8}$$

where the partial coefficient is 2 because $F_4 = F_6$, $F_3 = F_7$ and $F_2 = F_8$, based on symmetry. Similarly, F_1 can be obtained by Equation (9) when θ is negative.

$$F_1 = -F_5 - 2F_4 - 2F_3 - 2F_2 \tag{9}$$

The bending moment can be expressed as

$$M = \sum_{i=1}^{n} F_i R \cos \alpha_i \tag{10}$$

A simplified analytical model of the flange joint with a spigot based on the characteristic analysis of the FE and experimental methods has been proposed above.

4.4. Simulation

With the variation of the axial deformation of each section, the tangential force F_s of each section during the processes of loading and unloading is shown in Figure 16. If the tangential force of one section is positive, it means that the spigot is subject to tension; otherwise, it means compression. Due to the deformation of section 5 is 0 in these processes, the tangential force of section 5 is not present. The axial deformations of sections 1–4(6) decrease in sequence. It can be seen that sections 1, 2, and 8 experience the whole process, as shown in Figure 12c. Meanwhile, the curves of section 3(7) and section 4(6) coincide with those in Figure 12a,b, respectively. The different sections are in different contact statuses.

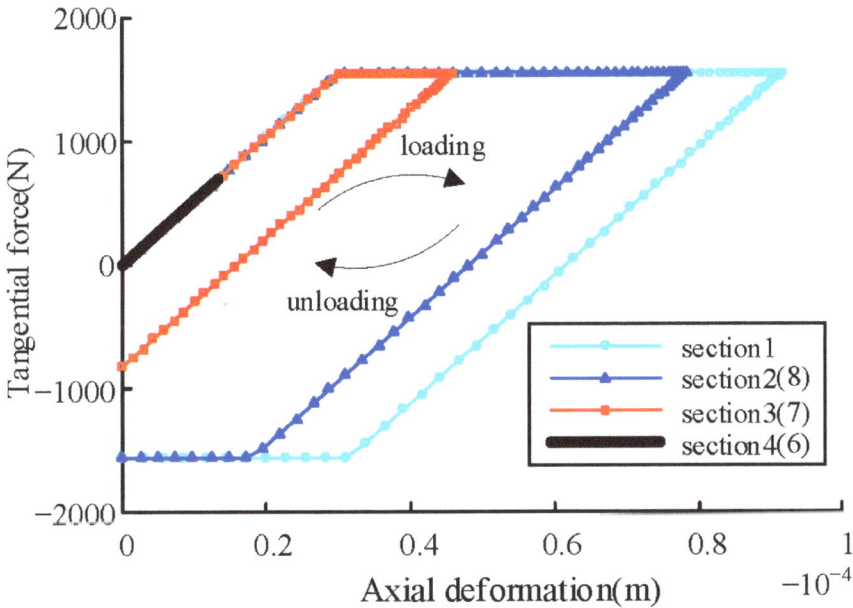

Figure 16. Tangential force variation with the deformation of each section.

Figure 17 presents the tangential force F_s of each section during the processes of loading and unloading with the angle of rotation. It can be observed that with the increase in the angle of rotation in the loading process, sections 1, 2(8), and 3(7) gradually enter the sliding condition from the sticking condition, and the corresponding tangential force first increases and then remains unchanged until entering the unloading process. In the unloading process, the switch between the sliding condition and the sticking condition occurs in the same sequence. These sections experience the process shown in Figure 12c. section 4(6) is close to section 5, and the deformation caused by the same angle of rotation is small. The tangential force cannot reach the condition of the stiffness decreasing suddenly, and no slip occurs in section 4(6), so the loading and unloading curves of section 4(6) coincide. It is noticeable that the tangential force of section 5 is large enough at the end of the unloading process that this section begins sliding. However, it is difficult to determine when section 5 begins sliding in the program; thus, the sliding of section 5 in the latter part of the unloading process is ignored. Since loading is reversed at the beginning of the next process, the spigot of section 5 is immediately set to the status of sliding. Due to the closure of the flange, section 5 cannot slip when subjected to a negative tangential force (compression direction). Only after the loading process is finished, section 5 begins to open and slip when subjected to a positive tangential force (tensile direction).

Figure 18 shows the comparison between the FE and simplified analytical models. The angle of rotation θ is taken as a variable in the calculation of the bending moment. The amplitude of θ is set as 0.0037 rad, which is equal to the angle of rotation caused by the bending moment of 1200 N·m in the FE method. Very similar curves can be observed, except that there are obvious fluctuations in the simplified analytical model curve near the zero point. This is because the sliding of section 5 in the latter part of the unloading process is ignored in the simplified analytical model mentioned above. The consistency of the curves shows that the simplified analytical model can simulate the stiffness characteristics of the flange joint, such as the bending stiffness decreasing suddenly and the hysteresis loop.

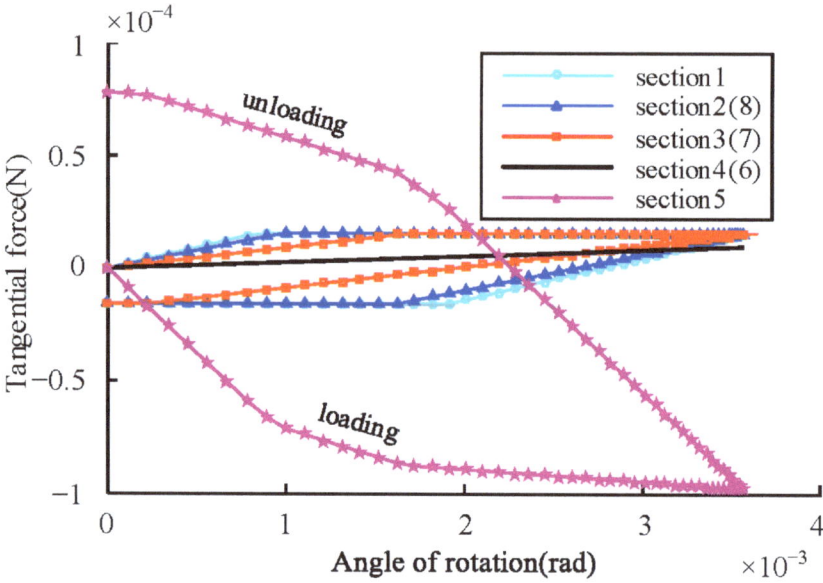

Figure 17. Tangential force variation with the angle of rotation.

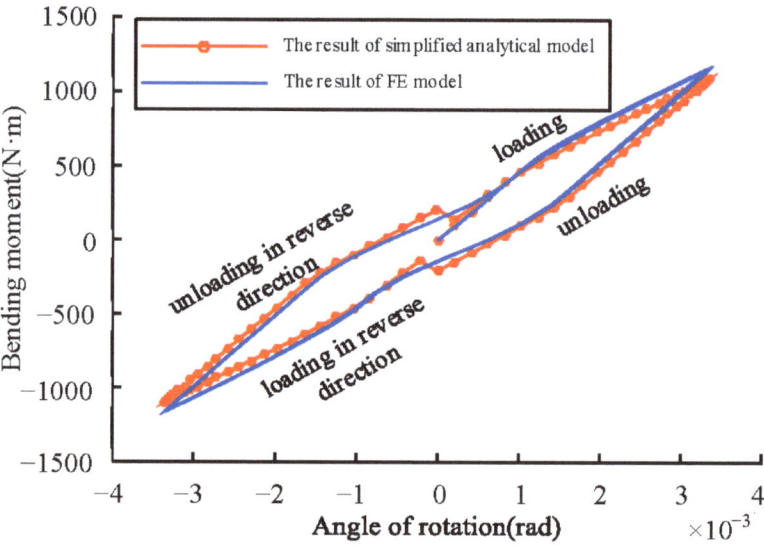

Figure 18. Angle of rotation curves of the FE and simplified analytical models.

In order to verify the applicability of the simplified analytical model, several additional loading conditions are applied, and the resulting curves are shown in Figure 19. The amplitudes of the angle of rotation are also given by the FE method; 0.001 rad, 0.0024 rad, and 0.0037 rad correspond to 600 N · m, 900 N · m, and 1200 N · m bending moments, respectively. All three curves in Figure 19 are consistent with the curves in Figure 5, except for the fluctuations near the zero point. Therefore, the proposed model's applicability is verified.

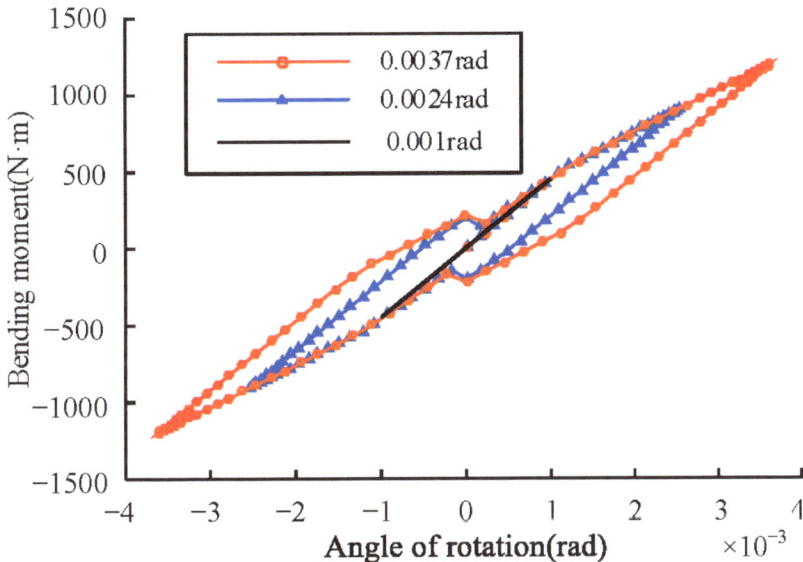

Figure 19. Angle of rotation–loading curve of the analytical model under different loads.

5. Conclusions

The stiffness characteristics of a flange joint with a spigot are studied based on FE and experimental methods. On this basis, a simplified analytical model is proposed and simulated. Within the parameters covered in this paper, the main conclusions can be summarized as follows:

1. When the load applied on the flange joint is large enough to cause the sliding of the contact surface of the spigot, hysteresis characteristics of the flange joint appear.
2. The experiment established in this paper verifies the validity of the stiffness characteristics obtained by FE analysis.
3. The proposed simplified analytical model can be utilized to simulate the deformation and sliding status of the contact surface of the spigot.
4. The hysteresis stiffness characteristics of the flange joint with a spigot can be obtained through the analytical model.

The simplified analytical model will be used in steady dynamic analysis in the following research, which requires a large amount of simulation, or in aero-engine dynamic characteristics analysis. Compared with the FE model for a fixed model parameter, the simplified analytical model is more convenient to use. Besides, it could effectively reduce the degrees of freedom of the system model. Certainly, the purpose of this paper is to provide a foundation for the research on the flange joint with a spigot, and further comprehensive and in-depth research is needed in the future to get closer to the physical model.

Author Contributions: Conceptualization, H.L., J.W. and Y.L. (Yu Liu); methodology, H.L. and Y.L. (Yu Liu); software, H.L. and Y.L. (Yu Liu); validation, J.W. and Y.L. (Yu Liu); investigation, J.W. and Y.L. (Yu Liu); resources, H.L., Z.W. and Y.L. (Yifu Long); data curation, H.L., Y.L. (Yu Liu), Z.W. and Y.L. (Yifu Long); writing—original draft preparation, H.L., Y.L. (Yu Liu) and Y.L. (Yifu Long); writing—review and editing, Y.L. (Yu Liu), J.W. and Y.L. (Yifu Long); visualization, H.L. and Y.L. (Yu Liu); supervision, J.W. and Y.L. (Yu Liu); project administration, J.W., Y.L. (Yu Liu) and Z.W.; funding acquisition, J.W and Y.L. (Yu Liu). All authors have read and agreed to the published version of the manuscript.

Funding: This work is financially supported by the National Natural Science Foundation of China (NSFC) under Grant No. 52205115.

Institutional Review Board Statement: Not applicable.

Informed Consent Statement: Not applicable.

Data Availability Statement: Not applicable.

Conflicts of Interest: The authors declare no conflict of interest.

Nomenclature

Symbols
k_1	Bending stiffness before a change in the initial loading stage
k_2	Bending stiffness after a change in the initial loading stage
k_s	Tangential contact stiffness of the spigot
k_t	Contact stiffness of the flange plate
δ	Displacement of the flange plate
F, F_s, F_t	Total tension force, tangential force of the spigot, tangential force of the flange plate
P	Contact pressure
μ	Sliding friction coefficient
δ_{max}	Peak value of displacement load
δ_c	Critical value of displacement load
θ	Angle of rotation for the flange plate
δ_i	Displacement of the ith section
$\dot{\delta}_i$	Velocity of the ith section
F_i	Tension force of the ith section
M	Bending moment

References

1. Sun, W.; Guan, Z.; Chen, Y.; Pan, J.; Zeng, Y. Modeling of Preload Bolted Flange Connection Structure for Loosening Analysis and Detection. *Shock Vib.* **2022**, *2022*, 7844875. [CrossRef]
2. Marek, P.; Pawlicki, J.; Mościcki, A. Tightness Problems at the Flange Connection in Transient Temperature and High Pressure Condition. *Eng. Fail. Anal.* **2022**, *133*, 105986. [CrossRef]
3. Zhu, L.; Bouzid, A.-H.; Hong, J. A Method to Reduce the Number of Assembly Tightening Passes in Bolted Flange Joints. *J. Manuf. Sci. Eng.* **2021**, *143*, 121006. [CrossRef]
4. El Masnaoui, W.; DaidiÉ, A.; Lachaud, F.; Paleczny, C. Semi-Analytical Model Development for Preliminary Study of 3D Woven Composite/Metallic Flange Bolted Assemblies. *Compos. Struct.* **2021**, *255*, 112906. [CrossRef]
5. Nassiraei, H. Local Joint Flexibility of CHS X-Joints Reinforced with Collar Plates in Jacket Structures Subjected to Axial Load. *Appl. Ocean Res.* **2019**, *93*, 101961. [CrossRef]
6. Nassiraei, H. Geometrical Effects on the LJF of Tubular T/Y-Joints with Doubler Plate in Offshore Wind Turbines. *Ships Offshore Struct.* **2022**, *17*, 481–491. [CrossRef]
7. Zacal, J.; Folta, Z.; Struz, J.; Trochta, M. Influence of Symmetry of Tightened Parts on the Force in a Bolted Joint. *Symmetry* **2023**, *15*, 276. [CrossRef]
8. Jamia, N.; Jalali, H.; Taghipour, J.; Friswell, M.I.; Haddad Khodaparast, H. An Equivalent Model of a Nonlinear Bolted Flange Joint. *Mech. Syst. Signal Process.* **2021**, *153*, 107507. [CrossRef]
9. Qin, Z.; Han, Q.; Chu, F. Bolt Loosening at Rotating Joint Interface and Its Influence on Rotor Dynamics. *Eng. Fail. Anal.* **2016**, *59*, 456–466. [CrossRef]
10. Song, Y. Modeling, Identification and Simulation of Dynamics of Structures with Joints and Interfaces. Ph.D. Thesis, University of Illinois, Urbana-Champaign, IL, USA, 2004.
11. He, K.; Zhu, W.D. Detecting Loosening of Bolted Connections in a Pipeline Using Changes in Natural Frequencies. *J. Vib. Acoust.* **2014**, *136*, 034503. [CrossRef]
12. Yao, X.; Wang, J.; Zhai, X. Research and Application of Improved Thin-Layer Element Method of Aero-Engine Bolted Joints. *Proc. Inst. Mech. Eng. Part G J. Aerosp. Eng.* **2017**, *231*, 823–839. [CrossRef]
13. Luan, Y.; Guan, Z.-Q.; Cheng, G.-D.; Liu, S. A Simplified Nonlinear Dynamic Model for the Analysis of Pipe Structures with Bolted Flange Joints. *J. Sound Vib.* **2012**, *331*, 325–344. [CrossRef]
14. Wang, C.; Zhang, D.; Zhu, X.; Hong, J. Study on the Stiffness Loss and the Dynamic Influence on Rotor System of the Bolted Flange Joint. In Proceedings of the Volume 7A: Structures and Dynamics, Düsseldorf, Germany, 16 June 2014; American Society of Mechanical Engineers: Düsseldorf, Germany, 2014; p. V07AT31A020.
15. Bouzid, A.-H.; Vafadar, A.K.; Ngô, A.D. On the Modeling of Anisotropic Fiber-Reinforced Polymer Flange Joints. *J. Press. Vessel Technol.* **2021**, *143*, 061506. [CrossRef]

16. Ahmadian, H.; Jalali, H. Identification of Bolted Lap Joints Parameters in Assembled Structures. *Mech. Syst. Signal Process.* **2007**, *21*, 1041–1050. [CrossRef]
17. Iranzad, M.; Ahmadian, H. Identification of Nonlinear Bolted Lap Joint Models. *Comput. Struct.* **2012**, *96–97*, 1–8. [CrossRef]
18. Bograd, S.; Reuss, P.; Schmidt, A.; Gaul, L.; Mayer, M. Modeling the Dynamics of Mechanical Joints. *Mech. Syst. Signal Process.* **2011**, *25*, 2801–2826. [CrossRef]
19. Oldfield, M.; Ouyang, H.; Mottershead, J.E. Simplified Models of Bolted Joints under Harmonic Loading. *Comput. Struct.* **2005**, *84*, 25–33. [CrossRef]
20. Ouyang, H.; Oldfield, M.J.; Mottershead, J.E. Experimental and Theoretical Studies of a Bolted Joint Excited by a Torsional Dynamic Load. *Int. J. Mech. Sci.* **2006**, *48*, 1447–1455. [CrossRef]
21. Van-Long, H.; Jean-Pierre, J.; Jean-François, D. Behaviour of Bolted Flange Joints in Tubular Structures under Monotonic, Repeated and Fatigue Loadings I: Experimental Tests. *J. Constr. Steel Res.* **2013**, *85*, 1–11. [CrossRef]
22. Firrone, C.M.; Battiato, G.; Epureanu, B.I. Modelling the Microslip in the Flange Joint and Its Effect on the Dynamics of a Multi-Stage Bladed Disk Assembly. In Proceedings of the Volume 7A: Structures and Dynamics, Seoul, Republic of Korea, 13 June 2016; American Society of Mechanical Engineers: Seoul, Republic of Korea, 2016; p. V07AT32A032.
23. Shi, W.; Zhang, Z. An Improved Contact Parameter Model with Elastoplastic Behavior for Bolted Joint Interfaces. *Compos. Struct.* **2022**, *300*, 116178. [CrossRef]
24. He, L.; Li, T. Undamped Non-Linear Vibration Mechanism of Bolted Flange Joint under Transverse Load. *J. Asian Archit. Build. Eng.* **2023**, *22*, 1507–1532. [CrossRef]
25. Liu, Y.; Wang, J.; Chen, L. Dynamic Characteristics of the Flange Joint with a Snap in Aero-Engine. *Int. J. Acoust. Vib.* **2018**, *23*, 168–174. [CrossRef]
26. Shuguo, L.; Yanhong, M.; Dayi, Z.; Jie, H. Studies on Dynamic Characteristics of the Joint in the Aero Engine Rotor System. *Mech. Syst. Signal Process.* **2012**, *29*, 120–136. [CrossRef]
27. Yu, P.; Li, L.; Chen, G.; Yang, M. Dynamic Modelling and Vibration Characteristics Analysis for the Bolted Joint with Spigot in the Rotor System. *Appl. Math. Model.* **2021**, *94*, 306–331. [CrossRef]
28. Liu, Y.; Zhao, D.; Guo, X.; Ai, Y. Stiffness and Geometry Characteristics of Flange Connection with Bolt Failure and Its Influence on Rotor Dynamics. *Proc. Inst. Mech. Eng. Part C J. Mech. Eng. Sci.* **2023**, 095440622211474. [CrossRef]

Disclaimer/Publisher's Note: The statements, opinions and data contained in all publications are solely those of the individual author(s) and contributor(s) and not of MDPI and/or the editor(s). MDPI and/or the editor(s) disclaim responsibility for any injury to people or property resulting from any ideas, methods, instructions or products referred to in the content.

Article

A Multi-Cavity Iterative Modeling Method for the Exhaust Systems of Altitude Ground Test Facilities

Keqiang Miao [1], Xi Wang [1], Meiyin Zhu [2,*], Song Zhang [3], Zhihong Dan [3], Jiashuai Liu [1], Shubo Yang [1], Xitong Pei [4], Xin Wang [3] and Louyue Zhang [1]

[1] School of Energy and Power Engineering, Beihang University, Beijing 100191, China; kqmiao@buaa.edu.cn (K.M.); xwang@buaa.edu.cn (X.W.); ljsbuaa@buaa.edu.cn (J.L.); yangshubo@buaa.edu.cn (S.Y.); sy2004112@buaa.edu.cn (L.Z.)
[2] Beihang Hangzhou Innovation Institute Yuhang, Hangzhou 310023, China
[3] Science and Technology on Altitude Simulation Laboratory, AECC Sichuan Gas Turbine Establishment, Mianyang 621703, China; goom2619@163.com (S.Z.); dzh798318_cym@163.com (Z.D.); tywangxin2012@163.com (X.W.)
[4] Research Institute of Aero-Engine, Beihang University, Beijing 100191, China; peixitong@buaa.edu.cn
* Correspondence: mecalzmy@buaa.edu.cn

Abstract: To solve the modeling problem of altitude ground test facility (AGTF) exhaust systems, which is caused by nonlinearity along the gas path and the difficulty of ejection factor calculation, a multi-cavity iterative modeling method is presented. The components of exhaust systems, such as the exhaust diffuser and cooler, are built with a series of volumes. It overcomes the disadvantage that traditional lumped-parameter models have, whereby they cannot calculate the dynamic parameters along the gas path. The exhaust system model is built with an iterative method based on multi-cavity components, and simulations are carried out under experimental conditions. The simulation results show that the maximum error of pressure is 2 kPa in the steady state and less than 6 kPa in the transient process compared with experimental data. Closed-loop simulations are also carried out to further verify the accuracy and effectiveness of the multi-cavity iterative exhaust system modeling method.

Keywords: altitude ground test facility; exhaust system; multi-cavity iterative model

1. Introduction

Altitude ground test facilities (AGTFs) can test the performances of aero engines over the flight envelope on the ground by simulating the inlet and outlet environment conditions of aero engines [1]. In order to reduce energy consumption in altitude ground test experiments with the aim of energy conservation and reducing expenses, the modeling of AGTFs has been a hot issue over the last decades. Moreover, the demand of the precise modeling of AGTF to design an advanced control algorithm with the purpose of realizing full mission profile flight trajectory continuous simulations for aero engines has been increasing for many years [2]. The exhaust system is an important component of AGTFs that is responsible for simulating air conditions after the engine nozzle. The modeling problem of the exhaust system results from the nonlinearity of components [3], especially the exhaust diffuser, which cools the gas exiting the engine nozzle by mixing it with air ejected from the test chamber and increases the static pressure of the mixed gas at the same time. The traditional lumped-parameter modeling method is insufficient to reflect the coupled dynamic of temperature and pressure inside. Moreover, due to the distance between the engine nozzle and the exhaust diffuser, the traditional method [4] cannot calculate the ejected secondary flow effectively. Research on exhaust diffuser simulations can be simplified with no induced secondary flow [5].

At the system level, the Arnold Engineering Development Complex (AEDC), which is located in America, has been conducting research on AGTF modeling and has been updating simulation software since 2000 [6–8]. A mathematical model of an AGTF in Canada was built by Boraira in 2006 [9]. In Germany, a digital simulation platform was built by Bierkamp et al. [10] and was developed to a semi-physical simulation platform by Weisser in 2013 [11]. In China, a system-level model of AGTF was established by Pei et al. in 2019 [12].

At present, the majority of research on AGTF modeling focuses on the lumped-parameter method. In the platform built by Pei et al. in reference [12], the exhaust system of an AGTF is considered single volume, which sacrifices the accuracy of the exhaust system model. Many algorithms are presented to compensate system simulation results by improving the accuracy of components. Zhu et al. proposed a coordinate positioning and regression algorithm to improve the accuracy of butterfly valves in exhaust systems [13]. Sosunov considered the thermal inertia of pipes and proposed an unsteady temperature calculation method [14]. In 2022, Liu et al. started to focus on the axial dynamic performance in AGTFs and built a quasi-one-dimensional flow model of a pipe [15] because Sheeley et al. proved that the accuracy of the lumped-parameter model is inadequate [16]. Moreover, robust control algorithms such as scheduled proportional–integral control [17], linear active disturbance rejection [18] and μ synthesis [19] are presented as a compromise to the inaccurate system model. It limits the development of the advanced control method, which maximizes the ability of AGTFs for more complicated experiments.

It can be concluded that there are increasing demands for an accurate model of exhaust systems that reflects the dynamic of the parameters along the symmetry components of the system. Therefore, a multi-cavity iterative modeling method for the exhaust systems of ATGFs is proposed in this paper. It separates exhaust systems into a series of volumes and calculates the parameters in each volume with the lumped-parameter method, because the exhaust system is composed of symmetry components, which means that the dynamic in the circumferential direction can be ignored. The dynamic of parameters in the axial direction can be reflected with lumped parameters in the ordered volumes. Moreover, the multi-cavity iterative modeling method introduces the initial guess values to ensure the smooth calculation of parameters. The iterative method takes the exhaust system as a whole and updates a series of system states at the same time, which increases the convergence of the model and speeds up the simulation process appropriately.

This paper is organized as follows. In Section 2, the multi-cavity components of the exhaust system are built. In Section 3, the exhaust system is built with a component model. The actual states of the system are achieved by solving residual differential equations with the Newton–Raphson iterative method. In Section 4, a simulation under different flight conditions is carried out, and the simulation results are compared with experimental data. Moreover, the simulation results are compared with the lumped-parameter model. Finally, conclusions are given in Section 5.

2. Component Model of the Exhaust System

The exhaust system is composed of a test chamber, exhaust diffuser, cooler, butterfly valves and pipe volume. Its structure is shown in Figure 1. The boundary conditions of the exhaust system consist of environment conditions, engine outlet conditions and Valve 3 outlet pressure conditions provided with the air extraction system. The environment condition includes temperature and pressure before Valve 1 and Valve 2. Since the outlet of the engine cannot be measured with sensors in the experiment directly, the mass flow, temperature, pressure and velocity, after the aero engine, are calculated with the inlet condition and states of aero engine, which consist of inlet mass W_0, inlet total temperature T_0, inlet total pressure P_0, power level angle (PLA), etc.

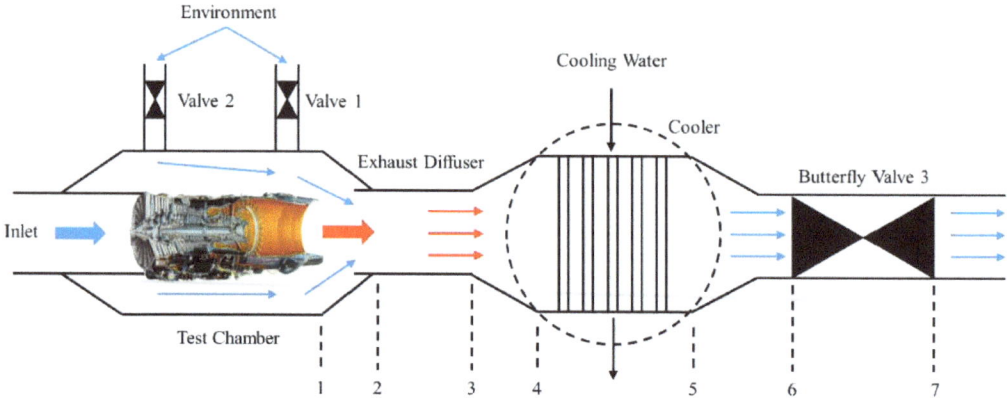

Figure 1. Structure of exhaust system.

The working medium in the exhaust system is a real gas that consists of the gas from the engine nozzle and air in the test chamber. However, because the pressure in the exhaust system is much lower than the atmospheric pressure (101 kPa) and the temperature is higher than 288.15 K, it is also reasonable to treat it as an ideal gas.

The units of the variables used in the modeling process are shown in Table 1.

Table 1. The units of variables.

Variable	Meaning	Units
p	Pressure	Pa
T	Temperature	K
A	Area	m^2
d	Diameter	m
V	Volume	m^3
v	Velocity	m/s
W	Mass flow	kg/s
m	Mass	kg
c	Specific heat	J/(kg*°C)
R	Gas constant	J/(kg*°C)
h_t	Unit enthalpy	J/kg
h	Heat transfer coefficient	W/(m²*°C)
E	Energy	J
U	Internal energy	J
$\dot{Q}$	Heat transfer rate	J/s

2.1. Exhaust Diffuser Model

The exhaust diffuser is used to reduce the temperature and velocity of the high-temperature and high-speed gas discharged from engine nozzle. Moreover, it reduces the load of the air extraction system by increasing the static pressure of the gas.

The exhaust diffuser model built with the multi-cavity method is divided into three parts, which are the ejection model, mixture model and expansion model, and into four sections, which are boundaries of the three models and are shown in Figure 2.

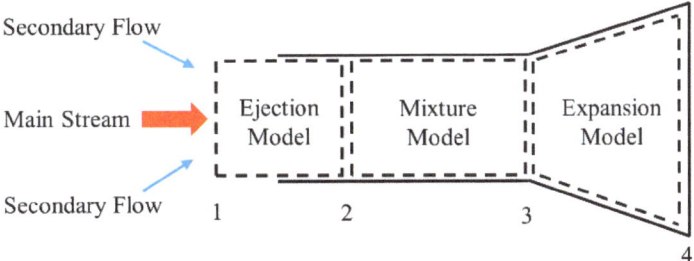

Figure 2. Multi-cavity exhaust diffuser model.

First, some theorems of aerodynamics need to be introduced.

Theorem 1. *Aerodynamic function $\pi(\lambda)$ describes the relationship between the static pressure p and the total pressure p_t. Moreover, it can be calculated with the specific heat ratio k and the velocity coefficient λ shown in Equation (1) [20].*

$$\pi(\lambda) = \frac{p}{p_t} = (1 - \frac{k-1}{k+1}\lambda^2)^{\frac{k}{k-1}} \qquad (1)$$

where, in $\lambda = \frac{v}{v_{cr}}$, v is the velocity of gas and v_{cr} is critical acoustic velocity.

Theorem 2. *The flow formula in Equation (2) describes the relationship between the mass flow W and the total parameters of the gas [20].*

$$W = K\frac{p_t}{\sqrt{T_t}}Aq(\lambda) \qquad (2)$$

where $K = \sqrt{\frac{k}{R}(\frac{2}{k+1})^{\frac{k+1}{k-1}}}$ is calculated with the specific heat ratio k and gas constant R, and the flow function $q(\lambda) = (\frac{k+1}{2})^{\frac{1}{k-1}}\lambda(1 - \frac{k-1}{k+1}\lambda^2)^{\frac{1}{k-1}}$ is calculated with the specific heat ratio k and the velocity coefficient λ. p_t is the gas total pressure, T_t is the gas total temperature and A is the flow area.

Theorem 3. *The impulse of a gas can be calculated with the mass flow, velocity, static pressure, flow area or total pressure, flow area, and flow function $f(\lambda)$. They are equal to each other, which is shown in Equation (3) [20].*

$$WV + pA = p_tAf(\lambda) \qquad (3)$$

where $f(\lambda) = (\frac{2}{k+1})^{\frac{1}{k-1}}q(\lambda)z(\lambda)$ and $z(\lambda) = \frac{1}{\lambda} + \lambda$.

Secondly, initial guess values need to be introduced into the calculation process of the exhaust diffuser model. They are listed in Table 2.

Table 2. The initial guess values of exhaust diffuser model.

Variable	Meaning
u	Ejection factor
P_{s2}	Static pressure at Section 2
P_{t3}	Total pressure at Section 3
λ_4	Velocity coefficient at Section 4

Then, the calculation process from Sections 1–4 is shown as follows.

2.1.1. Ejection Model

In the ejection process, the main stream expands slightly and increases the speed of the secondary flow in the ejection model shown in Figure 3. The ejection model calculates the mass of the secondary flow W_{1s} with the mass of the main stream W_{1m} and the ejection factor u, as shown in Equation (4). The flow area of the main stream and the secondary flow at Section 2, which are denoted as A_{2m} and A_{2s}, are also obtained.

$$W_{1s} = uW_{1m} \tag{4}$$

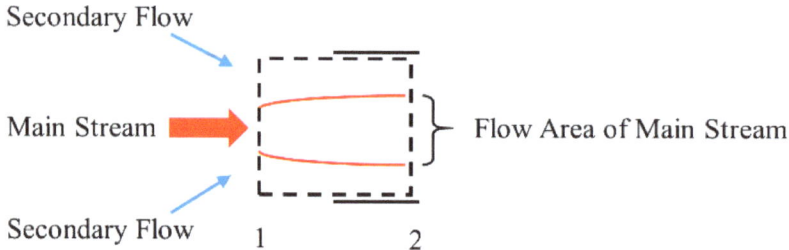

Figure 3. Diagram of ejection model.

The total pressure and total temperature of the main stream and secondary flow from Sections 1 and 2 are invariable because it is an isentropic expansion process. It is denoted as

$$p_{t2m} = p_{t1m}, \; p_{t2s} = p_{t1s} \tag{5}$$

$$T_{t2m} = T_{t1m}, \; T_{t2s} = T_{t1s} \tag{6}$$

where p_{t1m} is the total pressure of the main stream at Section 1 and p_{t2m} is the total pressure of the main stream at Section 2. T_{t1m} is the total temperature of the main stream at Section 1, and T_{t2m} is the total temperature of the main stream at Section 2. p_{t1s} is the total pressure of the secondary flow at Section 1, and p_{t2s} is the total pressure of the secondary flow at Section 2. T_{t1s} is the total temperature of the secondary flow at Section 1, and T_{t2s} is the total temperature of the secondary flow at Section 2.

Then, $\pi(\lambda)$ of the main stream at Section 2 can be calculated with the total pressure P_{s2m} and the initial guess value P_{s2}.

$$\pi(\lambda_{2m}) = \frac{p_{s2}}{p_{t2m}} \tag{7}$$

The velocity coefficient of the main stream can be obtained with the inverse function in Theorem 1. It is denoted as

$$\lambda_{2m} = \pi^{-1}(\lambda_{2m}) \tag{8}$$

Then, $q(\lambda)$ of the main stream at Section 2 can be obtained with

$$q(\lambda_{2m}) = \left(\frac{k+1}{2}\right)^{\frac{1}{k-1}} \lambda_{2m} \left(1 - \frac{k-1}{k+1}\lambda_{2m}^2\right)^{\frac{1}{k-1}} \tag{9}$$

Moreover, because of the flow conservation law, we have

$$W_{2m} = W_{1m} \tag{10}$$

$$W_{2s} = W_{1s} \tag{11}$$

The equivalent flow area of the main stream at Section 2 is defined as

$$A_{2m} = \frac{q(\lambda_{1m})}{q(\lambda_{2m})} A_{1m} \tag{12}$$

where A_{1m} is the output area of the engine.

Then, the equivalent flow area of secondary flow at Section 2 is defined as

$$A_{2s} = \frac{\pi d^2}{4} - A_{2m} \tag{13}$$

where d is the diameter of the exhaust diffuser.

2.1.2. Mixture Model

The mixture model calculates the temperature and pressure of the well-mixed gas by mixing the main stream and secondary flow at Section 2, as shown in Figure 4. The mixing process should meet the laws of flow conservation, energy conservation, impulse conservation and pressure balance.

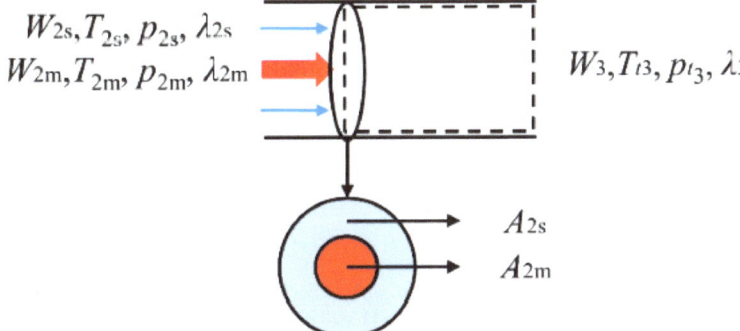

Figure 4. Diagram of mixture model.

The velocity coefficient λ_{2s} can be obtained with Theorem 2 because W_{2s}, T_{t2s}, P_{t2s} and A_{2s} are obtained from Equations (5), (6), (11) and (13). Then, $\pi(\lambda_{2s})$ can be obtained with Theorem 1.

The static pressure of the secondary flow at Section 2 can be calculated with $\pi(\lambda)$ in Equation (14).

$$p_{s2s} = p_{t2s} \pi(\lambda_{2s}) \tag{14}$$

The static pressure of the main stream and secondary flow should be balanced at Section 2. This means that $p_{s2m} = p_{s2s}$. The relative error of the static pressure at Section 2 is denoted with

$$e_1 = (p_{s2s} - p_{s2m})/p_{s2m} \tag{15}$$

The calculation process of static pressure balance, including Equations (7)–(15), is shown in Figure 5.

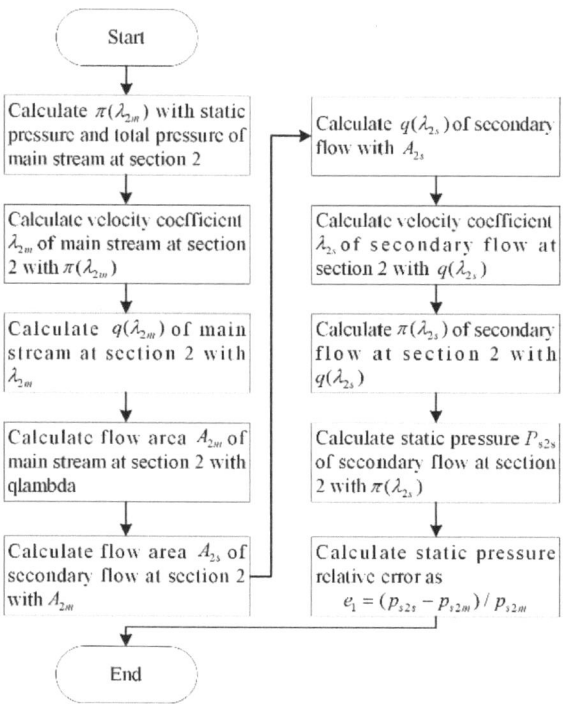

Figure 5. Calculation process of static pressure balance at Section 2.

The unit enthalpy at Section 3 can be calculated with the energy conservation law, as shown in Equation (16). The temperature is obtained with Equation (17).

$$h_{t3} = (h_{t2m}W_{2m} + h_{t2s}W_{2s})/W_3 \tag{16}$$

$$T_{t3} = h_{t3}/C_p \tag{17}$$

where $W_3 = W_{2s} + W_{2m}$, and h_t is unit enthalpy at each section.

The impulse of the gas at Sections 2 and 3 can be obtained with the right part of Equation (3) in Theorem 3. Because the mass flow at Sections 2 and 3 meets the impulse conservation law, we have

$$p_{t2m}A_{2m}f(\lambda_{2m}) + p_{t2s}A_{2s}f(\lambda_{2s}) = p_{t3}A_3f(\lambda_3) \tag{18}$$

The relative error of impulse is denoted as

$$e_2 = (p_{t2m}A_{2m}f(\lambda_{2m}) + p_{t2s}A_{2s}f(\lambda_{2s}) - p_{t3}A_3f(\lambda_3))/(p_{t2m}A_{2m}f(\lambda_{2m}) + p_{t2s}A_{2s}f(\lambda_{2s})) \tag{19}$$

2.1.3. Expansion Model

The inputs and outputs of the expansion model are shown in Figure 6. It is assumed that the process between Sections 3 and 4 is isentropic and adiabatic. Therefore, the total temperature and total pressure can be obtained with

$$p_{t4} = p_{t3} \tag{20}$$

$$T_{t4} = T_{t3} \tag{21}$$

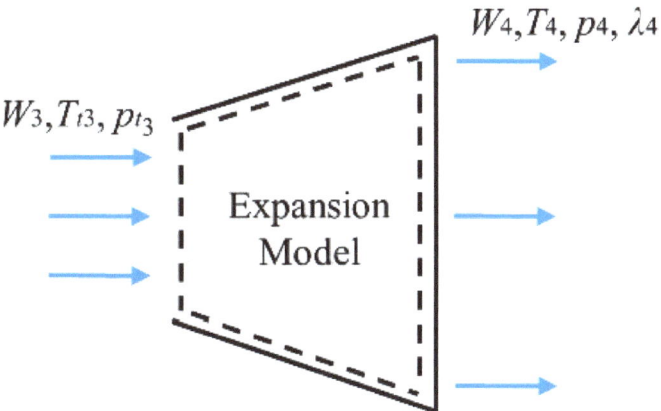

Figure 6. Diagram of expansion model.

Then, the mass flow at Section 4 can be calculated with Theorem 3 as

$$W_4 = K \frac{p_{t4}}{\sqrt{T_{t4}}} A_4 q(\lambda_4) \qquad (22)$$

where $A_4 = \frac{\pi D^2}{4}$ is the exit area of the expansion stage.

As a result of flow conservation, we have $W_4 = W_3$. Then, the relative error of the mass flow is defined as

$$e_3 = (W_4 - W_3)/W_3 \qquad (23)$$

The exit static pressure of the exhaust diffuser model is

$$p_4 = \pi(\lambda_4) p_{t4} = (1 - \frac{k-1}{k+1} \lambda_4^2)^{\frac{1}{k-1}} p_{t4} \qquad (24)$$

The exit static temperature of the exhaust diffuser model is

$$T_4 = \tau(\lambda_4) T_{t4} = (1 - \frac{k-1}{k+1} \lambda_4^2) T_{t4} \qquad (25)$$

2.2. Cooler Model

In the exhaust system, the cooler improves the gas flow capacity and reduces the aging of the equipment by reducing the gas temperature. The high-temperature gas passing through the cooling pipes with the cooling water inside is cooled to the allowable working temperature. The structure of the cooler in the exhaust system is shown in Figure 7. It is divided into n volumes, and it is assumed that the temperature in each volume between pipes is even. Then, the temperature decreases gradually from $T_{v,1}$ to $T_{v,n}$ in the volumes.

The heat transfer process of each pipe is shown in Figure 8. The heat transfers from the gas to the pipe and then to the cooling water. Finally, the heat is brought out of the system by the cooling water. The heat transfer rate between the gas and the pipe is denoted as

$$\dot{Q}_1 = h_1 A_1 (T_1 - T_m) \qquad (26)$$

where h_1 is heat transfer coefficient between the gas and the pipe, A_1 is the contact area, T_1 is the average temperature of gas and T_m is the average temperature of the pipe.

The heat transfer rate between the water and the pipe is denoted as

$$\dot{Q}_2 = h_2 A_2 (T_2 - T_m) \tag{27}$$

where h_2 is heat transfer coefficient between the gas and the pipe, A_2 is the contact area, T_2 is the average temperature of cooling water and T_m is the average temperature of the pipe.

The heat quantity brought out of the system by the cooling water is

$$\dot{Q}_3 = W_2 C_{p2} (T_2 - T_{2in}) \tag{28}$$

where W_2 is the mass of the cooling water in each pipe, C_{p2} is the specific heat of the cooling water, T_2 is the average temperature of the cooling water and T_{2in} is the original temperature of the cooling water.

Therefore, the energy change rate of the pipe is

$$\frac{dE_m}{dt} = \dot{Q}_1 + \dot{Q}_2 \tag{29}$$

Moreover, the energy of the pipe can be defined as

$$E_m = c_m m_m T_m \tag{30}$$

where c_m is the specific heat of the pipe, m_m is the mass of the pipe and T_m is the average temperature of the pipe.

Another definition of the energy change rate of the pipe is shown in Equation (31), which is obtained by differentiating Equation (30).

$$\frac{dE_m}{dt} = c_m m_m \frac{dT_m}{dt} \tag{31}$$

The differential equation of the pipe temperature is obtained by substituting Equation (29) into Equation (31). It is denoted as

$$\frac{dT_m}{dt} = \frac{h_1 A_1 (T_1 - T_m) + h_2 A_2 (T_2 - T_m)}{c_m m_m} \tag{32}$$

With the same calculation, we can obtain the differential equation of the gas temperature and the differential equation of the water temperature, as shown in Equations (33) and (34). The output temperature of the gas T_{1out} and the output temperature of the cooling water T_{2out} are shown in Equations (35) and (36).

$$\frac{dT_1}{dt} = -\frac{h_1 A_1 (T_1 - T_m)}{c_{p1} W_1} \tag{33}$$

$$\frac{dT_2}{dt} = \frac{-h_2 A_2 (T_2 - T_m) - W_2 C_{p2} (T_2 - T_{2in})}{c_{p2} W_2} \tag{34}$$

$$T_{1out} = T_1 + \frac{dT_1}{dt} \tag{35}$$

$$T_{2out} = T_{2in} + \frac{dT_2}{dt} \tag{36}$$

where c_{p1} is the specific heat of the gas, and W_1 is the mass of the gas flow through a pipe.

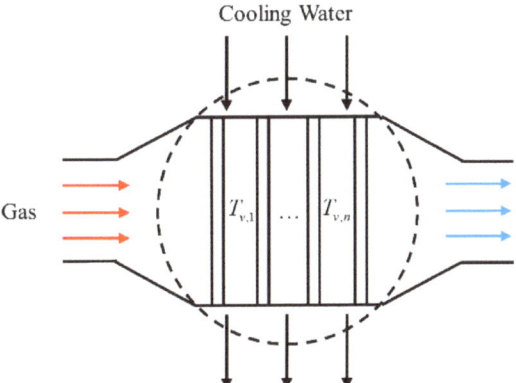

Figure 7. Structure of cooler.

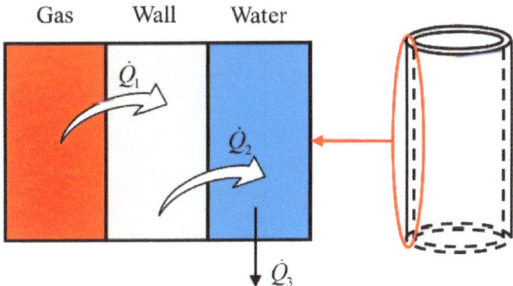

Figure 8. Diagram of cooling pipe heat transfer.

2.3. Butterfly Valve Model

A butterfly valve is used to regulate the pressure in the test chamber by controlling the mass flow exiting the exhaust system. The mass passing the butterfly valve depends on the pressure and temperature before the valve, the pressure after the valve and the opening of the valve. The structure of butterfly is shown in Figure 9.

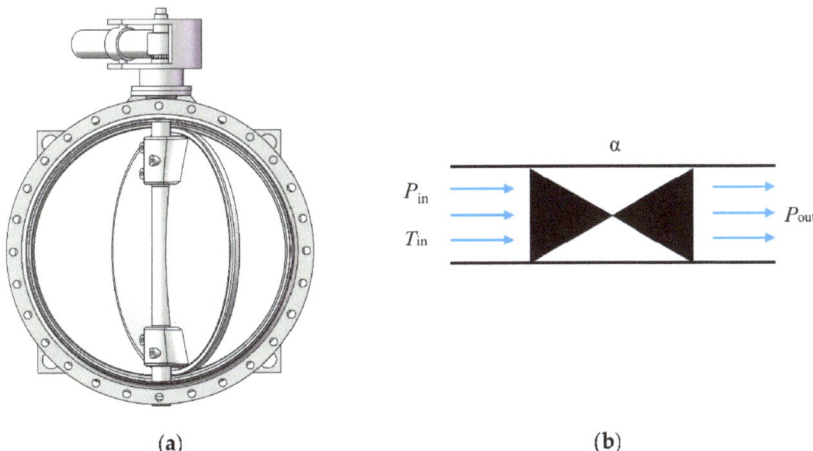

Figure 9. Model of butterfly valve: (**a**) Structure of butterfly valve; (**b**) Diagram of butterfly valve.

The butterfly valve is modeled in the form of a flow characteristic model with a flow coefficient map, as shown in Figure 10, and the flow function is denoted in Equation (37).

$$W = \phi \frac{\pi D^2}{4}\left(1 - \cos\frac{\alpha \pi}{180}\right) \cdot p_{in} \cdot \sqrt{\frac{2}{RT_{in}}} \qquad (37)$$

where W is the mass flow through the valve, and D is the diameter of the valve. The flow coefficient ϕ can be calculated with the valve pressure ratio $p_r = \frac{p_{out}}{p_{in}}$ and the valve opening α.

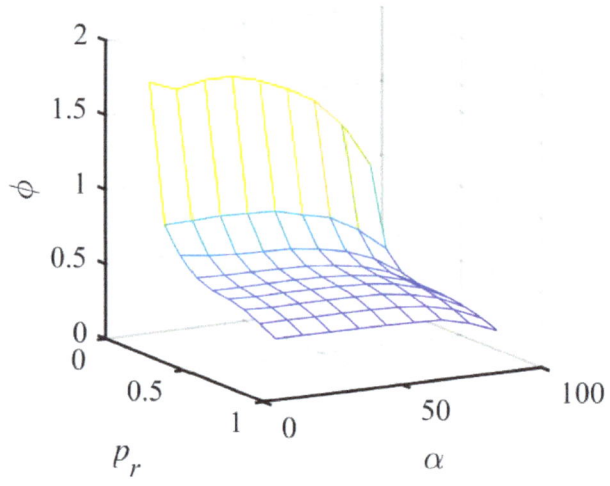

Figure 10. Flow coefficient map of butterfly valve.

2.4. Test Chamber Model and Pipe Volume Model

The test chamber is the test section of the exhaust system that simulates the exhaust environment of the engine at a high altitude by controlling the mass of the entering flow and exiting flow. The entering flow is controlled with Valve 1 and Valve 2 in this test chamber, which is the valve flow in Figure 11. The exiting flow is ejected by the engine exhaust gas, which is the secondary flow in Figure 11.

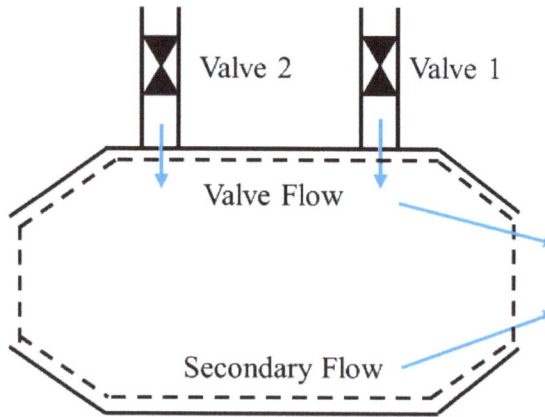

Figure 11. Structure of test chamber.

The internal energy equation of the air in the test chamber is defined as

$$U = WC_vT = W(C_p - R)T \tag{38}$$

where U is the internal energy of the air in the test chamber, W is the mass of the air in the test chamber, T is the temperature of the air in the test chamber, R is the gas constant, C_v is the specific heat of the gas at a constant volume and C_p is the specific heat of the gas at a constant pressure.

The energy differential equation in Equation (39) is obtained by differentiating Equation (38).

$$\frac{dU}{dt} = \frac{dW}{dt}(C_p - R)T + W(C_p - R)\frac{dT}{dt} \tag{39}$$

Moreover, the energy differential equation in Equation (39) can be deducted from the law of energy conservation.

$$\frac{dE}{dt} = W_{in}(C_pT_{in} + \frac{v_{in}^2}{2}) - W_{out}(C_pT_{out} + \frac{v_{out}^2}{2}) \tag{40}$$

where E is energy of the air in the test chamber, and W_{in} is the mass of the air entering the test chamber, which is the valve flow. T_{in} is the temperature of the air entering the test chamber. v_{in} is the velocity of the air entering the test chamber and is calculated with the mass of the air and the inlet area of the test chamber. W_{out} is the mass of the air exiting the test chamber, which is the secondary flow. T_{out} is the temperature of the air exiting the test chamber. v_{out} is the velocity of the air exiting the test chamber and is calculated with the mass of the air and the exiting area of the test chamber.

It is assumed that the air in the test chamber does not exchange heat with the wall. Therefore,

$$\frac{dE}{dt} = \frac{dU}{dt} \tag{41}$$

The change in the mass is denoted as

$$\frac{dW}{dt} = W_{in} - W_{out} \tag{42}$$

The ideal gas equation of state is shown in Equation (43).

$$PV = WRT \tag{43}$$

The temperature differential equation of the air in the test chamber, as shown in Equation (44), can be deduced with Equations (39)–(43).

$$\frac{dT}{dt} = \frac{RT}{PV(C_P - R)}\left[(C_P - R)T(W_{out} - W_{in}) + \left(C_PT_{in} + \frac{v_{in}^2}{2}\right)W_{in} - \left(C_PT_{out} + \frac{v_{out}^2}{2}\right)W_{out}\right] \tag{44}$$

The pressure differential equation of the ideal gas, as shown in Equation (45), can be obtained by differentiating the ideal gas equation of state in Equation (43).

$$\frac{dP}{dt} = \frac{RT}{V}\frac{dW}{dt} + \frac{P}{T}\frac{dT}{dt} \tag{45}$$

The pressure differential equation of the air in the test chamber, as shown in Equation (46), can be obtained by substituting Equations (43) and (44) into Equation (45).

$$\frac{dP}{dt} = \frac{R}{V(C_P - R)}\left[\left(C_PT_{in} + \frac{v_{in}^2}{2}\right)W_{in} - \left(C_PT_{out} + \frac{v_{out}^2}{2}\right)W_{out}\right] \tag{46}$$

The pressure and temperature inside the test chamber at time $t + \Delta t$ can be denoted with the pressure and temperature at time t and the variation within Δt.

$$P^+ = P + \frac{dP}{dt}\Delta t \tag{47}$$

$$T^+ = T + \frac{dT}{dt}\Delta t \tag{48}$$

The pipe volume model represents the pipe connecting the components above. In the exhaust system, it is the volume between the exhaust diffuser and Valve 3. It can be modeled the same as the test chamber model with one input and one output. However, the heat brought out of the volume by the cooler must be considered. Therefore, Equation (40) is augmented as Equation (49). The differential equations of temperature and pressure in Equations (44) and (46) are transformed into Equations (50) and (51). A diagram of the pipe volume model is shown in Figure 12.

$$\frac{dE}{dt} = W_{in}(C_p T_{in} + \frac{v_{in}^2}{2}) - W_{out}(C_p T_{out} + \frac{v_{out}^2}{2}) - \sum \dot{Q}_1 \tag{49}$$

where $\sum \dot{Q}_1$ is the total heat quantity brought out of the gas in the cooler.

$$\frac{dT}{dt} = \frac{RT}{PV(C_P - R)}\left[(C_P - R)T(W_{out} - W_{in}) + \left(C_P T_{in} + \frac{v_{in}^2}{2}\right)W_{in} - \left(C_P T_{out} + \frac{v_{out}^2}{2}\right)W_{out} + \sum \dot{Q}_1\right] \tag{50}$$

$$\frac{dP}{dt} = \frac{R}{V(C_P - R)}\left[\left(C_P T_{in} + \frac{v_{in}^2}{2}\right)W_{in} - \left(C_P T_{out} + \frac{v_{out}^2}{2}\right)W_{out} + \sum \dot{Q}_1\right] \tag{51}$$

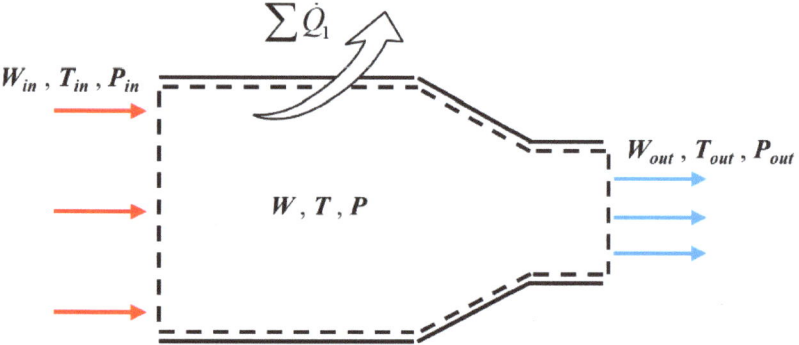

Figure 12. Diagram of pipe volume model.

3. Iterative Model of the Exhaust System

The exhaust system consists of two dynamic cavities whose temperature and pressure change when the flow balance of the inlet and outlet breaks. One of the cavities is the test chamber, and the other is the pipe volume. Since the main purpose of the exhaust system is to simulate the exhaust pressure of the engine, the pressure in the two dynamic cavities is chosen as the system states. The static pressure in the test chamber (p_{1s}) is mainly decided by the valve flow passing through Valve 1 and Valve 2 and the secondary flow ejected by the engine. The static pressure in the pipe volume (p_5) is mainly decided by the output mass of the exhaust diffuser and the mass passing through Valve 3.

In the calculation process of the exhaust diffuser, four initial guess values are introduced. This means that four residual equations are needed. By solving the residual equations with the iteration method, the initial guess values can be replaced with the true values of system gradually. The first three equations are shown in Equations (15), (19) and (23).

The fourth equation is the balance of the exhaust diffuser output pressure p_4 and the pipe volume pressure p_5, as shown in Equation (52).

$$e_4 = (p_5 - p_4)/p_5 \tag{52}$$

The relationship between the components is shown in Figure 13. The inlet conditions, system states and control values of the exhaust system model are listed in Table 3.

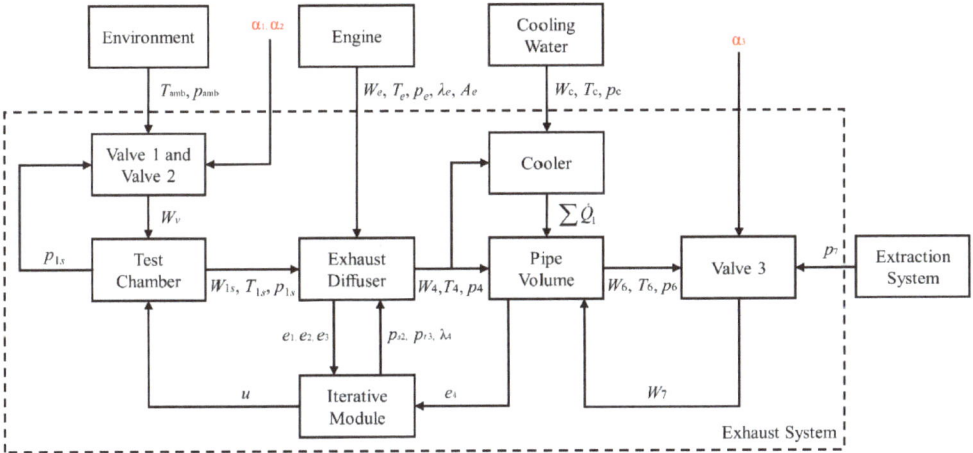

Figure 13. Logic diagram of exhaust system iterative model.

Table 3. Inlet conditions and control values of exhaust system model.

Type	Variable	Meaning
Inlet conditions	p_{amb}	Environment pressure
	T_{amb}	Environment temperature
	W_e	Mass flow of engine nozzle exit
	T_e	Temperature of engine nozzle exit
	p_e	Pressure of engine nozzle exit
	λ_e	Velocity coefficient of engine nozzle exit
	A_e	Flow area of engine nozzle exit
	W_c	Mass flow of cooling water
	T_c	Temperature of cooling water
	p_c	Pressure of cooling water
	p_7	Pressure after Valve 3
System states	p_{1s}	Static pressure in test chamber
	p_6	Static pressure in pipe volume
Control values	α_1	Opening of Valve 1
	α_2	Opening of Valve 2
	α_3	Opening of Valve 3

4. Model Simulation and Verification

In order to verify the accuracy of the exhaust system model, its simulation results should be compared with experimental data. The experimental data used in this paper were obtained in an experiment during the research of a turbofan engine at the AGTF of the AECC Sichuan Gas Turbine Establishment. It tests the performances of engines from idle to intermediate states at different flight heights, including 3 km, 5 km and 10 km. The experimental data were divided into four different types, which are mass flow, temperature, pressure and valve opening. They were obtained with four different kinds of sensors, which were orifice meters, temperature sensors, pressure sensors and linear displacement sensors.

As a result of the linear relationship between the displacement of the actuator and the opening of butterfly valve, as shown in Equation (53), the α_1, α_2 and α_3 could be calculated with the displacement measured with the linear displacement sensors. The arrangement of the sensors in the exhaust system is shown in Figure 14. Their manufacturers and accuracy are listed in Table 4. The measured experimental data are shown in Figure 15.

$$\alpha = \frac{90}{L_{\max}} L \tag{53}$$

where $L_{\max}$ is the maximum stroke of the hydraulic cylinder, and L is the current displacement measured by the sensor.

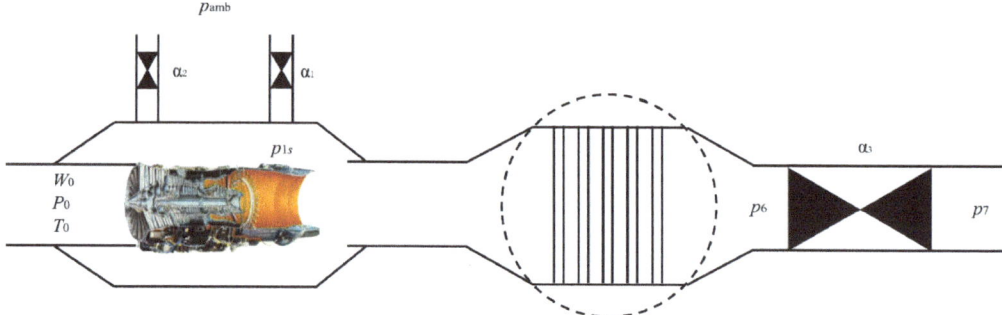

Figure 14. Arrangement of sensors.

Table 4. Type and accuracy of sensors.

Type	Manufacturer	Accuracy
Orifice meter	Chuanyi	±1%
Temperature sensor	Therncway	±0.5%
Pressure sensor	Honeywell	±0.1%
Linear displacement sensor	MTS	±0.1%

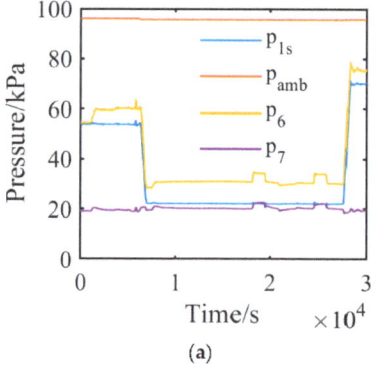
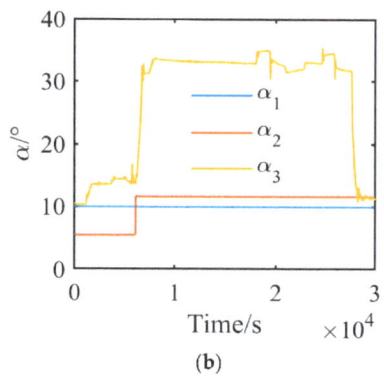

(a)   (b)

Figure 15. Experimental data: (**a**) Pressure of exhaust system; (**b**) Opening of valves in exhaust system.

4.1. Model Verification with Experimental Data and Observed Engine Parameters

The exhaust system model was built with Matlab in Simulink. The simulation platform is shown in Figure 16. It consisted of the experimental data module, engine data observer module, exhaust system controller module, actuator module and exhaust system module.

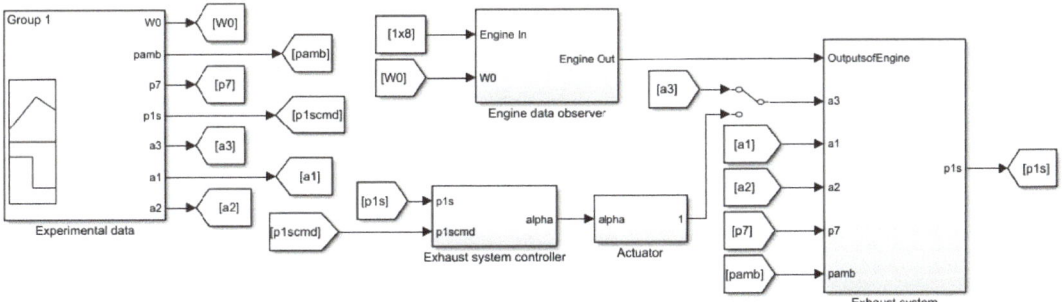

Figure 16. Simulation platform for exhaust system validation.

The experimental data module contained the data that are shown in Figure 15. The engine data observer module calculated the outlet mass flow W_e, pressure P_e, temperature T_e and nozzle throat area A_e of the engine based on the states of the engine because they were not measured directly. They are shown in Figure 17. The controller and the actuator were used to test the performance of the exhaust system built with the multi-cavity method in a closed-loop system. It was a PI controller with P = 0.001 and I = 0.1. The pressure in the test chamber was regulated to the desired value by adjusting the opening of Valve 3 with this PI controller by switching the manual switch module and activating the control loop. The transfer function of the valve actuator was assumed as $\frac{1}{0.1s+1}$.

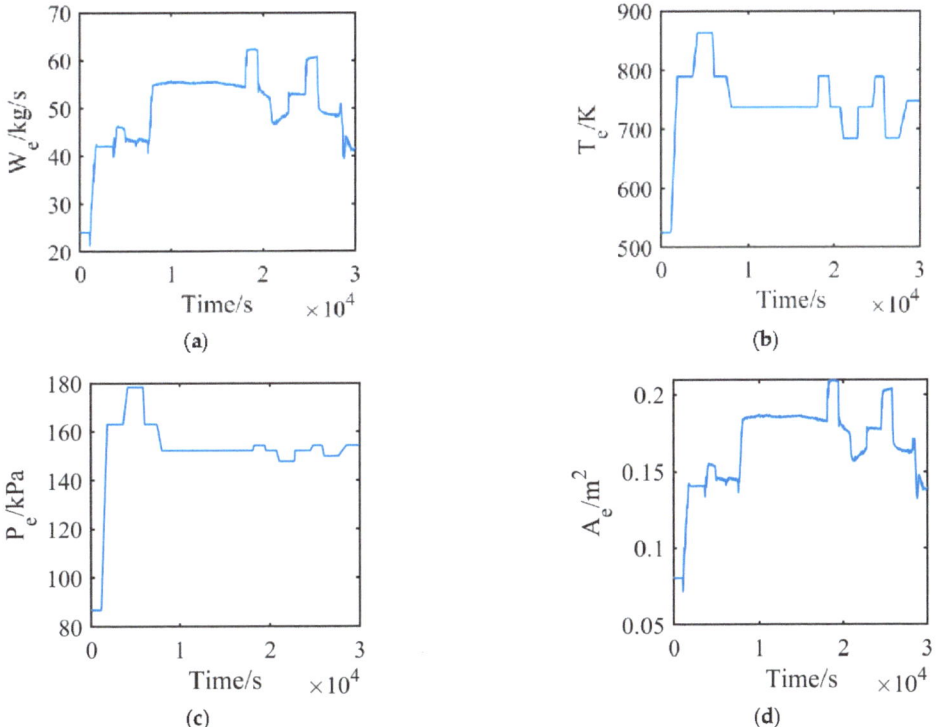

Figure 17. Observed engine parameters: (**a**) Mass flow of engine; (**b**) Total temperature after turbine; (**c**) Total pressure after turbine; (**d**) Throat area of engine nozzle.

The simulation platform was solved with the Heun method and a fixed step size of 0.01.

A simulation under the same working conditions as in the experiment was carried out. The inlet conditions and the control values of the exhaust system model that are shown in Table 2 were set with the experimental data module and the engine data observer module, as shown in Figure 16. Other parameters were set as constants. The temperature of the environment and cooling water was 288.15 K. The velocity coefficient λ_e at the nozzle throat was 1.

The simulation results compared with experimental data are shown in Figure 18. The pressure error between the simulation results and experimental data was no more than 2 kPa in the steady state and less than 6 kPa in the transient process.

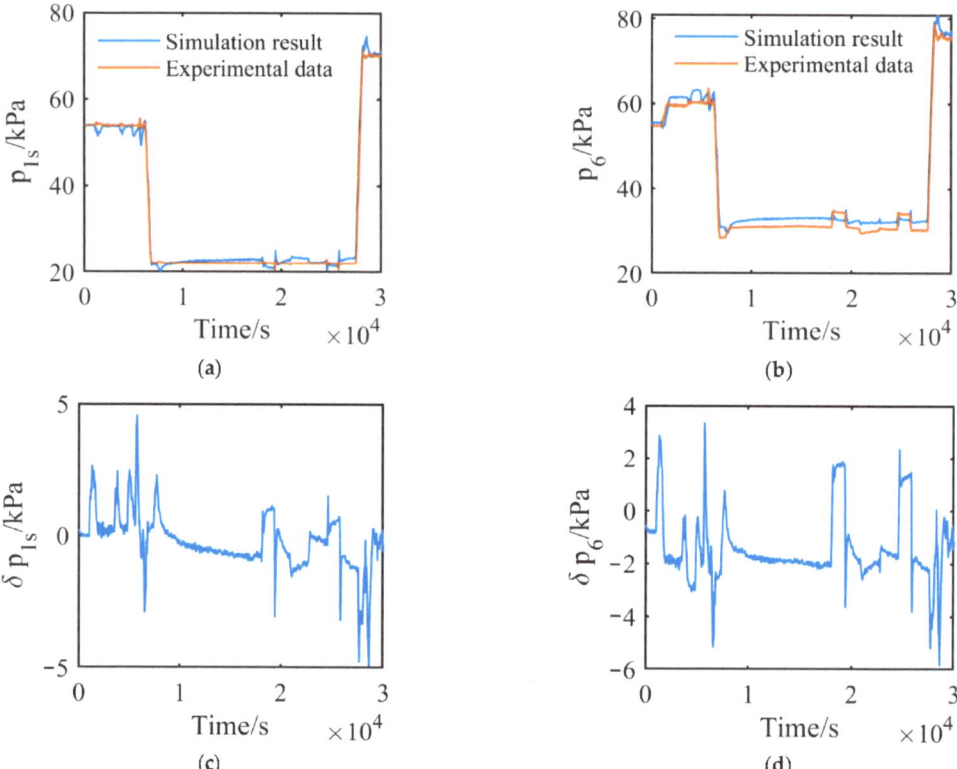

Figure 18. Simulation results of pressure: (**a**) Pressure in test chamber; (**b**) Pressure before Valve 3; (**c**) Error of pressure in test chamber; (**d**) Error of pressure before Valve 3.

4.2. Simulation Results Compared with Lumped-Parameter Model

It should be known that, in the modeling procedure of the test chamber and pipe volume model, the the main reason for the pressure dynamic in the volume is the inflow and outflow mass. Therefore, the mass calculation becomes an important part of the exhaust system model for both accurate calculations and dynamic performance analysis.

The simulation results of mass are shown in Figure 19. Taking the exhaust system shown in Figure 13 as a whole, the total mass flow into the system consists of the valve flow W_v, which is the sum of the mass that passes through Valve 1 and Valve 2, and the engine flow W_e. The mass flow out of the system includes mass that passes through Valve 3, which is denoted as W_7. They are equal to each other in the steady state, and both of them increase and decrease with the engine state as the mass flow W_e changes. It should be mentioned that W_7 increases around time 6000 s and decreases around time 27,600 s quickly, as shown in Figure 19a. It results from rapid opening change of Valve 3, as shown in Figure 15b, with the purpose of changing the pressure p_{1s} in the test chamber, as shown in Figure 15a. The dynamic performance of the test chamber, as shown in Figure 11, is defined with the input denoted as valve flow W_v and the output denoted as secondary flow W_{1s}. It is shown in Figure 19b that the valve flow W_v equals the secondary flow W_{1s} in the steady state, and the balance is broken in the transient process of the engine or exhaust system.

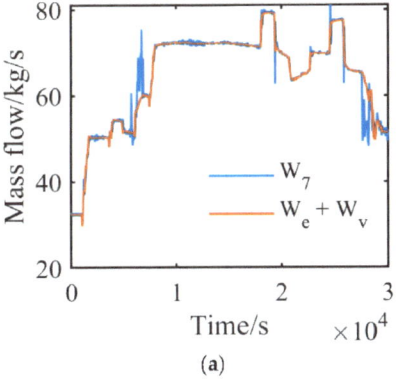
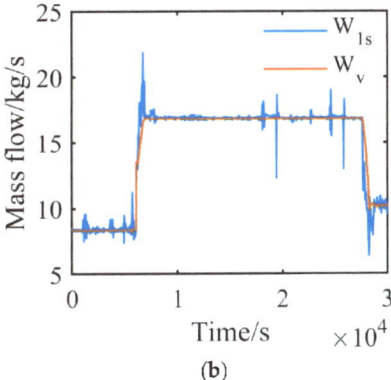

Figure 19. Simulation results of mass flow: (**a**) Mass flow of exhaust system; (**b**) Mass flow of test chamber.

The secondary flow of mass W_{1s}, which is ejected by the main stream of mass W_{1m} and defined in Equation (4), is the output of the test chamber model and one of inputs of the exhaust system volume. It is a bridge connecting the two main volumes, which is the test chamber and pipe volume, as shown in Figure 13, in the exhaust system. The multi-cavity iterative exhaust system model calculates secondary flow based on the exhaust parameters of the engine, the pressure in the test chamber and the pressure in the exhaust system by iteration. It makes the model have the advantage of calculating the pressure difference more accurately compared with the traditional lumped-parameter model. The lumped parameter model of AGTF exhaust systems was first proposed in reference [12], and its logic diagram is shown in Figure 20. It simplifies the exhaust system as a single volume without considering the pressure dynamic in the exhaust diffuser and the heat exchange in the cooler.

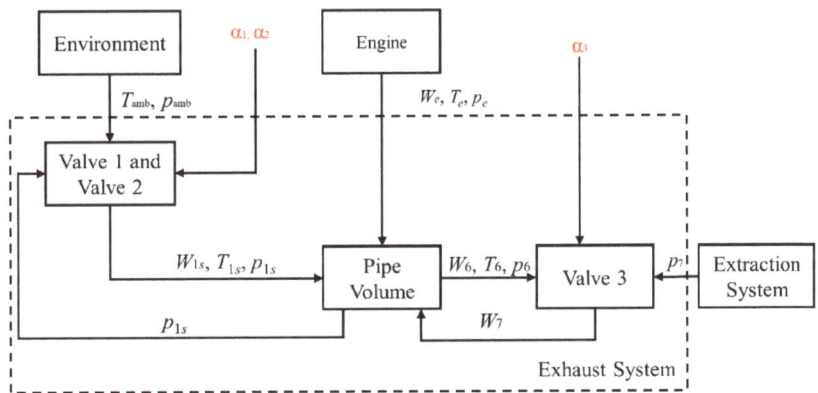

Figure 20. Logic diagram of exhaust system built with lumped parameters.

Therefore, to further verify the advantage of the exhaust system built with the multi-cavity iterative modeling method, it was compared with the traditional lumped-parameter model. The simulation was carried out under the same conditions shown in Figure 15. The simulation results are shown in Figure 21, where the experimental data are marked with a solid line, and the simulation results are marked with a dashed line.

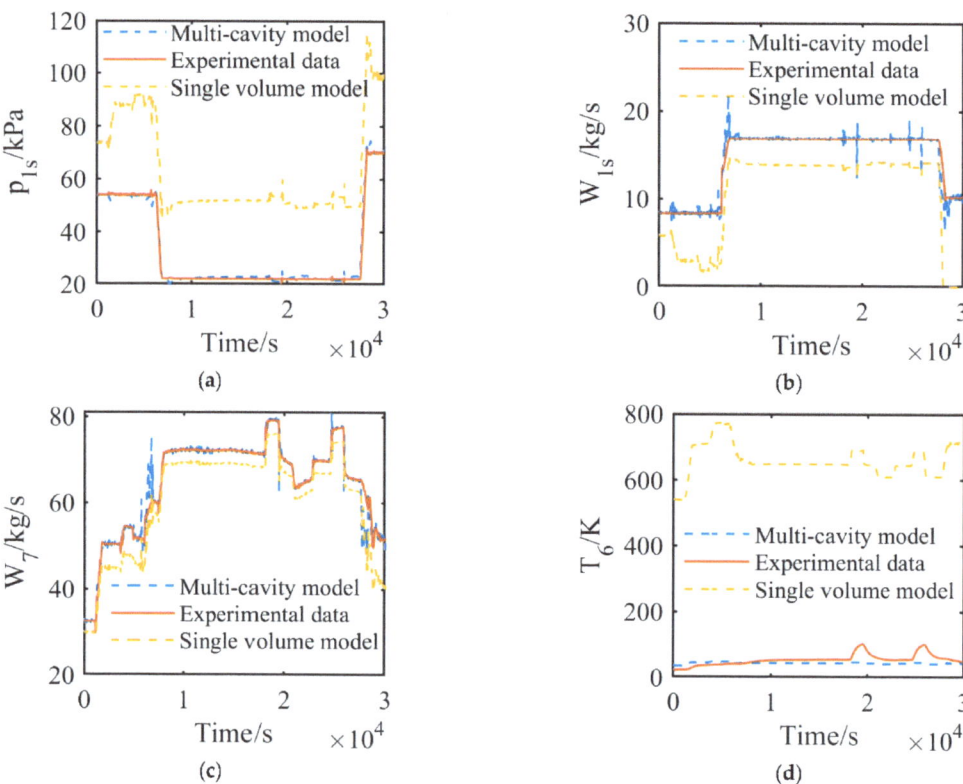

Figure 21. Parameter comparison between lumped-parameter model experimental data: (**a**) Pressure of exhaust system; (**b**) Valve flow of exhaust system; (**c**) Total flow of exhaust system; (**d**) Temperature before Valve 3.

It is shown in Figure 21a that the test chamber pressure obtained with the single-volume lumped-parameter model is much higher than that of the experimental data and the multi-cavity model simulation results. The secondary flow cannot be obtained with the single-volume model and is simplified as the valve flow. Due to the higher pressure in the test chamber, the valve flow W_v calculated with the single-volume model is smaller than that of the experimental data and the multi-cavity model simulation results. The difference between the experimental data and the multi-cavity model simulation results in Figure 21b was a result of replacing the experimental W_{1s} with W_v, because the secondary flow could not be measured directly in the experiment.

Moreover, the total mass flow of the exhaust system is shown in Figure 21c. The smaller total mass flow of the single-volume model results from the much higher temperature before Valve 3, because the lumped-parameter model does not consider the heat exchange in the cooler. The temperature before Valve 3 is shown in Figure 21d.

4.3. Closed-Loop Simulation Results

To verify the performance of the exhaust system built with the multi-cavity method in a closed-loop system, the manual switch module was switched, and the control loop was activated, as shown in Figure 16.

The simulation results are shown in Figure 22. The pressure in the test chamber fits the experimental data, and the difference between $α_3$ is relatively small, which further verifies the accuracy of the exhaust system model. It proves that the exhaust system model can be used to design and test the control algorithm. Moreover, the response of the mass flow, shown in Figure 23, is similar to the mass flow response, shown in Figure 19, with a smaller mass flow fluctuation.

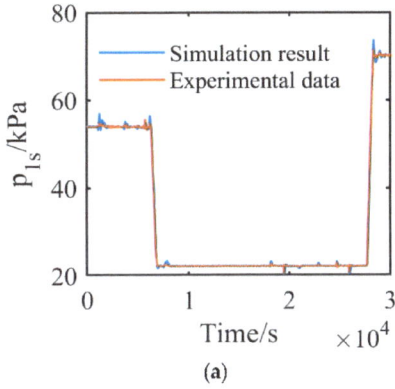
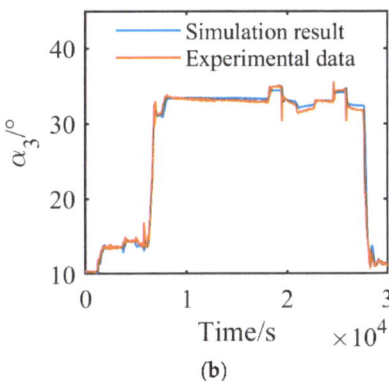

Figure 22. Comparison of closed-loop simulation results and experimental data: (**a**) Pressure of exhaust system; (**b**) Opening of Valve 3 in exhaust system.

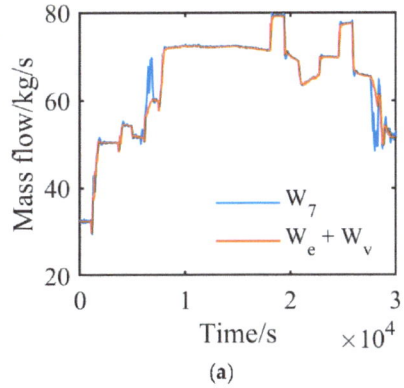
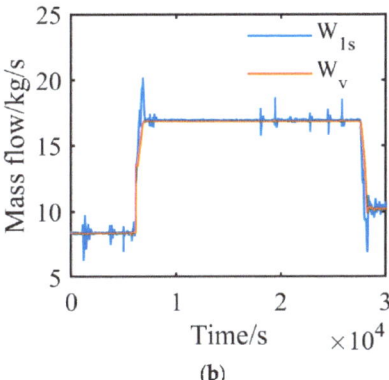

Figure 23. Closed-loop simulation results of mass flow: (**a**) Flow of exhaust system; (**b**) Flow of test chamber.

5. Conclusions

A multi-cavity iterative modeling method is proposed, which aims at improving accuracy and reflects the dynamic process of the exhaust system model. The conclusions that were obtained are listed as follows:

1. The multi-cavity iterative model is able to simulate the pressure in the test chamber and in the exhaust system. The simulation results show that the maximum error of pressure in the test chamber is 2 kPa in the steady state and 5 kPa in the transient process compared with the experimental data. The pressure error in the exhaust system is no more than 2 kPa

in the steady state and no more than 6 kPa in the transient process compared with the experimental data.

2. The simulation results of the multi-cavity iterative model were compared with the lumped-parameter model under the same working conditions. This showed that the model built with the proposed method has advantages of calculating the pressure and mass in the exhaust system accurately over the lumped-parameter model.

3. The multi-cavity iterative model was tested in a closed-loop system. The results show that the pressure in the test chamber can be controlled to the desired value, and the opening of the valve is only slightly different from the experimental data, which further verifies the accuracy of the exhaust system model and proves the model's capability of being used to design and test the control algorithm.

Impurities such as water vapor should be considered in the future, because engine emissions have been a hot issue for decades, and new AGTFs that are able to carry out research on the icing and noxious gases of engines are being built. It increases the demand of the exhaust system model, and the modeling method should be improved to support the research.

Author Contributions: Conceptualization, K.M. and X.W. (Xi Wang); methodology, K.M. and X.W. (Xi Wang); software, K.M., M.Z. and J.L.; validation, K.M., S.Z., Z.D. and X.W. (Xin Wang); writing—original draft preparation, K.M.; writing—review and editing, K.M., M.Z., S.Y. and X.P.; visualization, K.M. and L.Z.; supervision, M.Z.; project administration, X.W. (Xi Wang). All authors have read and agreed to the published version of the manuscript.

Funding: This research was funded by the AECC Sichuan Gas Turbine Establishment Stable Support Project, grant number GJCZ-0011-19; the National Science, Technology Major Project, grant number 2017-V-0015-0067; and the Technology Major Project, grant number 2019-V-0010-0104.

Data Availability Statement: Not applicable.

Conflicts of Interest: The authors declare no conflict of interest.

References

1. Hou, M.J. *Aero-Engine Altitude Simulating Test Techniquesl*; Aviation Industry Press: Beijing, China, 2014; p. 1.
2. Zhu, M. *Thermal Fluid Mechanism Analysis and the Research of Modern Control Methods for Altitude Ground Test Facilities*; Beihang University: Beijing, China, 2021.
3. Gerald, A.; Joseph, K. Numerical modeling of altitude diffusers. In Proceedings of the 19th Advanced Measurement and Ground Testing Technology Conference, New Orleans, LA, USA, 17–20 June 1996.
4. Li, G.; Zhang, J. *Aeronautic Ejector/Mixer*; National Defense Industry Press: Beijing, China, 2007; pp. 15–26.
5. Byung, H.; Jihwan, L.; Sunghyun, P.; Ji, H.; Woongsup, Y. Design and Analysis of a Second-Throat Exhaust Diffuser for Altitude Simulation. *J. Propuls. Power* **2012**, *28*, 1091–1104.
6. Montgomery, P.A.; Burdette, R.; Krupp, B. A Real-Time Turbine Engine Facility Model and Simulation for Test Operations Modernization and Integration. In Proceedings of the ASME Turbo Expo 2000: Power for Land Sea, and Air, Munich, Germany, 8–11 May 2000.
7. Montgomery, P.A.; Burdette, R.; Wilhite, L.; Salita, S. Modernization of a Turbine Engine Test Facility Utilizing a Real-Time Facility Model and Simulation. In Proceedings of the ASME Turbo Expo 2001: Power for Land Sea, and Air, New Orleans, LA, USA, 4–7 June 2001.
8. Davis, M.; Montgomery, P. A Flight Simulation Vision for Aeropropulsion Altitude Ground Test Facilities. In Proceedings of the ASME Turbo Expo 2002: Power for Land Sea, and Air, Amsterdam, The Netherlands, 3–6 June 2002.
9. Boraira, M.; Van Every, D. Design and Commissioning of a Multivariable Control System for a Gas Turbine Engine Test Facility. In Proceedings of the 25th AIAA Aerodynamic Measurement Technology and Ground Testing Conference, San Francisco, CA, USA, 5–8 June 2006.
10. Bierkamp, J.; Köcke, S.; Staudacher, S.; Fiola, R. Influence of ATF Dynamics and Controls on Jet Engine Performance. In Proceedings of the ASME Turbo Expo 2007: Power for Land Sea, and Air, Montreal, QC, Canada, 14–17 May 2007.
11. Weisser, M.; Bolk, S.; Staudacher, S. Hard-in-the-Loop-Simulation of a Feedforward Multivariable Controller for the Altitude Test Facility at the University of Stuttgart. 2013. Available online: https://www.dglr.de/publikationen/2013/301179 (accessed on 9 February 2022).
12. Pei, X.T.; Zhang, S.; Dan, Z.H.; Zhu, M.Y.; Qian, Q.M.; Wang, X. Study on Digital Modeling and Simulation of Altitude Test Facility Flight Environment Simulation System. *J. Propul. Technol.* **2019**, *40*, 1144–1152.

13. Zhu, M.; Zhang, S.; Dan, Z.; Pei, X.; Wang, W.; Wang, X. A coordinate positioning and regression algorithm for the flow characteristics of a large butterfly valve. *Gas Turbine Exp. Res.* **2017**, *30*, 39–44.
14. Sosunov, V. The history of aviation engine development in the USSR and the 60th anniversary of CIAM. In Proceedings of the 26th Joint Propulsion Conference, Orlando, FL, USA, 16–18 July 1990.
15. Pei, X.; Liu, J.; Wang, X.; Zhu, M.; Zhang, L.; Dan, Z. Quasi-One-Dimensional Flow Modeling for Flight Environment Simulation System of Altitude Ground Test Facilities. *Processes* **2022**, *10*, 377. [CrossRef]
16. Sheeley, J.M.; Sells, D.A.; Bates, L.B. Experiences with Coupling Facility Control Systems with Control Volume Facility Models. In Proceedings of the 42nd AIAA Aerospace Sciences Meeting and Exhibit, Reno, NE, USA, 5–8 January 2004.
17. Zhu, M.; Wang, X.; Zhang, S.; Dan, Z.; Pei, X.; Miao, K.; Jiang, Z. PI Gain Scheduling Control for Flight Environment Simulation System of Altitude Ground Test Facilities Based on LMI Pole Assignment. *J. Propuls. Technol.* **2019**, *40*, 2587–2597.
18. Qian, Q.; Dan, Z.; Zhang, S.; Pei, X.; Wang, W. Linear active disturbance rejection control method for intake pressure control in aero-engine transient test. *J. Aerosp. Power* **2019**, *34*, 2271–2279.
19. Zhu, M.; Wang, X.; Dan, Z.; Zhang, S.; Pei, X. Two freedom linear parameter varying u synthesis control for flight environment testbed. *Chin. J. Aeronaut.* **2018**, *32*, 1204–1214. [CrossRef]
20. Pan, J.; Shan, P. *Fundamentals of Gasdynamics*; National Defense Industry Press: Beijing, China, 2011; pp. 87–95.

Article

Generic Modeling Method of Quasi-One-Dimensional Flow for Aeropropulsion System Test Facility

Jiashuai Liu [1], Xi Wang [1], Xitong Pei [2,3], Meiyin Zhu [4], Louyue Zhang [1], Shubo Yang [1,*] and Song Zhang [3]

1. School of Energy and Power Engineering, Beihang University, Beijing 100191, China; ljsbuaa@buaa.edu.cn (J.L.); xwang@buaa.edu.cn (X.W.); sy2004112@buaa.edu.cn (L.Z.)
2. Research Institute of Aero-Engine, Beihang University, Beijing 100191, China; peixitong@buaa.edu.cn
3. Science and Technology on Altitude Simulation Laboratory, AECC Sichuan Gas Turbine Establishment, Mianyang 621703, China; 1510450119@student.cumtb.edu.cn
4. Beihang Hangzhou Innovation Institute Yuhang, Hangzhou 310023, China; mecalzmy@buaa.edu.cn
* Correspondence: yangshubo@buaa.edu.cn

Abstract: To support the advanced controller design and verification of the Aeropropulsion System Test Facility (ASTF), it is necessary to establish a mathematical model of ASTF with high precision and replace the current lumped parameter model. Therefore, a quasi-one-dimensional flow model of ASTF is established considering friction, localized losses, heat transfer, etc. Moreover, a generic modeling method is proposed for quasi-one-dimensional flow. With this method, all component models of ASTF are composed of staggered central control volume (CCV) and boundary control volume (BCV) and connected through virtual control volume. Thus, the properties of quasi-one-dimensional flow, such as spatial effect and time delay, can be easily addressed during the modeling process. The simulation results show that the quasi-one-dimensional flow model has higher accuracy than the lumped parameter model. Comparing the simulation results of the quasi-one-dimensional flow model with the test data, the relative errors of flow and pressure are less than 2.2% and 1.4%, respectively, further verifying the correctness of the proposed modeling method.

Keywords: quasi-one-dimensional flow; Aeropropulsion System Test Facility; modeling; virtual control volume; numerical simulation

Citation: Liu, J.; Wang, X.; Pei, X.; Zhu, M.; Zhang, L.; Yang, S.; Zhang, S. Generic Modeling Method of Quasi-One-Dimensional Flow for Aeropropulsion System Test Facility. Symmetry 2022, 14, 1161. https:// doi.org/10.3390/sym14061161

Academic Editors: Longfei Chen, Fatemeh Salehi, Zheng Xu, Guangze Li, Bin Zhang and Mihai Postolache

Received: 17 May 2022
Accepted: 3 June 2022
Published: 5 June 2022

Publisher's Note: MDPI stays neutral with regard to jurisdictional claims in published maps and institutional affiliations.

Copyright: © 2022 by the authors. Licensee MDPI, Basel, Switzerland. This article is an open access article distributed under the terms and conditions of the Creative Commons Attribution (CC BY) license (https:// creativecommons.org/licenses/by/ 4.0/).

1. Introduction

By providing airflow at pressures and temperatures experienced during flight, Aeropropulsion System Test Facility (ASTF) can simulate flight conditions to test aircraft engines [1,2]. To simulate more realistic flight conditions, a lot of research on the controller of ASTF is required, which relies on reliable numerical simulation verification. Hence, it is necessary to establish a high-precision numerical simulation model of ASTF.

In fact, ASTF is a symmetrical large-scale pipeline system, including pipe, valve, mixer, flow deflector, air source, etc. As early as 1998, Schmidt et al. [3] proposed a simplified volume model for the pipe, which was obtained from the mass and energy balance. After that, Montgomer et al. [4,5] and Sheeley et al. [6] used the lumped parameter method to develop the math models of test facilities at the Arnold Engineering Development Complex (AEDC), including tanks, pipes, burners, etc. Recently, considering the flow characteristics of the valve and the dynamic characteristics of the volume, Pei et al. [7] established a mathematical model of ASTF. To further improve the accuracy of the ASTF model, Zhu et al. [8] proposed a multi-volume modeling method, which considered the effects of many factors, such as pressure loss of pipeline and fluid–solid heat transfer. The above studies were essentially based on the lumped parameter method, which assumes that the airflow in pipeline is instantaneously mixed evenly and has uniform properties [6]. However, the lumped parameter model cannot directly account for spatial effects and any losses due to friction, bending, etc. [6,9]. It also cannot reflect some important characteristics of ASTF, such as time delay. In addition, the fluid network method [10–12] is often used in the research of complex pipeline system, such as natural gas transportation system, engine fuel system, compressed air supply system, etc. It mainly calculates the pressure of each node of the pipeline system and does not care about the flow process inside the pipeline. It is difficult to consider the heat transfer of pipes and other components.

While ensuring high computational efficiency, the one-dimensional (strictly, quasi-one-dimensional) flow method is gradually being studied to improve model accuracy. Pei et al. [13] built a model of the flight environment simulation system, which required that the cavity only be connected with the throttling element. To capture the transient response of the wind tunnel, which is a special pipeline system, Rennie et al. [14] divided the tunnel circuit into small sections, and applied the mass, momentum, and energy conservation equations for each section. However, this model was not suitable for the case that the pipeline has branches. Liu et al. [15,16] established the finite elements state-space model for one-dimensional compressible fluid flow but did not study the model of throttling elements (e.g., control valve). After that, Boylston [9] studied the quasi-one-dimensional flow modeling method for AEDC's four-foot transonic wind tunnel, but it was still limited to pipe modeling. Inspired by their work, this paper introduces the quasi-one-dimensional flow method into the modeling of ASTF. However, due to the complexity of ASTF's equipment, models of large valve, mixer, pipe junction, etc., also need to be established. More importantly, previous studies lack a generic modular modeling method for all ASTF's components. Thus, a generic quasi-one-dimensional flow modeling method of ASTF is proposed. This method makes all component models consist of control volumes. Because the control volume considers heat transfer, friction, local pressure losses and other factors, these factors can be easily introduced into all component models to improve them. In addition, these component models are symmetrical, which facilitates the construction of the system-level model.

The main contributions of this paper are as follows:

(1) The concepts of CCV, BCV and virtual control volume are proposed, and a generic quasi-one-dimensional flow modeling method is established.
(2) All the component models and system-level model of ASTF are established, and the limitation that the cavity can only be connected with the throttling element is eliminated.

This paper is organized as follows. In Section 2, two basic control volumes as well as the calculation methods of parameters are introduced, serving as the backbone for the modeling framework. In Section 3, through the staggered connection of CCV and BCV, the component models of ASTF are constructed, including the pipeline model, control valve model, multi-port junction model, flow source/sink model and pressure/temperature boundary model. To unify the boundary of the component models and build the system-level model, the virtual control volume is proposed as the connection of models. Section 4 presents the characteristics and establishment method of the system-level model, and then constructs a numerical simulation model of ASTF. Section 5 describes the differences between the quasi-one-dimensional flow model and lumped parameter model in steady-state and dynamic processes. By comparing the simulation results with test data, the correctness of the modeling method proposed in this paper is further verified. Section 6 provides concluding remarks.

2. Control Volume

Assumptions:

1. The fluid in control volume is instantaneously mixed evenly and has uniform properties.
2. The gravitational potential energy is ignored.

According to the difference of the concerned sections, control volumes are classified into two types: central control volume (CCV) and boundary control volume (BCV), as shown in Figure 1. The temperature, pressure, density and energy of the fluid are calculated in the state parameter cross section. The velocity, momentum and Mach number of the fluid are calculated in the motion parameter cross section. The boundary of the CCV is the motion parameter cross section, while the boundary of BCV is the state parameter cross section. The fluid moves only along the axial direction, and the radial boundary is closed. Although the parameters of each section can be calculated in different ways, the basic conservation equation is mostly used, which is briefly reviewed in the following.

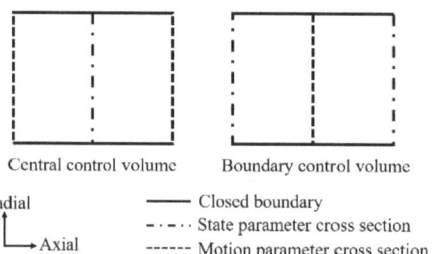

Figure 1. Differences between two types of control volumes.

2.1. Central Control Volume

As shown in Figure 2, it is assumed that the boundary parameters of CCV are known. According to the mature theory of unsteady flow of compressible fluid [9,17,18], the parameters of the CCV's state parameter cross section can be calculated as follows:

$$\frac{d(\rho_j V_j)}{dt} = \rho_{in} v_i A_i - \rho_{out} v_i A_i \quad (1)$$

$$\frac{d(E_j \rho_j V_j)}{dt} = \left(E_{in} + \frac{p_{in}}{\rho_{in}}\right)\rho_{in} v_i A_i - \left(E_{out} + \frac{p_{out}}{\rho_{out}}\right)\rho_{out} v_{i+1} A_{i+1} + \dot{Q}_j - \dot{\Psi} \quad (2)$$

where V_j is the volume of CCV. The subscript $i+1$ represents the downstream motion parameter section adjacent to motion parameter section i. E_{in}, p_{in}, ρ_{in}, E_{out}, p_{out} and ρ_{out} need to be calculated according to the parameters of adjacent upstream and downstream grids, such as taking the average value [9] or upwind scheme [19] (see Section 3.1 for details). Equation (1) represents the conservation of mass, that is,

$$\begin{pmatrix} \text{Time rate of change} \\ \text{of mass within the CCV} \end{pmatrix} = \begin{pmatrix} \text{Total mass flow rate} \\ \text{entering the CCV} \end{pmatrix} - \begin{pmatrix} \text{Total mass flow rate} \\ \text{leaving the CCV} \end{pmatrix}$$

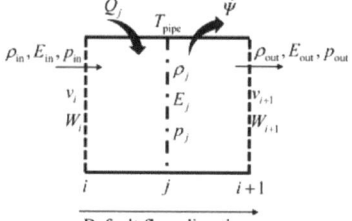

Figure 2. Schematic diagram of CCV.

Equation (2) represents the conservation of energy, that is,

$$\begin{pmatrix} \text{Time rate of change of} \\ \text{total energy within the CCV} \end{pmatrix} = \begin{pmatrix} \text{Rate of energy (transferred} \\ \text{by mass) entering the CCV} \end{pmatrix} \\ - \begin{pmatrix} \text{Rate of energy (transferred} \\ \text{by mass) leaving the CCV} \end{pmatrix} + \begin{pmatrix} \text{Rate of net} \\ \text{heat input} \end{pmatrix} - \begin{pmatrix} \text{Rate of net} \\ \text{work output} \end{pmatrix}$$

In Equation (2), $\dot{Q}_j = hA_s(T_{\text{pipe}} - T_j)$, and A_s is the surface area through which the convection heat transfer takes place, T_{pipe} is the wall temperature, and T_j is the temperature of the airflow in CCV [20,21]. Actually, h can be variable, and detailed calculation can be found in Refs. [9,22,23]. For ASTF, the fluid does not do shaft work, so $\Psi_s = 0$. Since the radial boundary of the control volume is a solid wall, where the velocity of the fluid is zero, the work done by the viscous force is zero. Then, $\Psi = \Psi_s + \Psi_v = 0$.

According to the state equation for ideal gas and the relationship between internal energy and state parameters, the pressure in the state parameter cross section can be obtained as [24]

$$p_j = (\gamma - 1)\rho_j \left[E_j - \frac{1}{2} v_j^2 \right] \tag{3}$$

where $v_j \approx \frac{1}{2}(v_i + v_{i+1})$.

2.2. Boundary Control Volume

As shown in Figure 3, it is assumed that the boundary parameters of BCV are known. According to the mature theory of the unsteady flow of compressible fluid [9,17,18], the parameters of BCV's motion parameter cross section can be calculated as follows:

$$\frac{dW_i}{dt} = (p_{j-1}A_{j-1} - p_jA_j + p_iA_j - p_iA_{j-1}) + \rho_{j-1}v_{\text{in}}^2 A_{j-1} - \rho_j v_{\text{out}}^2 A_j - \frac{V_i \rho_i f}{2D_i} v_i |v_i| - \frac{A_i \rho_i k}{2} v_i |v_i| \tag{4}$$

where v_{in} and v_{out} also need to be approximated (see Section 3.1 for details). The subscript j-1 represents the upstream state parameter section adjacent to state parameter section j. $p_i \approx \frac{1}{2}(p_{j-1} + p_j)$ and $\rho_i \approx \frac{1}{2}(\rho_{j-1} + \rho_j)$ are state parameters of the motion parameter cross section. Equation (4) represents the conservation of momentum, that is,

$$\begin{pmatrix} \text{Time rate of change of} \\ \text{momentum within the CCV} \end{pmatrix} = \begin{pmatrix} \text{Force on the fluid} \\ \text{due to pressure} \end{pmatrix} + \begin{pmatrix} \text{The rate of change of momentum} \\ \text{brought in by the incoming fluid} \end{pmatrix} \\ - \begin{pmatrix} \text{The rate of change of momentum} \\ \text{leaving with the outgoing fluid} \end{pmatrix} \\ - \begin{pmatrix} \text{Frictional force} \\ \text{acting against the flow} \end{pmatrix} - (\text{Force due to minor pressure losses})$$

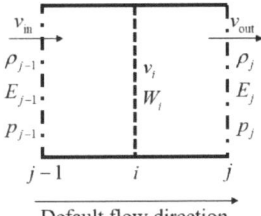

Figure 3. Schematic diagram of BCV.

By introducing f and k into Equation (4), the subsequent model can handle a variety of situations, such as friction, bends, flow obstructions, etc. For laminar flow (i.e., Re $\leq$ 2400), the friction factor f can be approximated by $64/\text{Re}$. When Re $>$ 2400, the friction factor f

can be approximated by $0.25 / \left[\log \left(\frac{\varepsilon}{3.7D} + \frac{5.74}{Re^{0.9}} \right) \right]^2$ [9]. The minor loss coefficient k is related to the structure of local obstacles and the actual pipeline, which is usually determined by experiments. The detailed description of k can be found in Refs. [9,25].

According to the definition of momentum, the velocity in the motion parameter cross section is obtained as

$$v_i = \frac{W_i}{\rho_i V_i} \qquad (5)$$

2.3. Outlook

CCV and BCV deal with parameters of different cross sections, and thus the one-dimensional flow process can be calculated by the staggered connection of CCV and BCV. Although this method is essentially similar to the one-dimensional flow in previous studies, this description method in this paper gives more convenience and universality to subsequent modeling. Any component and even the whole system can be described by the staggered connection of CCV and BCV, not limited to the pipeline.

The control volumes contain the parameters of different cross sections. In addition to the above calculation methods, for different components, their parameters are allowed to be calculated by different equations (see Section 3.2). Some parameters of the control volumes are unknown, which will be considered in the component modeling.

3. Component Models

As a pipeline system, the parameter distribution along the axial direction reflects the main characteristics of ASTF. Therefore, the difference of the radial parameters of airflow is ignored, and only the one-dimensional flow of the airflow in the pipeline system is considered. Based on this, the pipeline, control valve, multi-port junction, flow source/sink, pressure/temperature boundary and other components can be equivalent to the combination of CCV and BCV. Since the structure of each model is symmetrical, the direction of airflow does not affect the calculation of the model. Except for the flow source/sink model, which has a definite flow direction, the other models only need a default flow direction.

3.1. Pipeline Model

Based on the staggered grid, a one-dimensional flow model can be established [9,16,19,24]. Here, the CCV and BCV are staggered together to form a one-dimensional pipeline model, as shown in Figure 4. In the figure, the black triangle represents the input port, and the white triangle represents the output port. The subscript $j+1$ represents the downstream state parameter section adjacent to state parameter section j. The subscript $i-1$ represents the upstream motion parameter section adjacent to motion parameter section i.

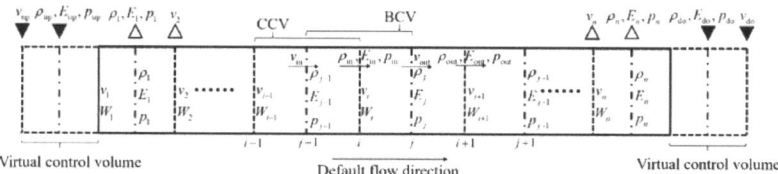

Figure 4. Schematic diagram of pipeline model.

The unknown state parameters can be calculated using the upwind scheme, and the unknown motion parameters can be approximated by the average value, as shown in Equations (6) and (7).

$$y_{\text{in}} = \begin{cases} y_{j-1}, v_i \geq 0 \\ y_j, v_i < 0 \end{cases}, \quad y_{\text{out}} = \begin{cases} y_j, v_i \geq 0 \\ y_{j+1}, v_i < 0 \end{cases} \qquad (6)$$

$$v_{\text{in}} = \frac{1}{2}(v_{i-1} + v_i) \,,\ v_{\text{out}} = \frac{1}{2}(v_i + v_{i+1}) \tag{7}$$

where $i = 1, 2, \cdots, n$ and $y \in \{\rho, E, p\}$.

Previous studies have carried out special treatment on the pipeline boundary, generally taking actual physical parameters as boundary conditions. In this paper, two virtual control volumes are set at both ends of the pipeline. The parameters of the virtual control volumes (i.e., the boundary conditions of the pipeline model) are provided by the upstream and downstream ports. Thus, the parameters of each cross section in the pipeline model can be directly calculated according to Equations (1)–(7). The calculation results of the virtual control volumes are returned as the output to the upstream and downstream ports, as shown in Figure 4.

3.2. Control Valve Model

ASTF's control valve is a large-size throttling component. The dynamic process of the airflow through the control valve directly affects the characteristics of ASTF. The presence of the valve plate makes the contact area between the valve body and airflow large so that the heat transfer process between the control valve and the airflow cannot be ignored. The control valve model can be established through the staggered connection of two CCVs and one BCV, as shown in Figure 5. The two CCVs are located on both sides of the throttling cross section of the control valve, which can reflect the dynamic process of the airflow. Moreover, the heat transfer process can be considered in the two CCVs to simulate the energy change of the airflow through the control valve.

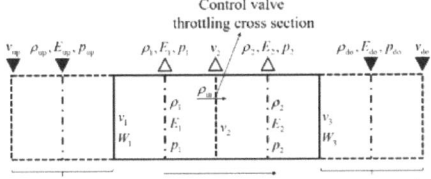

Figure 5. Schematic diagram of control valve model.

In ASTF, the structure of the large-size control valve is complicated, and it is difficult to establish its flow characteristic model with the traditional theoretical formula [26]. Actually, a large amount of data is retained during the operation of ASTF, which can be used to improve the accuracy of the model. The mass flow rate through the control valve can be written as [27]

$$\dot{m} = \varphi S \sqrt{2\rho_1 p_1} \tag{8}$$

where $\varphi = g(S, \pi)$ is the flow coefficient of control valve, and it is a function of pressure ratio $\pi = p_2/p_1$. g represents a complicated function. Through the fitting of test data and the calibration of the three-dimensional flow field simulation data, an interpolation table of the flow coefficient can be obtained to replace g [28].

The flow velocity at the throttling cross section can be calculated as

$$v_2 = \frac{\dot{m}}{\rho_{\text{in}} S} \tag{9}$$

where $\rho_{\text{in}} = \begin{cases} \rho_1, \dot{m} \geq 0 \\ \rho_2, \dot{m} < 0 \end{cases}$ is determined by the upwind scheme. The state parameter sections corresponding to subscripts 1 and 2 are shown in Figure 5.

The motion parameters at the throttle cross section are obtained, so Equation (4) is no longer needed. We set the virtual control volumes at both ends of the control valve, and the upstream and downstream ports provide the boundary conditions. According to Equations

(1)–(7), the parameters of each section (except the throttle cross section) can be calculated. Then the parameters of BCV are provided to the upstream and downstream ports.

3.3. Multi-Port Junction Model

The multi-port junction model is used for the intersection and branching of pipelines. The schematic diagram of the multi-port junction model with m inlets and k outlets is shown in Figure 6, and the virtual control volumes are also set for each port of the model.

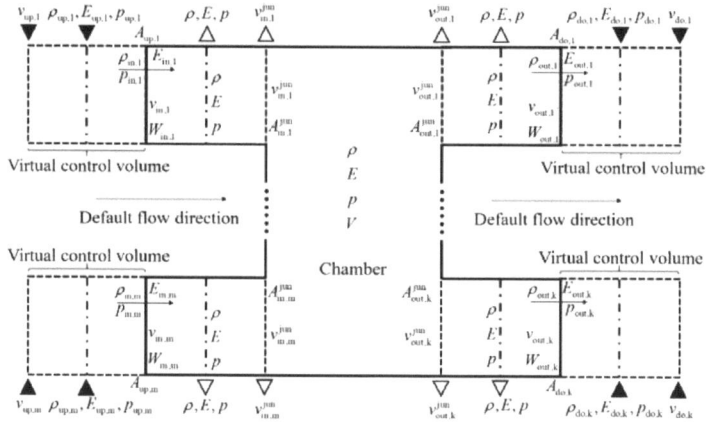

Figure 6. Schematic diagram of multi-port junction model.

Assuming that the airflow at the pipe junction is uniformly mixed instantaneously, according to Equations (1) and (2), the parameters in the multi-port junction can be calculated as

$$\frac{d(\rho V)}{dt} = \sum_{i=1}^{m} \rho_{in,i} v_{up,i} A_{up,i} - \sum_{j=1}^{k} \rho_{out,j} v_{do,j} A_{do,j} \tag{10}$$

$$\frac{d(E\rho V)}{dt} = \sum_{i=1}^{m} \left(E_{in,i} + \frac{p_{in,i}}{\rho_{in,i}} \right) \rho_{in,i} v_{in,i} A_{up,i} - \sum_{j=1}^{k} \left(E_{out,j} + \frac{p_{out,j}}{\rho_{out,j}} \right) \rho_{out,j} v_{out,j} A_{do,j} + \dot{Q} - \dot{\Psi} \tag{11}$$

From Equations (3), (10) and (11), the state parameters ρ, E, and p of the output ports can be determined. To unify the interface, it is still necessary to calculate the motion parameters of the output ports. Considering that the volume of multi-port junction is concentrated inside, and the control volumes at the inlet and outlet are relatively small, the compressibility of the air in these control volumes can be ignored. Therefore, the velocity of the inlet and outlet airflow can be expressed as follows according to the flow equilibrium,

$$v_{in,i}^{jun} = \frac{\rho_{in,i} v_{in,i} A_{up,i}}{\rho A_{up,i}^{jun}}, i = 1, 2, \cdots, m \tag{12}$$

$$v_{out,j}^{jun} = \frac{\rho_{out,j} v_{out,j} A_{do,j}}{\rho A_{out,j}^{jun}}, j = 1, 2, \cdots, k \tag{13}$$

In Equations (10)–(13), the state parameters such as $\rho_{in,i}$, $E_{in,i}$, $p_{in,i}$, $\rho_{out,j}$, $E_{out,j}$, and $p_{out,j}$, are calculated according to the upwind scheme shown in Equation (6), and the motion parameters $v_{in,i}$, $W_{in,i}$, $v_{out,j}$, and $W_{out,j}$, are calculated according to Equations (4) and (5) of BCV, for $i = 1, 2, \cdots m$ and $j = 1, 2, \cdots, k$.

3.4. Flow Source/Sink Model

The flow source model can be used to simulate the air supply unit of ASTF, where the mass flow rate and temperature are known. A single CCV is used as the flow source model, and a virtual control volume is set at its outlet, as shown in Figure 7. Since the virtual control volume is not set at the inlet of the flow source model, the parameters of the entrance section cannot be obtained directly by the upwind scheme. According to the mass flow rate, the entrance velocity is expressed as

$$v_1 = \frac{\dot{m}}{\rho_{in} A_{in}} \tag{14}$$

where the air flow density is approximated as $\rho_{in} \approx \rho_1$.

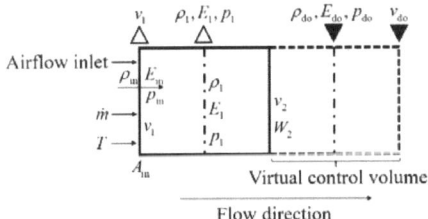

Figure 7. Schematic diagram of flow source model.

The airflow pressure therefore becomes $p_{in} = \rho_1 R T$. From Equation (3), the total energy E_{in} of inlet air flow can further be determined. Combining Equations (1)–(7), the remaining parameters of the control volume can be solved. The model output port parameters are shown in Figure 7.

The flow sink model can be used to simulate the intake air of an aircraft engine, and a virtual control volume is set at the inlet, as shown in Figure 8. Since the air flow direction is fixed, the state parameters of the exit section can be taken as

$$y_{out} = y_1 \tag{15}$$

where $y \in \{\rho, E, p\}$.

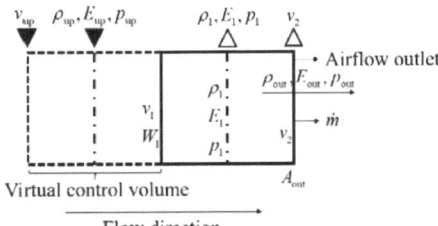

Figure 8. Schematic diagram of flow sink model.

The outlet flow rate is a known value, and then the outlet velocity is expressed as

$$v_2 = \frac{\dot{m}}{\rho_{out} A_{out}} \tag{16}$$

According to Equations (1)–(7), the parameters of the remaining sections can be obtained, and the parameters of output port are shown in Figure 8.

3.5. Pressure/Temperature Boundary Model

The pressure/temperature boundary model represents the known airflow boundary, such as atmospheric environment, etc. In addition to state parameters that can be output as known values, the model still needs to output motion parameters, as shown in Figure 9.

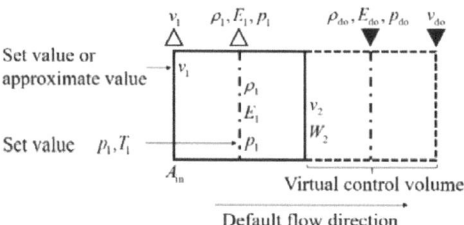

Figure 9. Schematic diagram of pressure/temperature boundary model.

For low-speed flow, kinetic energy accounts for a small proportion of the total energy of the air flow. Therefore, it is allowed to set a reasonable velocity based on experience without causing a large loss of accuracy. When the experience is insufficient or the speed is high, such as close to the air supply unit, it can be approximated as

$$v_1 \approx v_2 \qquad (17)$$

The state parameters p_1 and T_1 are known setting values, and the airflow density ρ_1 and total energy E_1 can be calculated according to the ideal gas state equation and Equation (3). In addition, a virtual control volume is added to the model, and the unknown motion parameters v_2 and W_2 are calculated according to Equations (4) and (5).

4. System-Level Model

4.1. Analysis of System-Level Modeling

The virtual control volume can conveniently handle model interfaces and boundaries. In Figure 10, the input/output ports of the upstream pipeline are respectively connected to the output/input ports of the downstream pipeline, so that the virtual control volumes essentially become CCVs of the connected pipelines. Then, Section I of the upstream pipeline and Section II of the downstream pipeline are seamlessly connected, and the parameters of the two sections remain the same. Although the pipelines connected in sections can be simplified into a single pipeline model, when the pipeline length reaches hundreds of meters, it is inconvenient to set all the pipeline parameters on one model. In addition, this method does not increase the computational complexity of the model.

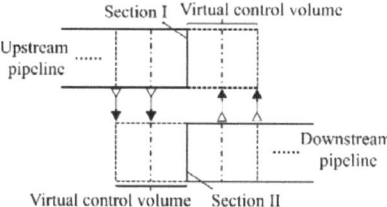

Figure 10. Interface connection diagram of pipeline model.

Figure 11 shows the connection between the pipeline and the control valve, in which the state parameter cross sections all follow the mass conservation equation and energy conservation equation, and only the calculation method of the motion parameter cross section is different. For the control valve, the motion parameters are calculated based on the mass flow rate through the control valve. For the pipeline, the motion parameters are calculated according to the momentum conservation equation.

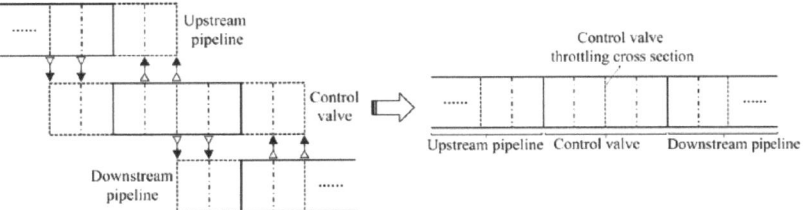

Figure 11. Connection of pipeline and control valve.

Therefore, for the system-level model, it is still in the form of complete staggered grids, in which the motion parameters can be combined with the characteristics of different components for targeted design and calculation. That also determines the flow characteristics of the airflow in ASTF. The characteristics of one-dimensional (strictly, quasi-one-dimensional) flow can be reflected in the whole system model. Due to the symmetry of the component models, the airflow direction in ASTF does not affect the modeling process.

Generally, due to the staggered grids and the average properties of some parameters (e.g., Equation (7)), the quasi-one-dimensional flow model cannot simulate supersonic flow stably and accurately [9]. Although reducing the grid size is a solution, the computational cost is very large because of the larger grid number. However, this does not affect the modeling of ASTF, because the airflow in ASTF is designed to be at a low Mach number.

4.2. ASTF Model

The ASTF structure is shown in Figure 12, which is mainly composed of pipelines, control valves, flow deflectors, mixer, buffer chamber, etc. The structure of ASTF is symmetrical. It regulates the flow of air source through the control valve and realizes the mixing of airflow with different temperatures and pressures in the mixer. After the mixed airflow passes through the buffer chamber and the special-shaped pipe, it is directly supplied to the aircraft engine. The buffer chamber can make the airflow mix more evenly. The special-shaped pipe makes the airflow more stable. A flow tube is used to measure the mass flow rate, temperature, pressure and other parameters. According to the actual structure and system working mechanism, the ASTF model is constructed as shown in Figure 13.

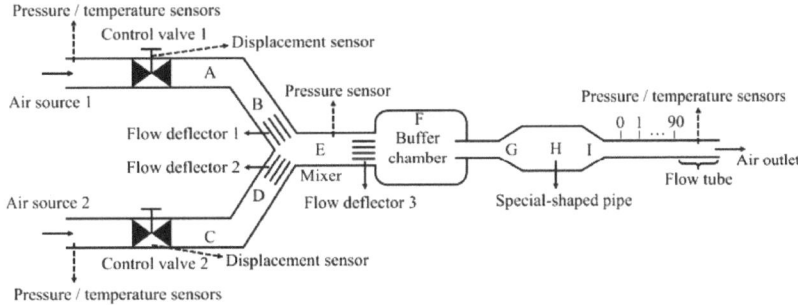

Figure 12. Schematic diagram of ASTF.

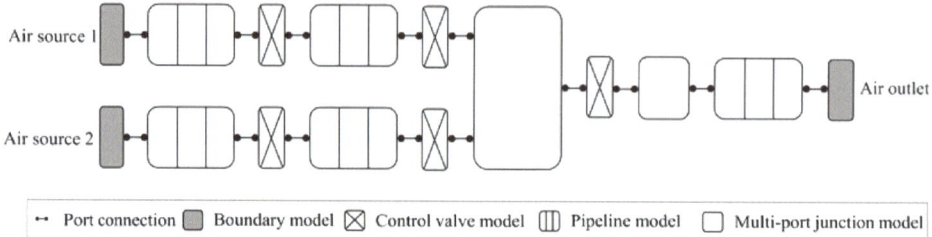

Figure 13. Model connection diagram.

In Figure 13, the boundary model represents the flow source/sink model or pressure/temperature boundary model, which needs to be selected according to the actual boundary type. As a throttling component, the flow deflector is represented by a control valve model with a fixed opening, and its flow coefficient is obtained by fitting three-dimensional flow field simulation data, as shown in Figure 14a. The flow coefficient of the control valve is obtained through test data and three-dimensional flow field simulation data, as shown in Figure 14b. The pressure ratio in Figure 14 represents the ratio of the airflow pressure after and before the control valve, and the area ratio represents the ratio of the valve opening area to the valve maximum cross-sectional area. The mixer has the characteristics of airflow mixing, while the buffer chamber is a typical large-scale volume, and both are modeled by the multi-port junction model. Since some component models are not strictly one-dimensional flow numerical models, the ASTF model is a quasi-one-dimensional flow model.

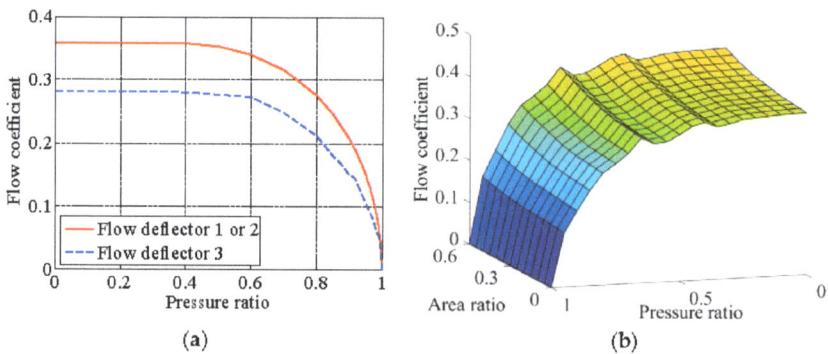

Figure 14. Flow coefficient. (**a**) Flow coefficient of flow deflector; (**b**) flow coefficient of control valve.

By setting the boundary model parameters (i.e., temperature/pressure or mass flow rate) and the opening of the control valves, the subsequent simulations can be carried out.

5. Simulation and Analysis

5.1. Difference between Quasi-One-Dimensional Flow Model and Lumped Parameter Model

According to Ref. [8], an ASTF model based on the multi-volume modeling method was established. This model is essentially a lumped parameter model. The control valve and flow deflector were used as throttling components to model the entire system as a combination of multiple volumes, in which the buffer chamber and the special-shaped pipe together formed a large volume, as shown in Figure 15. Compared with the lumped parameter model, the quasi-one-dimensional flow model takes into account factors such as friction and local losses (see Equation (4)), and the influence of these factors can be directly seen by the following simulation.

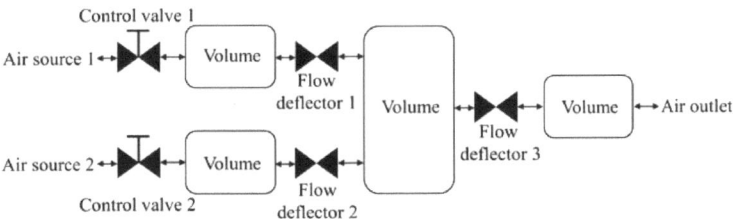

Figure 15. Connection diagram of lumped parameter model.

Boundary conditions:
1. The pressure and temperature of air source 1 were respectively 95.291 kPa and 282 K;
2. The pressure and temperature of air source 2 were respectively 147.053 kPa and 396 K;
3. The flow at air outlet was 223.2 kg/s.

The opening of control valve 1 was set to 72.6%. The opening of control valve 2 was maintained at 30% for 0~5 s (called stage 1) and stepped to 40% in the 5th second. The cross-section signs of ASTF are shown in Figure 12.

In stage 1, the pressure and temperature distribution of ASTF are listed in Tables 1 and 2. In fact, the lumped parameter model takes the pipeline between two throttling components as a volume. Therefore, the pressure and temperature of each cross section cannot be distinguished, and it is difficult to reflect the pressure and temperature gradients caused by expansion, bends, friction, and local obstacles. Conversely, the quasi-one-dimensional flow model can overcome these limitations and calculate the airflow parameters of the cross section where the real sensor is located. This lays the foundation for using sensor data to optimize the system-level model.

Table 1. Pressure at different sections.

Modeling Method	Pressure/kPa								
	A	B	C	D	E	F	G	H	I
Lumped parameter	88.4	88.4	90	90	86.4	76.6	76.6	76.6	76.6
Quasi-one-dimensional	88.6	87.8	91	89.6	85.9	78.6	71.6	74.4	72.4

Table 2. Temperature at different sections.

Modeling Method	Temperature/K								
	A	B	C	D	E	F	G	H	I
Lumped parameter	282.3	282.3	395.9	395.9	341.3	336.1	336.1	336.1	336.1
Quasi-one-dimensional	281.7	281.8	394.9	395.2	343	343.1	341	342.4	340.6

Select the equal-diameter straight pipeline at the outlet of ASTF and divide it into 91 nodes along the axial direction, as shown in Figure 12. When the opening of control valve 2 changed, the pressure and temperature of each node began to change with time. It can be seen from Figure 16 that the quasi-one-dimensional flow model can reflect the propagation process of the airflow. This means that ASTF has a time-delay characteristic, which provides a more realistic simulation environment for controller design and verification. However, for the lumped parameter model, the properties of the airflow are assumed to be instantaneously uniform in the pipeline, which cannot reflect the dynamic changes of the airflow parameters in space. When the pipeline is longer, it will cause a larger deviation. In addition, the flow characteristics of the airflow along the pipeline make the heat transfer process have spatial distribution characteristics, which can more accurately simulate the impact of the metal pipeline on the airflow energy.

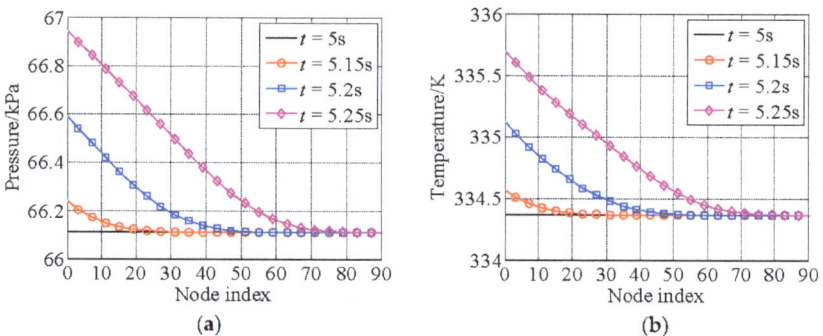

Figure 16. Pressure and temperature of each node. (**a**) Pressure change results; (**b**) temperature change results.

5.2. Model Verification Based on Test Data

To verify the accuracy of the quasi-one-dimensional flow model, the data measured by the sensors during the operation of ASTF were selected for simulation. The boundary conditions were the pressure and temperature measured by the sensors located at air sources 1, 2 and the air outlet. The valve opening measured by the displacement sensor was used as the input of the control valve model. The initial state of the model was set according to the ASTF system state at the beginning of the test data. The parameters to be compared were the mass flow (measured by the flow tube at the outlet) and the mixer pressure. The locations of these sensors are shown in Figure 12. In addition, for comparative analysis, the above simulation conditions were also applied to the lumped parameter model established according to Ref. [8], and the simulations were carried out.

The boundary conditions and valve opening are shown in Figure 17.

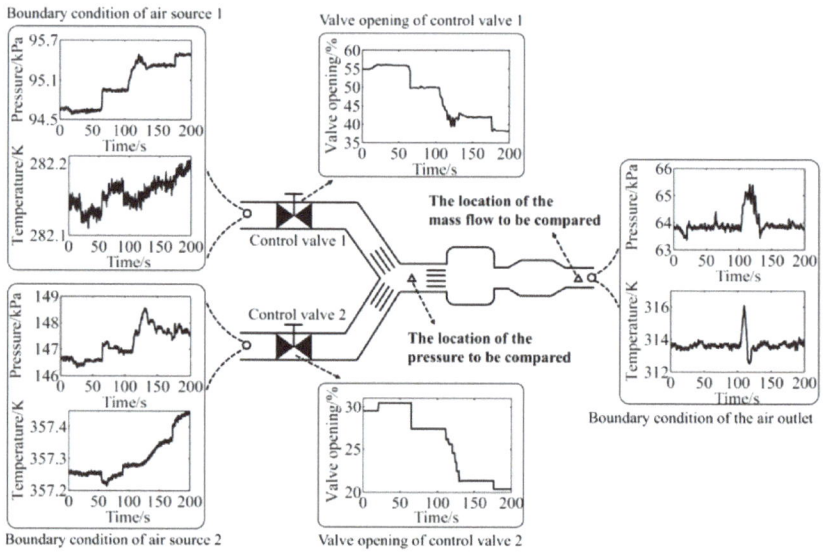

Figure 17. Boundary conditions and valve opening.

The comparison result of the mass flow rate at the air outlet is shown in Figure 18. Compared with the lumped parameter model, the simulation result of the quasi-one-dimensional flow model is more consistent with the test data in the steady state and dynamic process. In fact, the quasi-one-dimensional flow model has pressure losses caused by friction and cross-sectional area changes, resulting in different pressure distributions in ASTF. This makes the pressure ratio of the throttling component different in the two models, causing a difference in mass flow. It is worth noting that when studying the lumped parameter model in the past, this error between simulation data and test data is usually attributed to the inaccuracy of the control valve characteristic. However, the simulation result here reveals the defects of the lumped parameter model. Compared with the test data, the mass flow error results of the quasi-one-dimensional flow model are shown in Figure 19. The steady-state relative error is less than 1.5%, and the dynamic relative error is less than 2.2%.

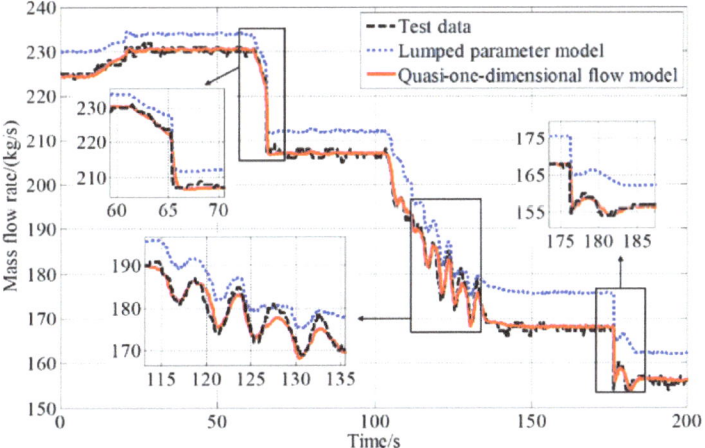

Figure 18. Mass flow comparison result.

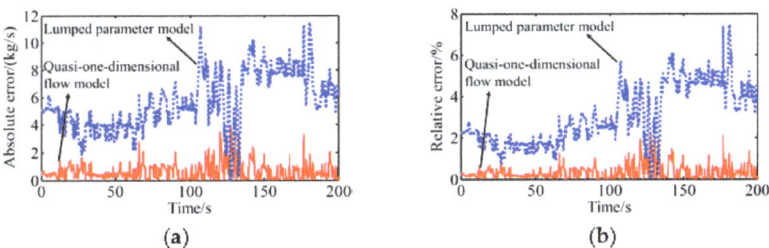

Figure 19. Mass flow error. (**a**) Absolute error of mass flow; (**b**) relative error of mass flow.

Figure 20 shows the pressure comparison result of the mixer. The pressure dynamics of the two models are basically consistent with the test data. Compared with the quasi-one-dimensional flow model, the pressure error of the lumped parameter model is larger, and the average pressure error is about 2 kPa, which is consistent with the simulation result of Ref. [8]. The pressure error results of the quasi-one-dimensional flow model are shown in Figure 21. In 20–65 s, the mass flow reaches the maximum, and the pressure error increases. At 65 s, the rapid action of the control valve causes a large dynamic pressure change, and the relative pressure error is less than 1.4%. In 65–200 s, the mass flow is relatively low, and the relative pressure error is less than 0.6%.

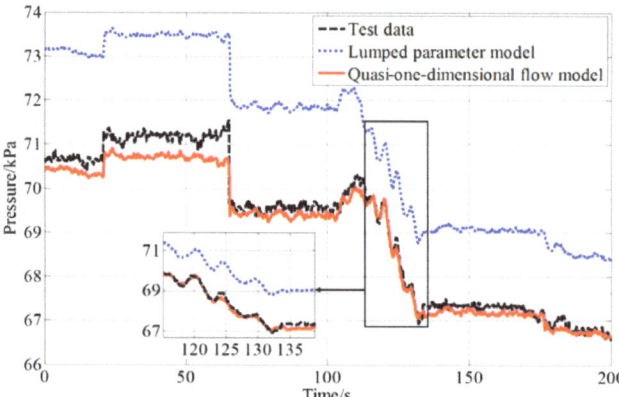

Figure 20. Comparison of pressure changes in mixer.

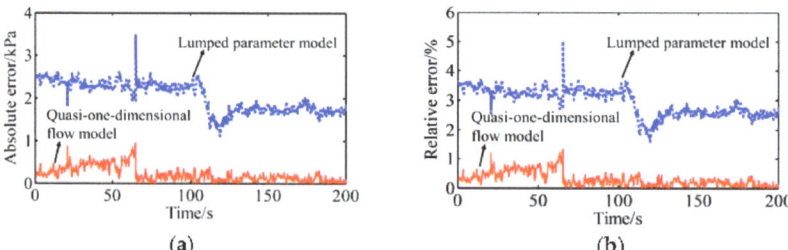

Figure 21. Pressure error of mixer. (**a**) Absolute error of pressure; (**b**) relative error of pressure.

6. Conclusions

In this study, a generic quasi-one-dimensional flow modeling method was used to establish a numerical simulation model of ASTF. Thereafter, the simulation and verification of the model were carried out. The main conclusions are summarized as follows.

(1) Compared with the lumped parameter model, the quasi-one-dimensional flow model can simulate spatial effect and time delay of the airflow.
(2) The simulation results of the quasi-one-dimensional flow model are basically consistent with the test data. The relative errors of mass flow and pressure are less than 2.2% and 1.4%, respectively, further verifying the correctness of the proposed modeling method.
(3) With the generic modeling method, the system-level model still has the form of a staggered grid, and the properties of quasi-one-dimensional flow, such as spatial effect and time delay can be easily addressed during the modeling process.

In the future, the friction factor and the minor loss coefficient will be calibrated through the test data to further improve the model accuracy.

Author Contributions: Conceptualization, J.L. and X.W.; methodology, J.L.; software, J.L. and X.P.; validation, J.L., M.Z. and S.Y.; formal analysis, S.Y.; investigation, L.Z. and S.Y.; resources, S.Z.; writing—original draft preparation, J.L.; writing—review and editing, X.W. All authors have read and agreed to the published version of the manuscript.

Funding: This work was funded by AECC Sichuan Gas Turbine Establishment Stable Support Project, grant number GJCZ-0011-19, and by National Science and Technology Major Project, grant number 2017-V-0015-0067.

Institutional Review Board Statement: Not applicable.

Informed Consent Statement: Not applicable.

Data Availability Statement: Not applicable.

Conflicts of Interest: The authors declare no conflict of interest.

Nomenclature

A	Flow area (m^2)
D	Diameter of pipe (m)
E	Total energy per unit mass of fluid consists of the kinetic, potential, and internal energies (J/kg)
f	Friction factor
h	Convection heat transfer coefficient (W/m^2/K)
k	Minor loss coefficient
$\dot{m}$	Mass flow rate (kg/s)
p	Pressure (Pa)
$\dot{Q}$	Heat flow rate (W)
Re	Reynolds number
S	Opening area of control valve (m^2)
t	Time (s)
T	Temperature (K)
u	Internal energy per unit mass of fluid (J/kg)
v	Flow velocity (m/s)
V	Volume (m^3)
W	Momentum (kg·m/s)
γ	Ratio of specific heats
ρ	Density (kg/m^3)
π	Pressure ratio
φ	Flow coefficient
ε	Surface roughness of the wall (m)
Ψ	Work output of the fluid (excluding xflow work) (J)
Ψ_s	Shaft work (J)
Ψ_v	Work done by viscous force (J)

Subscripts/Superscripts

do	Downstream port
i	Index to denote motion parameter cross section
in	Inlet property
j	Index to denote state parameter cross section
jun	Junction
k	Number of outlet ports
m	Number of inlet ports
n	Number of control volumes
out	Outlet property
up	Upstream port

References

1. Wang, X.; Zhu, M.Y.; Zhang, S.; Dan, Z.H.; Pei, X.T. Technology Development of Foreign Altitude Simulation Test Facilities Control System. *Gas Turb. Exp. Res.* **2017**, *30*, 49–55. Available online: https://kns.cnki.net/kcms/detail/detail.aspx?FileName=RQWL201706011&DbName=CJFQ2017 (accessed on 15 April 2022). (In Chinese)
2. Tian, J.H.; Dan, Z.H.; Zhang, S.; Wang, X. Environment Simulation Control Technology for Altitude Test Facility. *Aerosp. Power* **2021**, *3*, 64–68. Available online: https://kns.cnki.net/kcms/detail/detail.aspx?FileName=HKDL202103019&DbName=CJFQ2021 (accessed on 15 April 2022). (In Chinese)
3. Schmidt, K.J.; Merten, R.; Menrath, M.; Braig, W. Adaption of the Stuttgart University Altitude Test Facility for BR700 Core Demonstrator Engine Tests. In Proceedings of the ASME International Gas Turbine & Aeroengine Congress & Exhibition: The American Society of Mechanical Engineers, Stockholm, Sweden, 2–5 June 1998. [CrossRef]

4. Montgomery, P.A.; Burdette, R.; Krupp, B. A Real-Time Turbine Engine Facility Model and Simulation for Test Operations Modernization and Integration. In Proceedings of the ASME Turbo Expo 2000: Power for Land, Sea, and Air, Munich, Germany, 8–11 May 2000. [CrossRef]
5. Montgomery, P.A.; Burdette, R.; Wilhite, L.; Salita, S. Modernization of a Turbine Engine Test Facility Utilizing a Real-Time Facility Model and Simulation. In Proceedings of the ASME Turbo Expo 2001: Power for Land, Sea, and Air, New Orleans, LA, USA, 4–7 June 2001. [CrossRef]
6. Sheeley, J.M.; Sells, D.A.; Bates, L.B. Experiences with Coupling Facility Control Systems with Control Volume Facility Models. In Proceedings of the 42nd AIAA Aerospace Sciences Meeting and Exhibit, Reno, NE, USA, 5–8 January 2004. [CrossRef]
7. Pei, X.T.; Zhang, S.; Dan, Z.H.; Zhu, M.Y.; Qian, Q.M.; Wang, X. Study on Digital Modeling and Simulation of Altitude Test Facility Flight Environment Simulation System. *J. Propul. Technol.* **2019**, *40*, 1144–1152. (In Chinese) [CrossRef]
8. Zhu, M.Y.; Wang, X.; Pei, X.T.; Zhang, S.; Dan, Z.H.; Miao, K.Q.; Liu, J.S.; Jiang, Z. Multi-Volume Fluid-Solid Heat Transfer Modeling for Flight Environment Simulation System. *J. Propul. Technol.* **2020**, *41*, 2848–2859. (In Chinese) [CrossRef]
9. Boylston, B.M. Quasi-One-Dimensional Flow for Use in Real-Time Facility Simulations. Master's Thesis, University of Tennessee, Knoxville, TN, USA, 2011. Available online: https://trace.tennessee.edu/utk_gradthes/1058 (accessed on 9 February 2022).
10. Xu, Y.H. Research on Real-time Simulation Model and Algorithm for Incompressible Fluid Network Based on One-dimension Flow. Master's Thesis, Harbin Engineering University, Harbin, China, 2009. (In Chinese) [CrossRef]
11. Sun, K.; Liu, G.W.; Ma, C.T.; Niu, J.J. Calculation Model of Tube System and Value Based on 1D Network Model. *J. Eng. Therm.* **2017**, *38*, 1889–1895. Available online: https://kns.cnki.net/kcms/detail/detail.aspx?FileName=GCRB201709014&DbName=CJFQ2017 (accessed on 15 April 2022). (In Chinese)
12. Tao, Z.; Hou, S.P.; Han, S.J.; Ding, S.T.; Xu, G.Q.; Wu, H.W. Study on Application of Fluid Network into the Design of Air System in Engine. *J. Aerosp. Power* **2009**, *24*, 1–6. (In Chinese) [CrossRef]
13. Pei, X.T.; Liu, J.S.; Wang, X.; Zhu, M.Y.; Zhang, L.Y.; Dan, Z.H. Quasi-one-dimensional Flow Modeling for Flight Environment Simulation System of Altitude Ground Test Facilities. *Processes* **2022**, *10*, 377. (In Chinese) [CrossRef]
14. Rennie, R.M.; Sutcliffe, P.; Vorobiev, A.; Cain, A.B. Mathematical Modeling of Wind-Speed Transients in Wind Tunnels. In Proceedings of the 51st AIAA Aerospace Sciences Meeting Including the New Horizons Forum and Aerospace Exposition, Grapevine, TX, USA, 7–10 January 2013. [CrossRef]
15. Liu, K.; Cheng, M.S.; Zhang, Y.L. Dynamic Model of Priming Processes of Cryogenic Propellant Feed Lines. *J. Natl. Univ. Def. Technol.* **2003**, *25*, 1–5. (In Chinese) [CrossRef]
16. Cheng, M.S.; Liu, K.; Zhang, Y.L. Numerical Analysis of Pre-cooling and Priming Transients in Cryogenic Propellant Feed Systems. *J. Propul. Technol.* **2000**, *21*, 38–41. (In Chinese) [CrossRef]
17. Pan, J.S. *Fundamentals of Gasdynamics*; National Defense Industry Press: Beijing, China, 1989; pp. 40–63. (In Chinese)
18. Çengel, Y.A.; Boles, M.A. *Thermodynamics: An Engineering Approach*, 5th ed.; McGraw-Hill: New York, NY, USA, 2006; pp. 219–232.
19. Chen, Y.; Gao, F.; Zhang, Z.P.; Cai, G.B. Finite Volume Model for Quasi one-dimensional Compressible Transient Pipe Flow (I) Finite Volume Model of Flow Field. *J. Aerosp. Power* **2008**, *23*, 311–316. (In Chinese) [CrossRef]
20. Zhu, M.Y.; Wang, X. An Integral Type μ Synthesis Method for Temperature and Pressure Control of Flight Environment Simulation Volume. In Proceedings of the ASME turbo Expo 2017: Turbomachinery Technical Conference and Exposition, Charlotte, NC, USA, 26–30 June 2017. (In Chinese). [CrossRef]
21. Sutcliffe, P.; Vorobiev, A.; Rennie, R.M.; Cain, A.B. Control of Wind Tunnel Test Temperature Using a Mathematical Model. In Proceedings of the AIAA Ground Testing Conference, San Diego, CA, USA, 24–27 June 2013. [CrossRef]
22. Çengel, Y.A. *Heat Transfer: A Practical Approach*; McGraw-Hill: New York, NY, USA, 2010; pp. 11–25.
23. De Giorgi, M.G.; Fontanarosa, D. A Novel Quasi-one-dimensional Model for Performance Estimation of a Vaporizing Liquid Microthruster. *Aerosp. Sci. Technol.* **2019**, *84*, 1020–1034. [CrossRef]
24. Liu, K.; Zhang, Y.L. Finite Elements State-space Model for One-dimensional Compressible Fluid Flow. *J. Propul. Technol.* **1999**, *20*, 62–66. (In Chinese) [CrossRef]
25. Wang, X.Y. *Fundamentals of Aerodynamics*; Northwestern Polytechnical University Press: Xi'an, China, 2006; pp. 94–101. (In Chinese)
26. Wang, M.D. Research of Tank Pressurization System of Liquid Rocket Engine Test-bed. Master's Thesis, Shanghai Jiao Tong University, Shanghai, China, 2009. Available online: https://kns.cnki.net/kcms/detail/detail.aspx?FileName=2010200382.nh&DbName=CMFD2011 (accessed on 15 April 2022). (In Chinese)
27. Pei, X.T.; Zhu, M.Y.; Zhang, S.; Dan, Z.H.; Wang, X.; Wang, X. An Iterative Method of Empirical Formula for the Calculation of Special Valve Flow Characteristics. *Gas Turb. Exp. Res.* **2016**, *29*, 35–39. Available online: https://kns.cnki.net/kcms/detail/detail.aspx?FileName=RQWL201605009&DbName=CJFQ2016 (accessed on 9 February 2022). (In Chinese)
28. Wang, Y.L.; Wang, X.; Zhu, M.Y.; Pei, X.T.; Zhang, S.; Dan, Z.H. Comparative Study on Modeling Methods of Special Control Valves in Altitude Simulation Test Facility. *J. Propul. Technol.* **2019**, *40*, 1895–1901. (In Chinese) [CrossRef]

Article

Investigation on the Formation and Evolution Mechanism of Flow-Resistance-Increasing Vortex of Aero-Engine Labyrinth Based on Entropy Generation Analysis

Xiaojing Liu [1], Shuiting Ding [1,2], Longtao Shao [3], Shuai Zhao [1], Tian Qiu [1], Yu Zhou [1], Xiaozhe Zhang [4,5,*] and Guo Li [3]

1 Research Institute of Aero-Engine, Aircraft/Engine Integrated System Safety Beijing Key Laboratory, Beihang University, Beijing 100191, China; liuxiaojing@buaa.edu.cn (X.L.); dst@buaa.edu.cn (S.D.); zhaoshuai@buaa.edu.cn (S.Z.); qiutian@buaa.edu.cn (T.Q.); zybuaa@hotmail.com (Y.Z.)
2 Department of Aviation Engineering, Civil Aviation University of China, Tianjin 300300, China
3 School of Energy and Power Engineering, Beihang University, Beijing 100191, China; sltwhut@163.com (L.S.); lg666@buaa.edu.cn (G.L.)
4 Shenyang Engine Research Institute, Aero Engine Corporation of China, Shenyang 110015, China
5 Aero-Engine Thermal Environment and Structure Key Laboratory of Ministry of Industry and Information Technology, Nanjing University of Aeronautics and Astronautics, Nanjing 210016, China
* Correspondence: xzzmail@foxmail.com

Abstract: Labyrinth seals are widely employed in the air system of aircraft engines since they reduce the leakages occurring between blades and shrouds, which affect the entropy generation significantly. Excessive leakage flow of the labyrinth may be reduced the efficiency and performance of the engine. This paper proposes the concept of flow-resistance-increasing vortex (FRIV) on the top of the labyrinth that is based on the flow entropy generation mechanism of the stepped labyrinth and the main flow characteristics that lead to entropy generation. A three-dimensional simulation model of the labyrinth structure was established, and the model was compared and verified with the experimental data of the reference. The relative dissipation strength and vorticity distribution of the FRIV were theoretically analyzed. It was confirmed that the dissipative intensity distribution was the same as the vorticity distribution, and the correlation coefficient was larger in the labyrinth tip region. Therefore, a parametric study was conducted on design parameters related to the FRIV, including the teeth inclined angle, tooth crest width, step inclined angle, and other parameters. The results are beneficial for the construction of a stronger FRIV to reduce the leakage. This research is of great significance for the improvement of engine efficiency and for the reduction of fuel consumption in the future.

Keywords: labyrinth; flow characteristics; entropy generation; flow-resistance-increasing vortex; correlation coefficient; 3D simulation

1. Introduction

The labyrinth sealing technology is an important sealing technology in the air system of aero-engines, one that is a non-contact dynamic sealing composed of a rotor and a stator. There is a clearance between the tooth top and the static part, which directly affects the leakage flow of the labyrinth teeth. If the leakage flow rate of the labyrinth teeth is greater than the current design, it may directly reduce the efficiency and performance of the engine, and the maximum efficiency may even be reduced by 10% [1]. On the one hand, the kinetic energy of the fluid is consumed by the sudden expansion and contraction of the flow channel and converted to entropy production. On the other hand, when the rotating parts do work on the contact fluid, it will result in energy loss and entropy generation [2]. In other words, the main factors affecting entropy generation include the total pressure loss and the windage heating.

In the literature, a large number of studies on labyrinth seals can be found concerning experimental tests and numerical and analytical models. In terms of theoretical research, Martin first proposed the calculation method of the flow rate of the labyrinth teeth. The labyrinth tooth is regarded as a series of holes, and the fluid is assumed to be in an isothermal flow. However, this method is only suitable for incompressible streams, and the calculation error is large [3]. Then, Vermes revised the calculation formula proposed by Martin and added the correction term [4]. Zimmermann put forward a relational formula for calculating the carry factor or residual energy factor in the leakage flow rate of the labyrinth by means of experiments [5]. These theories are the basis for the research on the leakage of the labyrinth teeth. In terms of numerical research, the processing power of modern computers allows indeed for the simulation of more complex and detailed phenomena than that of the past years. On the basis of the experimental data of Waschka et al., Nayak et al. used a numerical simulation to study the effect of rotational speed on the sealing performance of the labyrinth teeth [6]. Lee et al. reviewed and summarized the literature on the grate seal structure since the 1960s, discussing the effect of sealing parameters on leakage flow [7]. Soemarwoto analyzed the flow mechanism of stepped teeth and stepped helical teeth by numerical simulation, and further studied the effect of tooth height on the vortex [8]. Rhode and Demko conducted a numerical study on the straight-through grate sealing structure by the finite difference method [9,10]. Stoff calculated the incompressible flow inside the straight-through labyrinth seal structure by the standard $k - \varepsilon$ turbulence model and analyzed the relationship between the leakage of the labyrinth structure and the pressure gradient [11]. Rhode and Sobolik et al. modified the standard $k - \varepsilon$ turbulence model by introducing the effective viscosity coefficient and studied the compressible flow in the straight-through labyrinth seal structure [12]. Rhode and Hibbs analyzed the effect of dimensional structure changes on the sealing performance of the labyrinth structure [13,14]. Demko and Morrison studied the effect of rotational speed on the flow of the straight-through labyrinth seal structure by numerical method [15]. Rapisarda et al. studied the influence of step position and tooth edge on leakage coefficient by using the numerical method and obtained the influence law of windage heating and vortices on leakage [16,17]. Kali charan Nvayak et al. studied the windage heating characteristics of the grooved stepped tooth-honeycomb bushing labyrinth seal structure by numerical method [18,19]. Kaliraj et al. analyzed static and rotational effects of labyrinth seals at various flows and geometrical dimensions and optimized leak flow by straight and steeped seal configurations [20]. Desando and Rapisarda focused on the implementation of a numerical model for rotating stepped labyrinth seals installed in low-pressure turbines [21]. Ganine presented a simplified coupled transient analysis methodology that allows for the assessment of the aerothermal and thermomechanical responses of engine components together with cooling air mass flow and pressure and temperature distributions in an automatic fully integrated way [22]. In the experimental research, Child provided an overview of the development of vortices in labyrinth seals and their main effects on self-excited instability [23]. Wittig et al. measured the leakage characteristics of the through-grate plane model under different pressure ratios and Reynolds through experiments, showing the influence of basic aerodynamic parameters on the through-grate flow coefficient [24]. Denecke et al. analyzed the influence of the inlet swirl on the windage heating and swirl development of the simple grate segment, which compared the experimental results with the calculation results McGreehan and Ko [25]. The results show that the local circumferential velocity and the development of the swirl in the grate flow channel are the key factors affecting the accuracy of the temperature rise calculation [26]. Willenboge et al. studied the influence of Reynolds number and inlet and outlet pressure ratio on the leakage characteristics of stepped grate structure through experiments, which showed that at a high Reynolds number, the flow coefficient may have a local maximum value [27]. Stocker and Rhode et al. studied the effect of the relative position of the wear groove and the axial direction of the labyrinth on the leakage characteristics of the labyrinth seal through experiments [28,29]. Millward and Edward studied the variation law of the heat generated by the wind resistance with the leakage amount and the pressure ratio under different labyrinth tooth structures [30]. Woschka et al. focused on analyzing the effect of rotational speed on head clearance, which showed that rotational speed has a great influence on the head clearance of labyrinth teeth [31]. Braun et al. accurately measured the tooth tip clearance with a static plane labyrinth tooth test bench and showed the effects of pressure ratio and tip clearance on the leakage characteristics of straight and stepped teeth [32]. Min et al. evaluated the leakage characteristics of a stepped labyrinth seal by experiments and computational fluid dynamics [33]. Khan et al. investigated the physical behaviors of heat and mass transfer flow with entropy generation through the effects of embedded parameters [34]. Ramzan et al. reached the conclusion that the entropy generation is increased with the enhancement of radiation parameter, Eckert number, Lewis number, temperature difference parameter, dimensionless constant parameter, Curie temperature, Prandtl number, and concentration difference parameter [35]. Although a large amount of work has been done on the research of labyrinth teeth, hardly anyone has investigated the characteristics of the labyrinth teeth region by the theoretical of entropy generation, which shows energy distribution accurately [36].

Labyrinth sealing technology is becoming increasingly more important in improving the performance of the engine with the advancement of materials and process technology. In practical engineering applications, the flow in the labyrinth seal area is very complicated and there is energy loss. Using the entropy production theory to conduct in-depth research on the flow characteristics of the stepped grate, the energy in this area can be quantitatively analyzed. In this work, flow analysis and reconstruction are performed on the grate region. A method is proposed to construct stronger flow resistance, increasing vortices and reducing leakage by enhancing the labyrinth tip velocity gradient and throttling effect. This research can serve as an important reference for the design and application of aero-engine labyrinth and plays an important role in reducing the kerosene consumption and improving thermal efficiency. The rest of this paper is organized as follows. In Section 2, the flow-resistance-increasing vortex (FRIV) on the top of the labyrinth tooth is introduced, and the leakage characteristics of the labyrinth tooth region are studied on the basis of the theory of entropy production. In Section 3, the labyrinth simulation model is established and checked. In Section 4, the simulation results are analyzed and compared with numerical calculations, which shows that the method proposed in this paper to reduce the leakage of the labyrinth is effective. The last section presents the conclusions of this research.

2. The Formation Mechanism of the FRIV at the Tip of the Labyrinth Tooth

2.1. Definition of Flow-Resistance-Increasing Vortices

The labyrinth structure is widely used in the air system of the engine, which is mainly composed of the turntable and the stator casing. The working principle of a labyrinth seal consists of separating regions at different pressures through the clearance, i.e., a gap placed between rotating and stationary components [37]. In this gap, the conversion of pressure energy into kinetic energy occurs. The latter can then be dissipated into thermal energy into the cavity downstream of the clearance or transferred to the next gap through the kinetic carryover [38]. This labyrinth seal adopted in the present work is a stepped type, with five labyrinth teeth [39]. It can be seen from Figure 1 that there is a vortex in the area behind the labyrinth teeth and another vortex structure in the labyrinth cavity. The vortex can hinder the airflow of the labyrinth tip, and we define the vortex as a flow-resistance-increasing vortices. The labyrinth geometric parameters and their representative symbols are shown in Table 1.

2.2. Theoretical of Entropy Generation

In this work, we focused on the sealing characteristics of the labyrinth; therefore, the system is partially simplified. The flow in the labyrinth is assumed to be steady, viscous, adiabatic, and compressible, and thermal conductivity λ and rotational speed n are not considered. Theoretically, the labyrinth structure of an aero-engine can be regarded as an adiabatic continuous power system. The theoretical deduction process of the entropy generation about aero-engine system is shown as follows, which includes viscous dissipation, heat transfer with temperature difference, metal friction, and mixing.

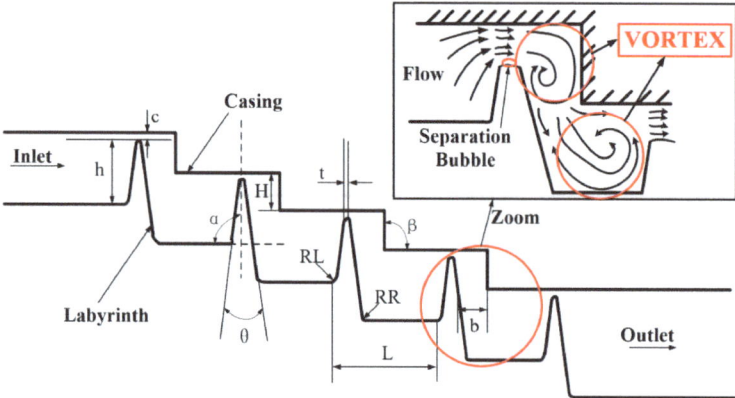

Figure 1. Geometrical dimension drawing of labyrinth seal structure.

Table 1. Symbols and initial values of labyrinth geometric parameters.

Geometric Parameters	Symbol	Initial Value	Unit
Step height	H	2	mm
Teeth pitch	L	7	mm
Tooth height	h	5	mm
Tooth tip width	t	0.3	mm
Labyrinth inclination angle	α	90	degrees
Angle between face teeth	θ	15	degrees
Seal clearance	c	0.36	mm
Tooth to step distance	b	2	mm
Step inclined angle	β	90	degrees
Root rounding of steps	R1	0	mm
Root rounding on pressure side	RL	0.5	mm
Fin tip fillet	RR	0.5	mm

The entropy equation of the open system [40] is as follows:

$$\delta S_g = dS_{CV} - \frac{\delta Q}{T_r} + s_2 \delta m_2 - s_1 \delta m_1 \qquad (1)$$

where S_g is the entropy generation; T_r is the wall temperature; Q is the amount of heat exchange; s_2, s_1 is the specific entropy; m_1, m_2 is the mass flow; and S_{CV} is the total entropy of the stable flow system.

For steady flow, adiabatic systems are as follows:

$$dS_{CV} = 0, \frac{\delta Q}{T_r} = 0 \qquad (2)$$

For the unit mass working medium, introduce Equation (2) into Equation (1), and the entropy generation equation is simplified to

$$s_g = s_2 - s_1 \qquad (3)$$

The energy equation is as follows [41]:

$$-w_s = h_2 - h_1 + \frac{V_2^2 - V_1^2}{2} = h_2^* - h_1^* = c_p(T_2^* - T_1^*) \quad (4)$$

where h_1 denotes import specific enthalpy, h_2 denotes export specific enthalpy, h_1^* denotes total import enthalpy, h_2^* denotes total export enthalpy, T_1^* denotes total inlet temperature, T_2^* denotes total outlet temperature, and c_p denotes constant pressure specific heat.

For a labyrinth in an air system, the entropy generation of the system can be expressed as follows [42]:

$$s_g = s_2 - s_1 = c_p \ln \frac{T_2}{T_1} - R \ln \frac{P_2}{P_1} = c_p \ln \frac{T_2^*}{T_1^*} - R \ln \frac{P_2^*}{P_1^*} \quad (5)$$

$$s_g = c_p \ln \frac{T_2^*}{T_1^*} - R \ln \pi \quad (6)$$

where T_1 denotes inlet static temperature, T_2 denotes outlet static temperature, P_1 denotes inlet static pressure, P_2 denotes outlet static pressure, P_1^* denotes inlet total pressure, P_2^* denotes outlet total pressure, T_1^* denotes inlet total temperature, T_2^* denotes outlet total temperature, c_p denotes constant pressure specific heat, and R denotes general gas constant.

For the windage heating, it is mainly caused by friction:

$$w_s = M\omega \quad (7)$$

where M denotes the frictional force, ω denotes the rotational angular velocity, and w_s denotes the frictional power.

The dimensionless parameters such as the discharge coefficient and windage heating can be expressed as follows [43]:

$$C_m = \frac{M}{0.5\rho\omega^2 r^5} \quad (8)$$

$$w_s = 0.5 C_m \rho \omega^3 r^5 \quad (9)$$

$$w_s = \dot{m} c_p \Delta T^* \quad (10)$$

$$\Delta T^* = \frac{C_m \rho \omega^3 r^5}{2\dot{m} c_p} \quad (11)$$

$$\frac{\dot{m}\sqrt{T^*}}{P^*} = f(\pi) \quad (12)$$

$$\Delta T^* = T_2^* - T_1^* = \frac{C_m \rho \omega^3 r^5 \sqrt{T_1^*}}{2 c p f(\pi) P_1^*} \quad (13)$$

The dissipative entropy generation in the flow originates from the dissipation generated by the viscosity of the fluid, resulting in the irreversible conversion of mechanical energy to internal energy. The equation of the relative dissipation intensity of the fluid micro-element is as follows [44]:

$$\phi = \left[2\left(\frac{\partial V_x}{\partial x}\right)^2 + 2\left(\frac{\partial V_y}{\partial y}\right)^2 + 2\left(\frac{\partial V_z}{\partial z}\right)^2 + \left(\frac{\partial V_y}{\partial x} + \frac{\partial V_x}{\partial y}\right)^2 + \left(\frac{\partial V_z}{\partial y} + \frac{\partial V_y}{\partial z}\right)^2 + \left(\frac{\partial V_x}{\partial z} + \frac{\partial V_z}{\partial x}\right)^2\right] - \frac{2}{3}\left(\frac{\partial V_x}{\partial x} + \frac{\partial V_y}{\partial y} + \frac{\partial V_z}{\partial z}\right)^2 \quad (14)$$

Flow vorticity can be written as

$$\begin{cases} \omega_x = \frac{1}{2}\left(\frac{\partial V_z}{\partial y} - \frac{\partial V_y}{\partial z}\right) \\ \omega_y = \frac{1}{2}\left(\frac{\partial V_x}{\partial z} - \frac{\partial V_z}{\partial x}\right) \\ \omega_z = \frac{1}{2}\left(\frac{\partial V_y}{\partial x} - \frac{\partial V_x}{\partial y}\right) \end{cases} \quad (15)$$

where V_x, V_y, V_z are the flow velocities in the x, y, and z directions, respectively.

In order to investigate the correlation between the vorticity and the relative dissipation intensity distribution, the correlation coefficient τ is defined as follows [45]:

$$\tau = \omega \cdot \phi \quad (16)$$

Introduce the vorticity transport equation [46]:

$$\frac{Dw}{Dt} = (\omega \bullet \nabla)V - \omega(\nabla \bullet V) + \nabla \rho \times \frac{\nabla p}{\rho^2} + v\nabla^2 \omega \quad (17)$$

where $(\omega \bullet \nabla)V$ denotes the stretching and bending of the vortex line caused by the velocity gradient of the flow field, which causes the velocity gradient in the flow field to change significantly and results in a change in the absolute value of the vorticity; $\omega(\nabla \bullet V)$ denotes the change in the magnitude of the vorticity caused by the volume change of the fluid micelle, which causes the absolute value of the vorticity to change; and $\nabla \rho \times \nabla p / \rho^2$ denotes the effect of baroclinic moment on vorticity due to non-parallel pressure gradients and density gradients. For a positive pressure fluid, whose density is only a function of pressure, and thus the density and pressure terms have the same gradient, this term is equal to 0, where $v\nabla^2 \omega$ denotes the viscous diffusion effect of vorticity. It can be seen that the first two items in the equation are the main reasons for the generation of the FRIV.

In conclusion, the local entropy generation of the labyrinth can be expressed as

$$s_g = c_p \ln\left(1 + \frac{C_m \rho \omega^3 r^5}{2c_p f(\pi) P_1^* \sqrt{T_1^*}}\right) - R \ln \pi \quad (18)$$

It can be seen that the major local entropy increase comes from the windage heating and the total pressure loss in the flow of the labyrinth.

3. Simulation Model and Verification of Labyrinth

In this section, the labyrinth model is established for simulation and was checked with the known experimental data. In the following sections, the simulation results and theoretical calculations are compared and analyzed.

3.1. Numerical Approach and Computational Meshing

In this paper, UG parametric modeling was adopted, and the step labyrinth and bushing are both axisymmetric. ANSYS CFX (ver. 19.0, ANSYS Inc., Canonsburg, PA, USA, 2018) [47], a commercial software program, was used for CFD analysis. Figure 2 shows the diagram of meshes, Figure 3 shows the diagram of local mesh refinement in the tip of the labyrinth that is critical for investigation [48]. Nevertheless, we ensured that the 3D was practically close to the 2D domain by setting the smallest width as far as we could and applied symmetry conditions to lateral faces. In order to reduce the amount of calculation, a two-dimensional axisymmetric (rotational) model was adopted; this method is recommended for 2D calculations according to the CFX manual [49]. ANSYS ICEM is used for unstructured meshing. ANSYS Fluent is used for the calculation, the inlet and outlet set as the pressure control, and the wall set as non-slip adiabatic. The 3D unsteady compressible N-S equation is solved by the finite volume method. The turbulence is modeled with the $k - \varepsilon$ model with a second-order upwind discretization scheme in time and space.

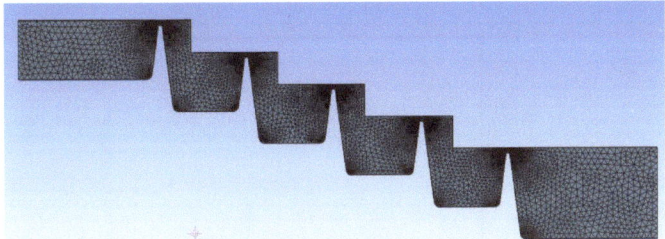

Figure 2. Diagram of meshes.

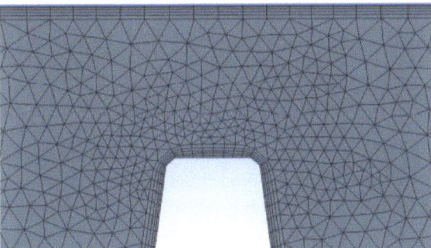

Figure 3. Diagram of local mesh refinement.

3.2. Boundary Conditions and Mesh Independent

Inlet temperature and outlet static pressure and temperature were used as boundary conditions to simulate the operating conditions of the test. Table 2 shows the boundary conditions.

Table 2. Boundary conditions.

Boundary Conditions	Value	Unit
Outlet pressure	101,325	Pa
Import and export pressure ratio	1.5	
Inlet temperature	300	K
Outlet temperature	300	K

Define the inlet and outlet pressure ratio as

$$\pi = \frac{P_2^*}{P_1^*} \tag{19}$$

where converted flow is calculated as follows:

$$\varphi = \frac{m\sqrt{T_1^*}}{P_1^* A} \tag{20}$$

There are two parameters that affect the number of grids and the calculation results in the mesh settings: global minimum mesh size and the minimum mesh size for local refinement of tooth tips. The mesh-independent calculation will be performed for different values of the parameters.

The global minimum mesh sizes were set as 2, 0.6, and 0.2 mm for three working conditions, and the minimum mesh size for local refinement of tooth tips was set as 0.03 mm. The grid numbers corresponding to the three working conditions were 180,000, 190,000, and 360,000. It can be seen from Figure 4 that the flow rate of the grate remained stable with global minimum mesh size varying from 0.2 to 2 mm. In this research, the global minimum mesh size was set as 0.6 mm.

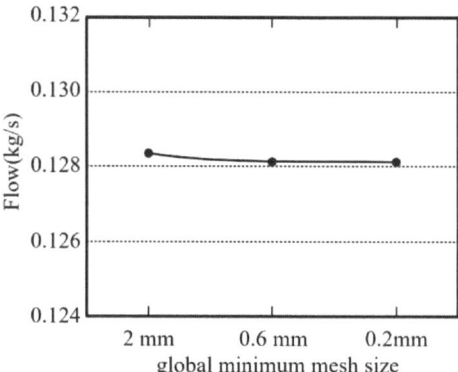

Figure 4. Global minimum mesh independent computation.

The minimum mesh size for local refinement of tooth tips was set as 0.1, 0.03, 0.02, and 0.01 mm for four working conditions, and the global minimum mesh size was set as 0.6 mm. The grid numbers corresponding to the four working conditions were 180,000, 200,000, 320,000, and 710,000. It can be seen from Figure 5, that when the minimum size of the refined mesh of the tooth tip was reduced to 0.03 mm, the flow rate of the labyrinth did not change significantly. Therefore, the minimum size of the local refined mesh of the tooth tip region in the CFD model was set as 0.03 mm, and the number of grids was around 200,000.

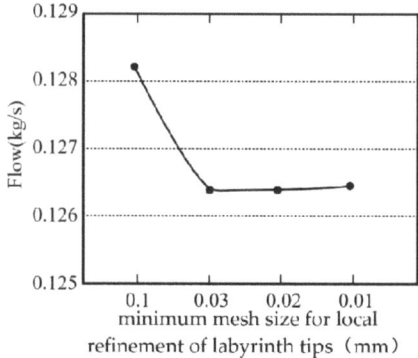

Figure 5. The minimum mesh size for local refinement independent computation.

3.3. Model Checking

In order to verify the accuracy of the parametric model, the calculation results of the simulation model in this paper were compared with the experimental data provided by known references [49]. The geometrical parameters of the labyrinth provided by the reference [50] are shown in Table 3. The data were input into the simulation model for calculating the leakage flow of the labyrinth under different inlet and outlet pressure ratios, and then we converted the leakage into the converted flow for comparison. The comparison results are shown in Table 4.

Table 3. Geometric parameters of the labyrinth.

Geometric Parameters	Symbol	Value	Unit
Seal clearance	c	0.36	mm
Tooth height	h	6	mm
Teeth pitch	L	6	mm
Tooth tip width	t	0.25	mm

Table 4. Comparison results.

SE. NU.	Pressure Ratio	Test Results	Converted Flow $\varphi \times 10^2$					
			Standard $k-\varepsilon$	Deviation	RNG $k-\varepsilon$	Deviation	SST $k-\omega$	Deviation
1	1.0357	0.7491	0.7813	4.29%	0.9567	27.71%	0.6659	−11.11%
2	1.0886	1.1621	1.2088	4.02%	1.4731	26.77%	1.0246	−11.83%
3	1.1598	1.5018	1.5467	2.99%	1.8720	24.65%	1.3265	−11.67%
4	1.2391	1.7480	1.7983	2.88%	2.1557	23.32%	1.5739	−9.96%
5	1.3878	2.0326	2.1025	3.44%	2.4773	21.88%	1.8981	−6.62%
6	1.4717	2.1616	2.2168	2.55%	2.5940	20.00%	2.0339	−5.91%
7	1.6290	2.3515	2.3816	1.28%	2.7448	16.72%	2.2314	−5.11%
8	1.7805	2.4116	2.4515	1.65%	2.8365	17.62%	2.3565	−2.28%
9	1.8974	2.4495	2.5106	2.50%	2.8872	17.87%	2.4371	−0.50%

In the present research, we briefly discussed the influence of various boundary conditions on the converted flow. It can be seen from Figure 6 that the converted flow was gradually increased with the increase of the pressure ratio. In addition, we can quantify that the conversion flow increased with the inlet temperature increase, and the conversion flow increased with the inlet pressure decrease from Equation (2).

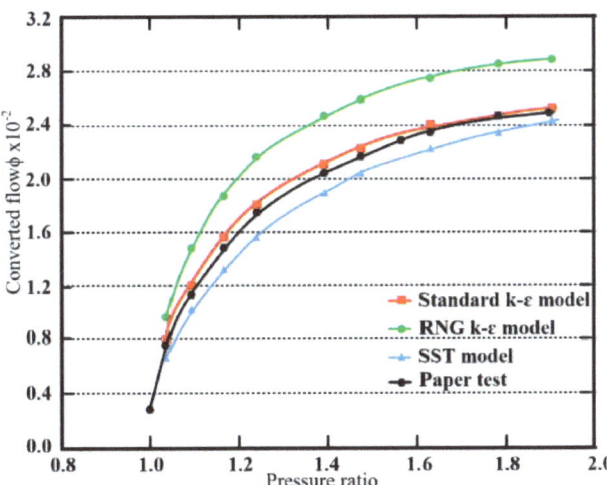

Figure 6. Comparison of turbulence models.

It can be seen from Table 4 that there were some differences in the calculation results obtained using the various turbulence models. In the calculated operating conditions, the maximum relative error of the converted flow was 4.29% which used the standard $k - \varepsilon$ turbulence model. Therefore, the standard $k - \varepsilon$ turbulence model was selected as the turbulence model for subsequent related calculations.

The results show that the calculation results of this paper are in fantastic agreement with the experimental results of the reference (Figure 7). In the working conditions, the maximum relative error of the converted flow rate was no more than 5%. The simulation model of the labyrinth established in this paper has high reliability.

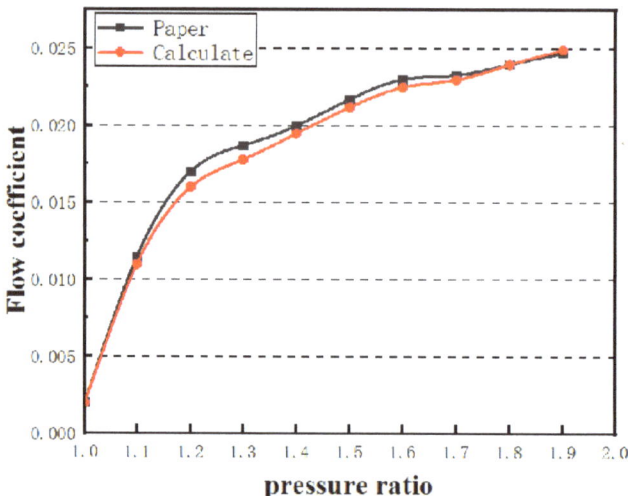

Figure 7. Comparison of calculation and reference test results.

4. Results and Discussion

In this section, the relationship between the FRIV and the relative dissipation intensity was first analyzed, being based on the theoretical research and simulation results. Then, the influence of the labyrinth tooth inclination angle, tooth tip width, step shape, and other parameters on the FRIV entropy generation were investigated. Finally, a more effective method to utilize the FRIV was proposed that is based on the research results.

4.1. The Formation Mechanism of FRIV

The flow field in the labyrinth region is shown in Figure 8. It can be seen that a high-speed airflow was formed in labyrinth region when the fluid flowed through the labyrinth teeth, which was mainly because of the strong throttling effect of the incoming flow at the tip. When the airflow passed through the labyrinth teeth, it suddenly expanded and was spread at a certain angle. Under the action of the flow resistance of the fluid in the labyrinth cavity and the frictional resistance of the wall surface, a vortex was formed in the labyrinth cavity.

Figure 9 shows the vorticity cloud map of the labyrinth, and Figure 10 shows the relative dissipation intensity distribution cloud map. Comparing and analyzing the two figures, we see that the distribution of the relative dissipative intensity was the same as the vorticity distribution, and it was mainly distributed on the wall and the tooth tip.

In order to express the correlation of the vorticity with the relative dissipative intensity distribution, we illustrated in Figure 11 the correlation coefficient cloud diagram, which was obtained by simulation. From the simulation results, it can be seen that there was a large correlation coefficient in the labyrinth tip. This result is consistent with the conclusion obtained from Equation (16), which indicates that there was a larger vortex. This vortex can be explained by the first term in Equation (17), which impeded the airflow at the tip and reduced leakage. When we input the vortex intensity ω obtained by simulation into Equation (18), it can be seen that the entropy generation increased exponentially.

4.2. Influence of Labyrinth Tip width Variation on Entropy Generation

The labyrinth teeth with the labyrinth tip widths of 0.2, 0.4, 0.6, and 0.8 mm were selected for comparative research. Figure 12 shows the flow field for different labyrinth tip widths. Figure 13 shows the distribution of entropy generation for different labyrinth tip widths.

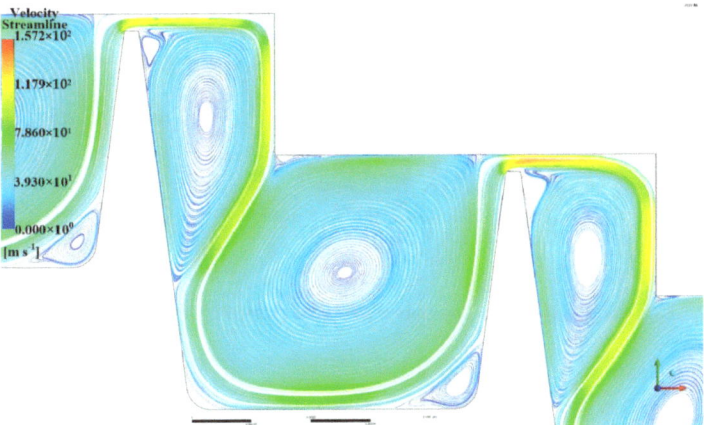

Figure 8. Formation diagram of the FRIV.

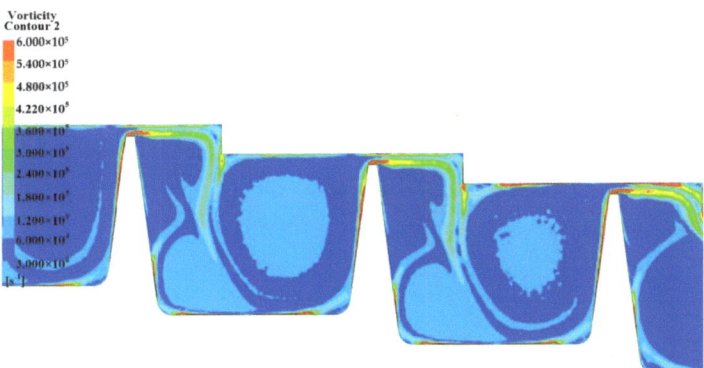

Figure 9. Cloud map of vorticity distribution.

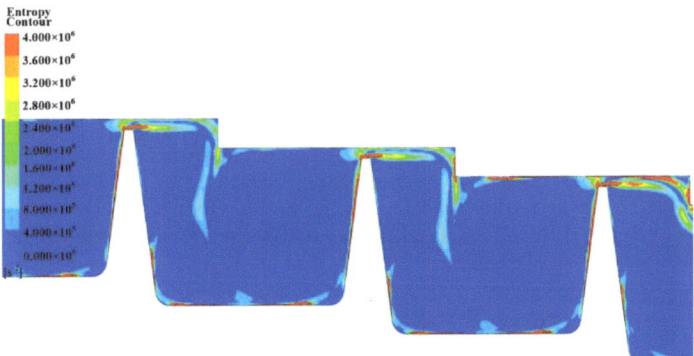

Figure 10. Cloud map of relative dissipative intensity distribution.

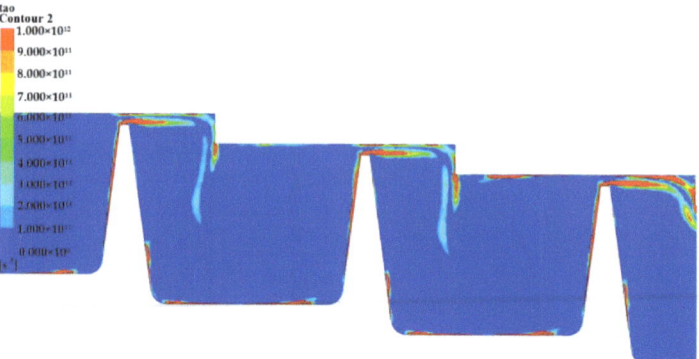

Figure 11. Correlation coefficient distribution.

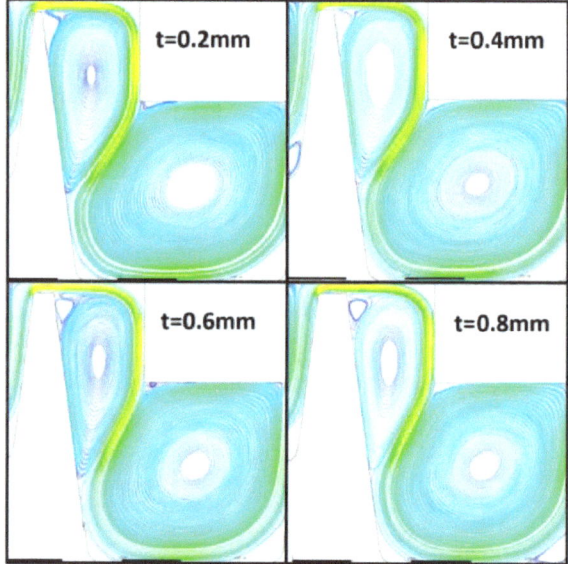

Figure 12. Flow field for different tip widths.

It can be seen from Figure 12 that with the increase of labyrinth width, the direction of airflow velocity was gradually reversed, and the strength of the FRIV was weakened. It can be seen from Equation (14) that ϕ may be decreased with the increase of the labyrinth tip width, and the relative dissipative strength decreased with the increase of the tip width. It can be seen from Equation (15) that the vortex strength decreased with the increase of the tip width and the influence of the vortex became weaker when the tip width was larger. When the tip width was small, the gas flow direction changed rapidly after the gas in the upper labyrinth cavity flowed through the tip, where the velocity gradient was large. This result was consistent with the conclusion in Equation (17) that the vortex was stretched and bent due to the velocity gradient of the flow field, which led to changes in the magnitude and direction of the vorticity. In addition, the major flow in the tip area may be affected by the extrusion of the vortex on the rear side of the labyrinth tooth tip. When the tooth tip width of the labyrinth was smaller, the vortex on the rear side of the tooth tip was stronger, which strengthened the influence on the flow in the tooth tip area and strengthened the compression effect of the tooth tip gap. Further, the flow field data (as shown in Figure 12) obtained by the simulation was output and substituted into the entropy generation of Equations (15) and (18). The relationship between the entropy generation and the airflow position can be obtained. It can be seen from Figure 14 that the results are consistent with the variation of entropy generation with tooth tip width (as shown in Figure 13).

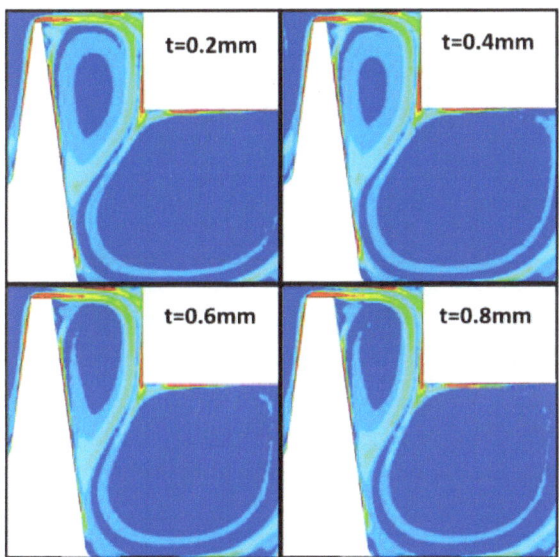

Figure 13. The distribution of entropy generation with different tip widths.

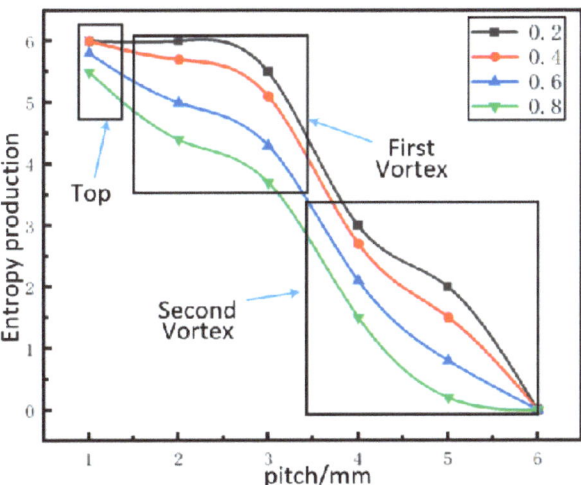

Figure 14. Variation of entropy generation with airflow position.

In conclusion, in the labyrinth tip, the entropy generation was the largest; with the increase of tooth tip width, the entropy generation of the FRIV gradually decreased.

4.3. Influence of Labyrinth Inclination on Entropy Generation

The labyrinth inclination angle α from 50° to 120° is divided into 8 groups for comparative research. Figure 15 shows the effect of different Labyrinth inclination angles on the vortex strength, Figure 16 shows the effect of different labyrinth inclination angles on the flow field.

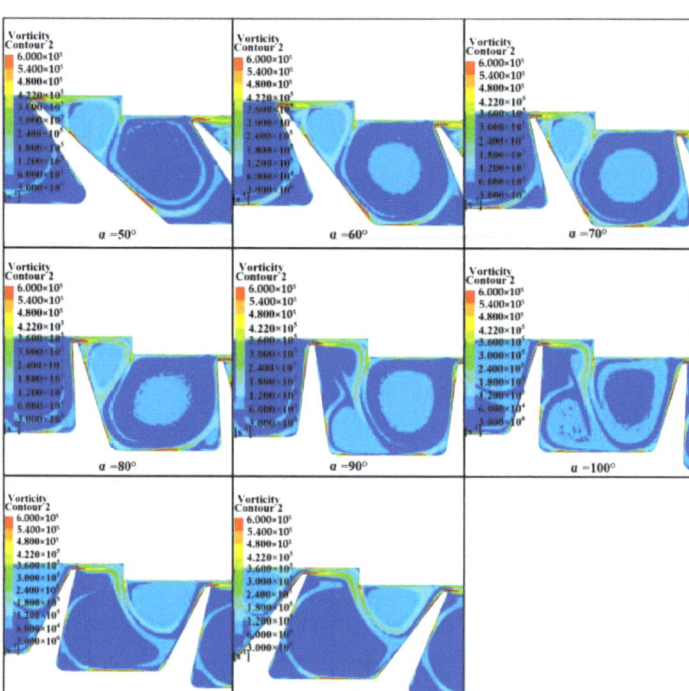

Figure 15. Distribution of the vortex with the change of the labyrinth inclination angle.

It can be seen from Figure 16 that when the labyrinth teeth were inclined forward, the FRIV on the top of the teeth increased significantly compared to the vertical position of the labyrinth teeth. The flow-resistance-increasing vortices almost occupied the clearance, which is beneficial to the sealing of the labyrinth. When the labyrinth teeth were inclined backward, the FRIV on the top of the labyrinth gradually weakened, being transferred to the low radius of the labyrinth back, forming two vortexes in the labyrinth cavity. In terms of the cloud image data shown in Figures 15 and 16, the trend diagram of the vortex intensity with the labyrinth inclination angle was obtained, as shown in Figure 17.

Figure 17 illustrates the labyrinth vortex and the thickness of vortex as a function of labyrinth inclination. It can be seen from Figure 17 that with the labyrinth inclination decreasing, the vortex increased, and the thickness of the vortex increased, indicating that the FRIV of the labyrinth was gradually strengthened. The FRIV occupied the radial clearance of the labyrinth and hindered the flow. When the labyrinth inclination angle was less than 60°, the growth trend of the vortex slowed down, and the thickness of the vortex remained stable. This result is consistent with the conclusion from the qualitative analysis of Equation (17). When the inclination angle became smaller, the velocity and direction of the flow field changed more greatly, which caused the velocity gradient in the flow field to change significantly and led to the change of the absolute value of the vorticity. As a result, the FRIV effect was enhanced accordingly.

4.4. Influence of Step Inclination Angle Change on Entropy Generation

Taking the labyrinth inclination angle of 60 degrees, the labyrinth teeth with step inclination angles of 60°, 70°, 80°, and 90° were selected for comparative research. Figures 18 and 19 show the flow field diagram and the entropy generation diagram in the labyrinth with different step inclination angles, respectively.

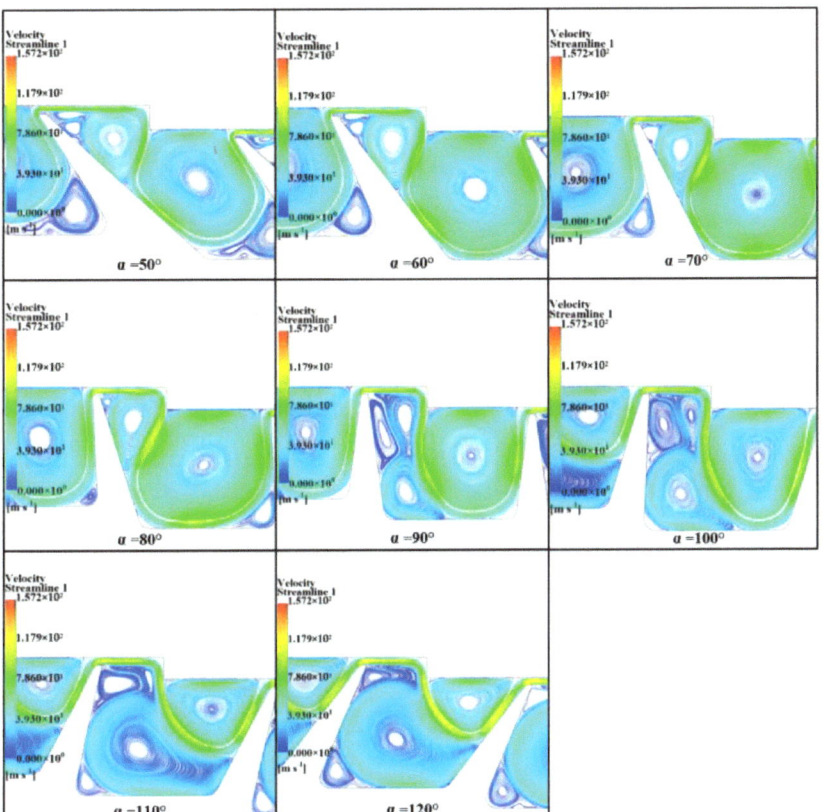

Figure 16. Flow field with the change of labyrinth inclination angle.

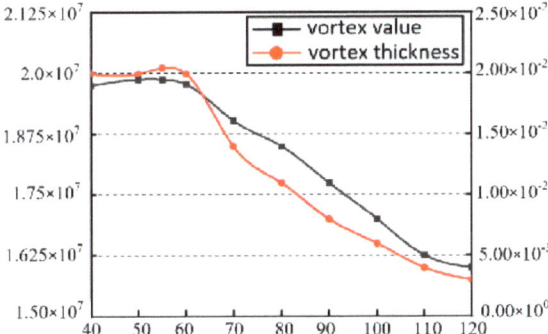

Figure 17. The trend of vortex with labyrinth inclination change.

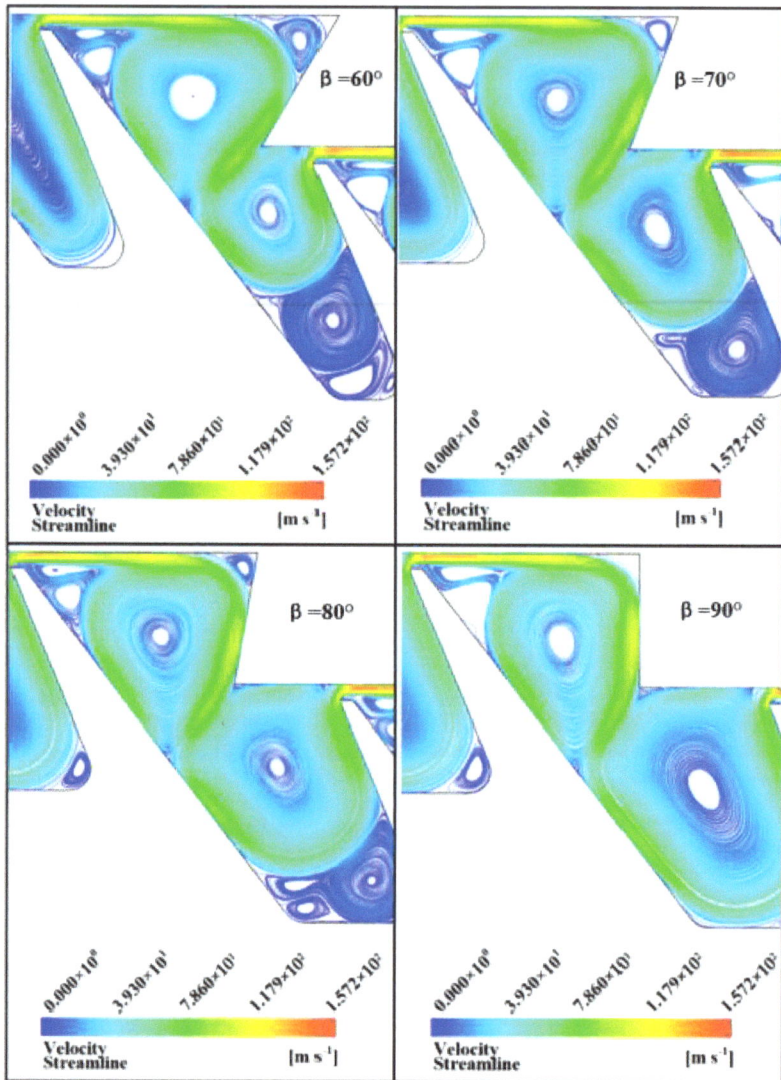

Figure 18. Flow field diagram with different step inclinations.

It can be seen from Figure 18 that the incoming flow through the labyrinth tip directly hit the step wall and then changed the flow direction, which flowed to the labyrinth cavity along the inclined angle of the step. Then, the air flow was divided into two parts, and two opposite vortices were formed at the inner step of the labyrinth and in the tooth cavity. As the inclination angle of the step decreased, the vortex at the inner step of the labyrinth was squeezed toward the labyrinth tip and increased. This result is consistent with the conclusion drawn from the second item in Equation (17). Due to the decrease of the inclination angle, the volume change rate of the airflow in the tooth tip became larger, which caused the absolute value of the vorticity to increase. Further, the flow field data in Figure 18 were substituted into the entropy generation Equations (15) and (18), wherein the relationship between the entropy generation and the airflow position as shown in Figure 20 can be obtained. It can be seen from Figure 20 that this result was consistent with the result

(Figure 19) of the simulation. In other words, with the increase of the inclination angle of the step, the entropy generation of the FRIV gradually decreased.

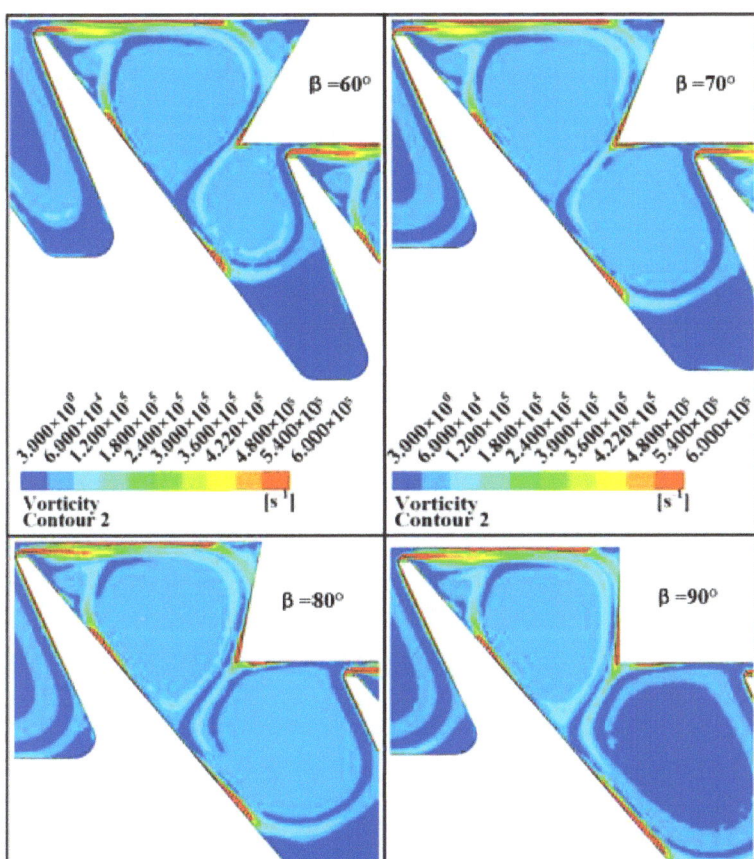

Figure 19. Distribution of the entropy generation with different step inclinations.

4.5. Influence of FRIV Strength on Leakage

In this paper, a stronger FRIV was constructed on the basis of the influence of geometric parameter changes on entropy production [51,52], which is defined as the best; the geometric parameters are shown in Table 5. The best entropy generation was studied and compared with the initial FRIV. Figure 21 shows the comparison flow field between the best and the initial. Figure 22 shows the comparison distribution of FRIV entropy generation between the best and the initial.

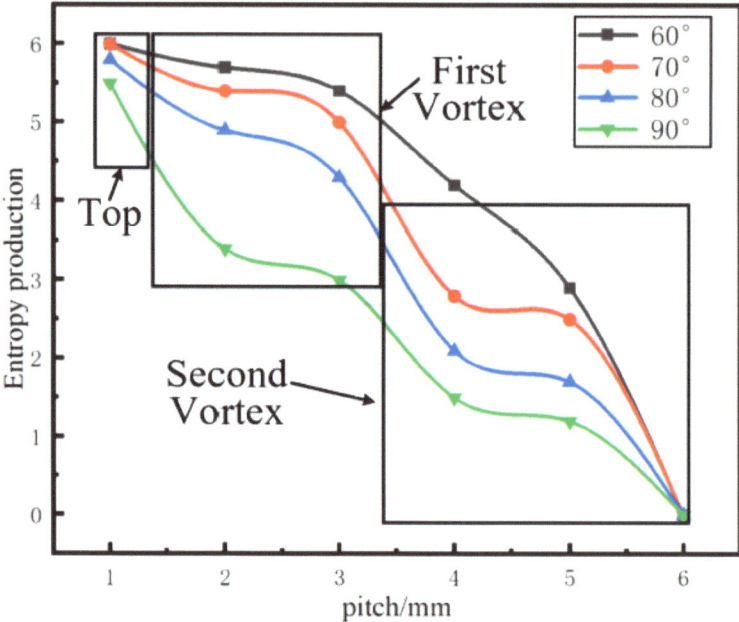

Figure 20. Entropy generation with step inclination change.

Table 5. The best labyrinth geometry parameters.

Geometrical Parameters	Symbol	Initial Value	Unit
Step height	H	2	mm
Tooth spacing	L	7	mm
Tooth height	h	5	mm
Tooth tip width	t	0.2	mm
Labyrinth inclination angle	α	60	degrees
Angle between face teeth	θ	15	degrees
Seal clearance	c	0.36	mm
Tooth to step distance	b	2	mm
Step inclined angle	β	60	degrees
Root rounding of steps	R1	0	mm
Root rounding on pressure side	RL	0.5	mm
Fin tip fillet	RR	0.5	mm

It can be seen from Figure 22 that the best entropy production was larger. From the calculation results in Table 6, the best flow rate through the top of the labyrinth teeth was reduced by 22%, which verifies the fact that the sealing effect of the labyrinth structure became stronger. The calculation results of entropy generation and leakage flow rate illustrate the fact that the method proposed in this study is effective in reducing the leakage of labyrinth teeth.

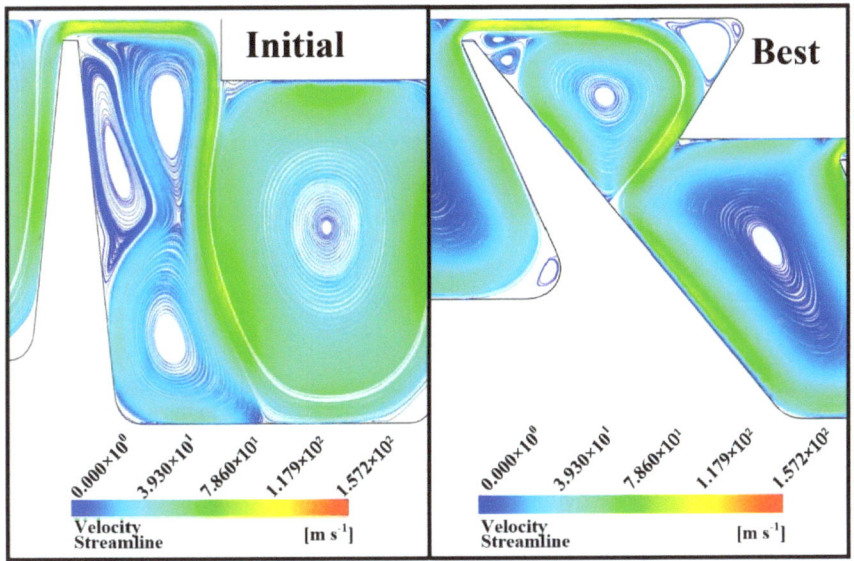

Figure 21. Flow field comparison diagram.

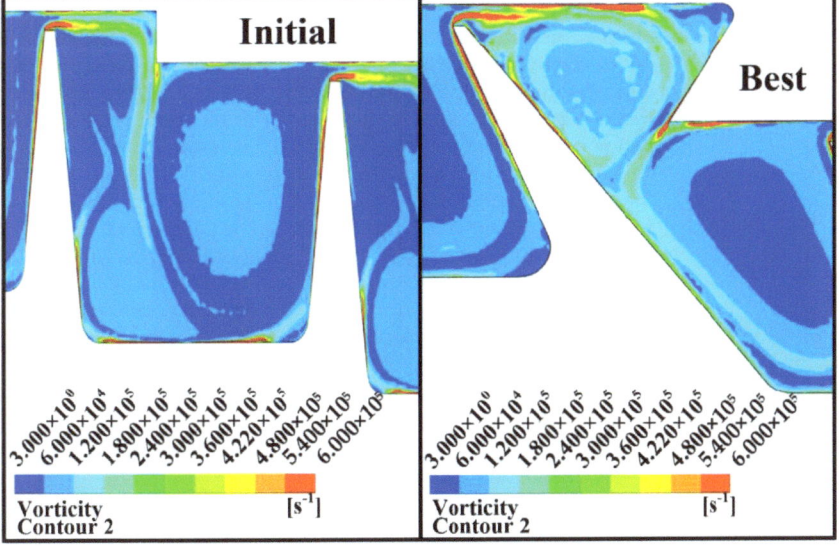

Figure 22. Comparison of entropy generation distribution.

Table 6. Calculation results.

Serial Number	Pressure Ratio	Standard $k-\varepsilon$	Best	Degree of Reduction
1	1.0357	0.7813	0.6124	21.6%
2	1.0886	1.2088	0.9546	21.02%
3	1.1598	1.5467	1.2565	18.76%
4	1.2391	1.7983	1.5239	15.25%
5	1.3878	2.1025	1.8981	12.09%
6	1.4717	2.2168	2.0339	10.50%
7	1.6290	2.3816	2.2314	8.4%
8	1.7805	2.4515	2.3565	5.91%
9	1.8974	2.5106	2.4371	4.91%

5. Conclusions

In this paper, the modeling and simulation of the aero-engine labyrinth structure was carried out, and the model was checked according to the known experimental data. The concept of FRIV was proposed, and the influence of related geometric parameters on the entropy generation of FRIV was studied.

(1) The relative dissipative intensity distribution was the same as the vorticity distribution, and there was a region with a large correlation coefficient at the top of the tooth, where the vortex was stronger and the dissipation effect was stronger.

(2) When the labyrinth width was small, the flow direction changed rapidly after the gas in the upper labyrinth cavity flowed into the labyrinth tip, which formed a FRIV at the labyrinth tip. With the increase of tooth width, the direction of fluid velocity was gradually reversed. In addition, the entropy generation and the resistance increase effect were weakened. When the labyrinth teeth were inclined forward, the FRIV on the top of the tooth increased significantly, and the entropy generation increased, almost occupying the entire labyrinth clearance. When the labyrinth teeth were inclined backward, the FRIV on the top of the tooth gradually weakened. As the inclination angle of the step decreased, the vortex at the inner step of the labyrinth squeezed toward the tooth tip and increased. In addition, the FRIV was enhanced, the entropy generation increased, and the sealing effect was strengthened.

(3) By selecting reasonable labyrinth geometric parameters, the labyrinth tip velocity gradient and throttling effect can be enhanced, and the $(\omega \cdot \nabla)V$ and $\omega(\nabla \cdot V)$ terms in the vorticity transport equation can be increased, which can strengthen FRIV and reduce leakage. The results of this research show that the reasonable labyrinth parameters can reduce leakage by 22%. Therefore, the method of reducing leakage proposed in this paper is effective.

Author Contributions: Methodology, X.L.; validation, S.D. and L.S.; formal analysis, X.Z.; investigation, S.Z. and G.L.; resources, T.Q.; writing—original draft preparation, Y.Z. All authors have read and agreed to the published version of the manuscript.

Funding: This work was funded by the Basic Research Program of the National Nature Science Foundation of China, grant number 51775025 and 51775013, China Key Research and Development Plan (No.2017YFB0102102, 2018YFB0104100).

Institutional Review Board Statement: Not applicable.

Informed Consent Statement: Not applicable.

Data Availability Statement: The data used to support the findings of this study are included within the article.

Acknowledgments: The Basic Research Program of the National Nature Science Foundation of China, grant number 51775025 and 51775013, China Key Research and Development Plan (No.2017YFB0102102, 2018YFB0104100).

Conflicts of Interest: The authors declare no conflict of interest.

References

1. Ludwig, L.P.; Johnson, R.L. *Sealing Technology for Aircraft Gas Turbine Engines*; AIAA: Reston, VA, USA, 2000.
2. Guo, H.; Feng, Q. Experiment on flow characteristic in rotating labyrinth with consideration of clearance change. *J. Aerosp. Power* **2018**, *33*, 1779–1786.
3. Hanzlik, H.J. Labyrinth Packing. US1831242A, 10 November 1931.
4. Vermes, G.Z. A Fluid Mechanics Approach to the Labyrinth Seal Leakage Problem. *J. Eng. Gas Turbines Power* **1960**, *83*, 161. [CrossRef]
5. Zimmermann, H.; Wolff, K.H. Air System Correlations: Part 1—Labyrinth Seals. In Proceedings of the Asme International Gas Turbine & Aeroengine Congress & Exhibition, Stockholm, Sweden, 2 June 1998.
6. Nayak, K.C. Effect of Rotation on Leakage and Windage Heating in Labyrinth Seals With Honeycomb Lands. *J. Eng. Gas Turbines Power* **2020**, *142*, 081001. [CrossRef]
7. Lee, S.I.; Kang, Y.J. Basic Research Trends on Labyrinth Seal of Gas Turbine. *KSFM J. Fluid Mach.* **2020**, *23*, 32–39. [CrossRef]
8. Soemarwoto, B.I.; Kok, J.C. Performance Evaluation of Gas Turbine Labyrinth Seals Using Computational Fluid Dynamics. In Proceedings of the ASME Turbo Expo 2007: Power for Land, Sea, and Air, Montreal, QC, Canada, 14–17 May 2007; Volume 4.
9. Rhode, D.L.; Demko, J.A. Prediction of Incompressible Flow in Labyrinth Seals. *J. Fluids Eng.* **1986**, *108*, 19–25. [CrossRef]
10. Demko, J.A.; Morrison, G.L.; Rhode, D.L. The Prediction and Measurement of Incompressible Flow in a Labyrinth Seal. *J. Eng. Gas Turbines Power* **1989**, *111*, 189–195. [CrossRef]
11. Stoff, H. Incompressible flow in a labyrinth seal. *J. Fluid Mech.* **2006**, *100*, 817–829. [CrossRef]
12. Rhode, D.L.; Sobolik, S.R. Simulation of Subsonic Flow Through a Generic Labyrinth Seal. *J. Eng. Gas Turbines Power* **1985**, *108*, 429–437. [CrossRef]
13. Rhode, D.L.; Hibbs, R.I. New model for flow over open cavities. I-Model development. *J. Propuls. Power* **1992**, *8*, 392–397. [CrossRef]
14. Rhode, D.L.; Hibbs, R.I. New model for flow over open cavities. II-Assessment for seal leakage. *J. Propuls. Power* **1971**, *8*, 398–402. [CrossRef]
15. Demko, J.A.; Morrison, G.L. Effect of shaft rotation on the incompressible flow in a labyrinth seal. *J. Propuls. Power* **1990**, *6*, 171–176. [CrossRef]
16. Rapisarda, A.; Desando, A. Rounded Fin Edge and Step Position Effects on Discharge Coefficient in Rotating Labyrinth Seals. *J. Turbomach.* **2016**, *138*, 011005. [CrossRef]
17. Scherer, T.; Waschka, W. Numerical Predictions of High-Speed Rotating Labyrinth Seal Performance: Influence of Rotation on Power Dissipation and Temperature Rise. In *International Symposium on Heat Transfer in Turbomachinery*; Begell House Inc.: Danbury, CT, USA, 1994.
18. Nayak, K.; Ansari, A. The Effect of Rub-Grooves on Leakage and Windage Heating in Labyrinth Seals with Honeycomb Lands. In Proceedings of the Aiaa/Asme/Sae/Asee Joint Propulsion Conference & Exhibit, San Jose, CA, USA, 14–17 July 2013.
19. Musthafa, K.C. The Effects of Tooth Tip Wear and its Axial Displacement in Rub-Grooves on Leakage and Windage Heating of Labyrinth Seals with Honeycomb Lands. In Proceedings of the 43rd AIAA/ASME/SAE/ASEE Joint Propulsion Conference & Exhibit (AIAA), Nashville, TN, USA, 8–11 July 2007.
20. Kaliraj, K.R.; Yepuri, G.B. Parametric Studies on Gas Turbine Labyrinth Seal for the Secondary Air Flow Optimization at Static and Rotating Conditions. In Proceedings of the ASME 2019 Gas Turbine India Conference, Chennai, India, 5–6 December 2019.
21. Desando, A.; Rapisarda, A. Numerical Analysis of Honeycomb Labyrinth Seals:Cell Geometry and Fin Tip Thickness Impact on the Discharge Coefficient. In Proceedings of the ASME Turbo Expo 2015, Montreal, QC, Canada, 15–19 June 2015.
22. Ganine, V.; Chew, J.W. Transient Aero-Thermo-Mechanical Multidimensional Analysis of a High Pressure Turbine Assembly Through a Square Cycle. *J. Eng. Gas Turbines Power* **2021**, *143*, 081008. [CrossRef]
23. Childs, D.W. Turbomachinery Rotordynamics: Phenomena, Modeling, and Analysis. In *Turbomachinery Rotordynamics Phenomena Modeling and Analysis*; John Wiley and Sons: Hoboken, NJ, USA, 1993.
24. Wittig, S.L.; Dörr, K.L. Scaling Effects on Leakage Losses in Labyrinth Seals. *J. Eng. Gas Turbines Power* **1983**, *105*, 305–309. [CrossRef]
25. Denecke, J.; Dullenkopf, K. Experimental Investigation of the Total Temperature Increase and Swirl Development in Rotating Labyrinth Seals. In Proceedings of the Asme Turbo Expo: Power for Land, Sea, & Air, Reno, NV, USA, 6–9 June 2005.
26. Mcgreehan, W.F.; Ko, S.H. Power Dissipation in Smooth and Honeycomb Labyrinth Seals. In Proceedings of the ASME 1989 International Gas Turbine and Aeroengine Congress and Exposition, Toronto, ON, Canada, 4–8 June 1989.
27. Willenborg, K.; Kim, S. Effects of Reynolds Number and Pressure Ratio on Leakage Loss and Heat Transfer in a Stepped Labyrinth Seal. *J. Turbomach.* **2001**, *123*, 815. [CrossRef]
28. Stocker, H.L.; Cox, D.M. *Aerodynamic Performance of Conventional and Advanced Design Labyrinth Seals with Solid-Smooth Abradable, and Honeycomb Lands*; NASA: Washington, DC, USA, 1977; pp. 1–272.
29. Rhode, D.L.; Adams, R.G. Rub-Groove Width and Depth Effects on Flow Predictions for StraightThrough Labyrinth Seals. *J. Tribol.* **2004**, *126*, 781–787. [CrossRef]

30. Millward, J.A.; Edwards, M.F. Windage Heating of Air Passing Through Labyrinth Seals. *J. Turbomach.* **1996**, *118*, 414–419. [CrossRef]
31. Waschka, W.; Wittig, S. Influence of High Rotational Speeds on the Heat Transfer and Discharge Coefficients in Labyrinth Seals. *J. Turbomach.* **1990**, *114*, 462–468. [CrossRef]
32. Braun, E.; Dullenkopf, K. Optimization of Labyrinth Seal Performance Combining Experimental, Numerical and Data Mining Methods. In Proceedings of the Asme Turbo Expo: Turbine Technical Conference & Exposition, Copenhagen, Denmark, 11–15 June 2012.
33. Min, S.H.; Soo, I.L. Effect of clearance and Cavity Geometries on Leakage Performance of a Stepped Labyrinth Seal. *Processes* **2020**, *8*, 1496.
34. Khan, N.S.; Shah, Q. Mechanical aspects of Maxwell nanofluid in dynamic system with irreversible analysis. *ZAMM J. Appl. Math. Mech.* **2021**, *101*, e202000212. [CrossRef]
35. Ramzan, M.; Khan, N.S. Mechanical analysis of non-Newtonian nanofluid past a thin needle with dipole effect and entropic characteristics. *Sci. Rep.* **2021**, *11*, 19378. [CrossRef] [PubMed]
36. Yu, Z.; Lifeng, H. Investigation on transient dynamics of rotor system in air turbine starterbased on magnetic reduction gear. *J. Adv. Manuf. Sci. Technol.* **2021**, *1*, 2021009. [CrossRef]
37. Jiang, W.; Bin, W. A Novel Blade Tip Clearance Measurement Method Based on Event Capture Technique. *Mech. Syst. Signal Process.* **2020**, *139*, 106626.
38. Weinberger, T.; Dullenkopf, K. Influence of Honeycomb Facings on the Temperature Distribution of Labyrinth Seals. In Proceedings of the ASME Turbo Expo 2010: Power for Land, Sea, and Air, Glasgow, UK, 14–18 June 2010.
39. Shuiting, D.; Ziyao, W. Probabilistic failure risk assessment for aeroengine disks considering a transient process. *Aerosp. Sci. Technol.* **2018**, *78*, 696–707.
40. Ding, S.; Che, W. Application of entropy equation in the judgement of flow direction in transient air system. *J. Aerosp. Power* **2017**, *32*, 2305–2313.
41. Bertin, J.; Cummings, R. *eBook Instant Access—for Aerodynamics for Engineers, International Edition*; Pearson: London, UK, 2013.
42. Allahverdyan, A.; Nieuwenhuizen, T. Steady adiabatic state: Its thermodynamics, entropy production, energy dissipation, and violation of Onsager relations. *Phys. Rev. E* **2000**, *62*, 845. [CrossRef]
43. Denecke, J.; Frber, J. Dimensional Analysis and Scaling of Rotating Seals. In Proceedings of the ASME Turbo Expo 2005: Power for Land, Sea, and Air. 2005, Reno, NV, USA, 6–9 June 2005.
44. Dagan, A.; Arieli, R. Solutions of the vorticity transport equation at high Reynolds numbers. In Proceedings of the Thirteenth International Conference on Numerical Methods in Fluid Dynamics, Rome, Italy, 6–10 July 1992.
45. Koh, Y.M. Vorticity and viscous dissipation in an incompressible flow. *KSME J.* **1994**, *8*, 35–42. [CrossRef]
46. Barati, R. The numerical solution of the vorticity transport equation. In Proceedings of the Third International Conference on Numerical Methods in Fluid Mechanics, Paris, France, 7 July 1972.
47. ANSYS Inc. *ANSYS CFX 19.0*; ANSYS Inc.: Canonsburg, PA, USA, 2018.
48. Yu, Z.; Longtao, S. Numerical and Experimental Investigation on Dynamic performance of Bump Foil Journal Bearing Based on Journal Orbit. *Chin. J. Aeronaut.* **2021**, *34*, 586–600.
49. ANSYS Inc. *ANSYS CFX-Solver Modeling Guide*; ANSYS Inc.: Canonsburg, PA, USA, 2011.
50. Prasad, B.V.S.S.; Manavalan, V.S. Computational and Experimental Investigations of Straight-Through Labyrinth Seals. In Proceedings of the Asme International Gas Turbine & Aeroengine Congress & Exhibition, Orlando, FL, USA, 2 June 1997.
51. Yu, Z.; Tong, X. Digital-twin-driven geometric optimization of centrifugal impeller with free-form blades for five-axis flank milling. *J. Manuf. Syst.* **2021**, *58*, 22–35.
52. Yu, Z.; Yue, S. Parametric Modeling Method for Integrated Design and Manufacturing of Radial Compressor Impeller. *Int. J. Adv. Manuf. Technol.* **2020**, *10*, 1178.

Article

An Experimental Investigation into the Thermal Characteristics of Bump Foil Journal Bearings

Yu Zhou [1], Longtao Shao [2], Shuai Zhao [1], Kun Zhu [3], Shuiting Ding [1,4], Farong Du [1] and Zheng Xu [5,*]

[1] Research Institute of Aero-Engine, Aircraft/Engine Integrated System Safety Beijing Key Laboratory, Beihang University, Beijing 100191, China; zybuaa@hotmail.com (Y.Z.); zhaoshuai@buaa.edu.cn (S.Z.); dst@buaa.edu.cn (S.D.); dufrong@hotmail.com (F.D.)
[2] School of Energy and Power Engineering, Beihang University, Beijing 100191, China; sltwhut@163.com
[3] Aero Engine Academy of China, Aero Engine Corporation of China, Beijing 100191, China; zkjason1213@163.com
[4] Department of Aviation Engineering, Civil Aviation University of China, Tianjin 300300, China
[5] Laboratory of Aeroengine Power System, Beihang Hangzhou Innovation Institute Yuhang, Hangzhou 310023, China
* Correspondence: zheng.xu@buaa.edu.cn

Citation: Zhou, Y.; Shao, L.; Zhao, S.; Zhu, K.; Ding, S.; Du, F.; Xu, Z. An Experimental Investigation into the Thermal Characteristics of Bump Foil Journal Bearings. *Symmetry* **2022**, *14*, 878. https://doi.org/10.3390/sym14050878

Academic Editor: Jan Awrejcewicz

Received: 18 March 2022
Accepted: 19 April 2022
Published: 25 April 2022

Publisher's Note: MDPI stays neutral with regard to jurisdictional claims in published maps and institutional affiliations.

Copyright: © 2022 by the authors. Licensee MDPI, Basel, Switzerland. This article is an open access article distributed under the terms and conditions of the Creative Commons Attribution (CC BY) license (https://creativecommons.org/licenses/by/4.0/).

Abstract: Bump foil journal bearings (BFJBs) are widely used in the superchargers of aviation piston engines (APEs). This paper proposes a method to evaluate the operating state of superchargers by monitoring the bearing temperature. A numerical model with a repeating symmetrical structure in the axial direction is established based on a certain type of supercharger, which solves the temperature field of BFJBs with the non-isothermal Reynolds equation and energy equation. It can be used to analyze the effect of thermal expansion on lift-off speed and stop-contact speed. A new test rig and six various BFJBs were designed to check the temperature characteristics of the BFJBs with variable load and speed. By comparing the numerical results with the experimental results, it was shown that the air film temperature increased almost linearly with the increase in bearing load and speed. However, the temperature increase caused by the rotation speed was significantly greater than the load. The structural parameters of the BFJB affected the bearing support stiffness, which had a nonlinear effect on the lift-off speed and air film temperature. Therefore, the proposed method to evaluate the state of superchargers with BFJBs was effective. These thermal characteristics can be used to guide BFJB design and predict the life cycle of BFJBs.

Keywords: bump foil journal bearing (BFJB); thermal characteristics; energy equation; test bench; lift-off speed; numeral calculations; experimental investigation; supercharger

1. Introduction

Bump foil journal bearings (BFJB) are compliant, self-acting hydrodynamic air bearings, with a thin hydrodynamic air film between the rotor and the top foil. The bump foil elastic support structure of BFJBs is a repeating symmetrical structure in the axial direction [1]. Compared with oil-lubricated bearings, BFJBs have distinct advantages of power density, which is conducive to improving the power-to-weight ratio of aviation piston engines (APEs). Some studies have shown that BFJBs can reduce the weight of small aviation engines by 15% and improve reliability by 10 times [2–4]. Moreover, BFJBs have the distinct advantages of low friction loss, long average life, high precision, and high efficiency, in addition to requiring no lubrication [5]. The hydrodynamic lubrication method can avoid the failure of superchargers caused by oil leakages and overcome extremely low-temperature environments at high altitudes [6,7], resulting in substantial benefits to the aerospace community.

With improvements to the flight envelope of APEs, the requirements for superchargers are also increasing [8,9]. BFJBs are selected for the turbocharger rotor system of APEs due to their high-speed stability and reliability [10,11]. The main parameters for evaluating BFJBs include lift-off speed, friction torque, and temperature characteristics, which significantly affect the performance of BFJBs. Although some parameters of BFJBs can be solved by numerical calculations, it is impossible to accurately calculate their parameters in real operating conditions due to simplifying assumptions. The programming and debugging processes of these algorithms are very complicated, and not suitable for engineering applications [12].

The performance of BFJBs is usually measured indirectly by experiments, which are costly [13–15]. If the working state of the bearing can be quickly evaluated directly through some parameters, the reliability of the bearing can be monitored and the life of the bearing can be predicted [16,17]. Since it is difficult to directly monitor the lift-off time and frictional resistance torque of BFJBs in practical applications, it is almost impossible to evaluate the operating status of BFJBs in real time. The method of directly measuring the temperature of BFJBs is practical and effective, and proper temperature management is beneficial to improving the capacity of bearings at high speed [18].

Many analytical and experimental investigations have been implemented to explore the thermal characteristics of BFJBs. NASA and the U.S. Air Force have developed multi-blade models that have been successfully used in military and commercial aircraft such as the Boeing 767, Boeing 757, and DC-10 gas foil bearings (GFBs) [19,20]. DellaCorte et al. established a high-temperature test bench for GFBs to measure the bearing capacity under high temperature and speed conditions, and concluded that the bearing capacity decreased with the increase in temperature [21]. Braun built a thrust foil air bearing test bench with a maximum test speed of 80,000 r/min, which was used by NASA to develop a future non-lubricated turbine engine [22]. The team of Feng K, from Hunan University, designed and built a GFB to measure the bearing temperature characteristics and rotor dynamic response under different rotational speed and load conditions, and successfully applied it to oil-free turbochargers and high-speed air compressors for fuel cells [23,24]. Pattnayak et al. developed a new bore for high-performance behaviors of self-acting GFBs based on numerical investigations, and compared it with the results of conventional bump foil/rigid bore aerodynamic bearings [25]. Ganaiet et al. proposed a method to reduce friction and improve the dynamic performance of a self-acting air foil journal bearing, using a texture with rectangular dimples [26]. Samanta et al. reviewed the key technological innovations in structural foil designs that leverage the stiffness and damping of foil structure for high-speed and high-load applications [27]. Liu X. et al. built a test rig for thrust gas bearings and measured the temperature under different working conditions [28]. Kumar et al. presented an overview of the work completed in the past few decades regarding the development of numerical models. They listed the efforts of several researchers around the world to conduct experimental investigations for predicting and analyzing the thermohydrodynamic behavior of gas foil bearings at different operating conditions [29].

Theoretical approaches used to investigate the performance of BFJBs have been discussed in various references. Salehi et al. explored the static performance and temperature field of a GFB with the Reynolds equation, the air film energy equation and the simple elastic base model [30]. Peng and Khonsari proposed a THD model to investigate the static performance of foil air bearings. The bearing structure adopts a simple elastic base model, and the field distributions of the film pressure and temperature are obtained by coupling the Reynolds equation and the energy equation [31]. Zhou Y. et al. simplified the interaction force in the bearing into the form of a spring, realizing the decoupling of the complex rotor system [32]. Feng K's team presented a numerical model for the 3D thermo-hydrodynamic analysis of sparse mesh bump-type foil bearings on air films, taking into account thermal convection in the cooling air, the thermal expansion of the bearing components, and the changes in material properties due to thermal variation. The lubrication model of the thrust bearing has been simultaneously coupled with the generalized Reynolds equation, film thickness equation, energy equation, viscosity equation, solid heat conduction equation, and solid thermal expansion equation [33,34]. Andres et al. developed a theoretical model that considers thermoelastic deformation, thermal expansion, and the centrifugal growth of the rotor [35,36]. Lai T. et al. analyzed foil bearing stability with the Timoshenko beam element, which is an accurate and fast method [37].

There is dry friction between the rotor shaft and the top foil of the BFJB before the turbocharger rotor lift-off, which causes the temperature of the rotor shaft and the elastic support foil to rise sharply. The elastic support foil will fail due to excessive temperature [38]; this can easily cause unnecessary losses and even lead to accidents involving machine crashes and fatalities [39]. Although many investigations have been conducted on the temperature characteristics of BFJBs, there is no suitable method to evaluate the state of oil-free turbochargers in real time. Considering these analyses above, this paper proposes a method which evaluates the performance and predicts the life cycle of BFJBs using temperature characteristics. The novelty of this work lies in the exploration of BFJB thermohydrodynamic characteristics as the key element of the proposed experimental investigation. A new BFJB test rig and various test BFJBs were designed to monitor the temperature field. The experimental results and the numerical calculations were compared to verify the method. The presented method can be used to guide BFJB design and predict the life of BFJBs. The rest of this paper is organized as follows: in Section 2, a numerical model of the BFJB considering the thermal deformation of foil is established. In Section 3, a new BFJB test rig and various test BFJBs are designed to measure the thermal characteristics. Section 4 presents the experimental results and compares them with numerical calculations to verify the feasibility of the proposed experimental method. The last section presents the conclusions of this paper.

2. Numerical Model of BFJBs

2.1. Energy Equation

A large amount of heat is generated when the supercharger rotor is running at high speed, and the prediction of the gas film temperature in the BFJB requires a simultaneous solution to the Reynolds equation and the energy equation. The Reynolds equation and the energy equation are both functions of pressure and temperature. It is very complicated to solve the full energy equation of the temperature field in the bearing [40]. Under non-isothermal conditions, the steady-state Reynolds equation for the gas film can be presented as [41]:

$$\frac{1}{R^2}\frac{\partial}{\partial \theta}\left(\frac{ph^3}{12\mu T}\frac{\partial p}{\partial \theta}\right) + \frac{\partial}{\partial y}\left(\frac{ph^3}{12\mu T}\frac{\partial p}{\partial y}\right) = \frac{\omega}{2}\frac{\partial}{\partial \theta}\left(\frac{ph}{T}\right) \quad (1)$$

where μ denotes the gas viscosity (Pa·s), T denotes the gas film temperature (°C), x denotes the circumferential direction, y denotes the bearing width direction; z denotes the gas film thickness direction; and ω denotes the rotor angular velocity (rad/s).

There is a certain relationship between the viscosity of the gas and the temperature [29]. This can be presented as:

$$\mu = a(T + T_{ref}) \quad (2)$$

where $a = 4 \times 10^{-8}$ and $T_{ref} = 458.75$ °C.

The heat transfer of the gas film is described by the energy transfer equation, and the process is partially simplified according to the actual heat transfer [42]:

(1) The thermal conductivity of the BFJB in the x direction is negligible compared with the convective heat transfer;
(2) The thickness of the gas film is of micrometer-scale, so the main form of energy transfer along the z direction is heat conduction. The simplified energy equation is provided as:

$$\rho c_p \left(u\frac{\partial T}{\partial x} + v\frac{\partial T}{\partial y} + \omega\frac{\partial T}{\partial z} \right) = k\left(\frac{\partial^2 T}{\partial y^2} + \frac{\partial^2 T}{\partial z^2} \right) + u\frac{\partial p}{\partial x} + v\frac{\partial p}{\partial y} + \mu\left(\left(\frac{\partial u}{\partial z}\right)^2 + \left(\frac{\partial v}{\partial z}\right)^2 \right) \quad (3)$$

where c_p (J/(kg·K)) denotes the specific heat capacity of the gas at constant pressure, k (W/(m·K)) denotes the thermal conductivity of the gas, u (m/s) denotes the velocity in the x direction of the gas film, v (m/s) denotes the velocity in the y direction of the gas film, and ω (m/s) denotes the velocity in the z direction of the gas film. The schematic diagram of the BFJB is shown in Figure 1.

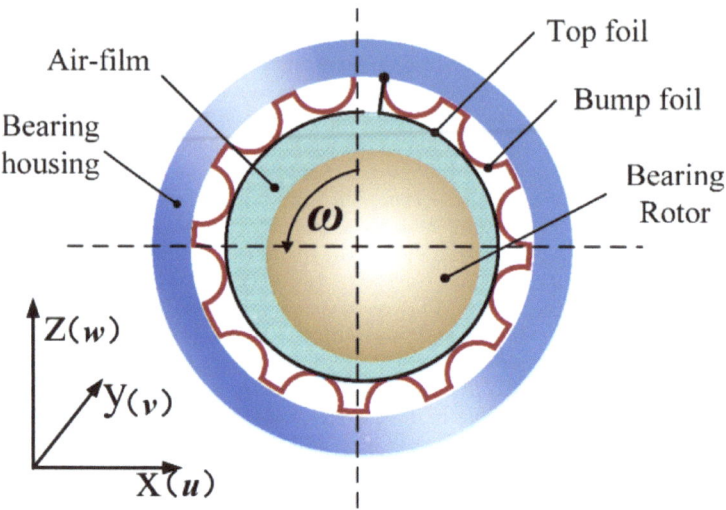

Figure 1. Schematic diagram of the BFJB.

The rotor growth at high temperature and high rotational speed significantly influences the air film because of its thinness. To reasonably predict the temperature field, the thermal and centrifugal effect on rotor growth cannot be ignored. The film thickness can be presented as:

$$h = c(1 + \varepsilon \cos) + (p - p_a) \cdot s / k_b - \Delta C_t - \delta_c \tag{4}$$

where c (m) denotes the initial air-film gap; s (m) denotes the foil pitch; k_b (N/m^2) denotes the axial unit foil length stiffness; and δ_c (m) denotes the centrifugal growth of the rotor.

The thermal expansion of the bearing system is represented by ΔC_t (m):

$$\Delta C_t = \delta_s + \delta_f - \delta_h \tag{5}$$

where δ_s denotes the thermal expansion of the rotor; δ_f denotes the thermal expansion of the bump foil structure; and δ_h denotes the thermal expansion of the bearing top foil.

The centrifugal growth for the rotor due to high-speed rotation is calculated as:

$$\delta_c = \frac{\rho R^3 \omega^2}{4E}(1 - v) \tag{6}$$

where E denotes the elasticity modulus; ρ denotes the material density; denotes the Poisson ratio; and R denotes the rotor radius.

The gas film energy boundary conditions are [43]:

$$\begin{aligned} T|_{z=h_{\min}} &= T_{\text{rotor}} \\ T|_{z=h_{\max}} &= T_{\text{top}} \\ T|_{\text{steady}} &= T_{\text{air}} \end{aligned} \tag{7}$$

where T_{rotor} denotes the shaft temperature; T_{top} denotes the top foil temperature; and T_{air} denotes the air gap temperature.

In this paper, the finite difference method and the Newton–Raphson method are combined to solve the model. It can be seen from Equations (1)–(7) that the non-isothermal Reynolds equation, the gas film energy equation, and the gas film thickness equation are coupled with each other; therefore, ignoring the influence of assumptions, a steady-state convergent solution can be obtained in an iterative manner [44].

2.2. Thermohydrodynamic (THD) Model of BFJBs

In this section, Figure 2 presents a schematic of the THD model for the BFJB consisting of the bearing shell, bump foil, top foil, and rotating shaft [45]. According to the working principle of BFJBs, the heat generated from the viscous shearing in the air film is partly converted into the heat energy of the air film, among other components; moreover, these other parts diffuse into the atmosphere in the form of heat energy through the top foil, bump foil, bearing housing, and rotor. Compared with the bearing structure size, the thickness of the air film can be ignored, so the proportion of heat diffused into the environment by the air film is very small. Because the top foil is very thin, the heat transfer area in the lateral direction is negligible; therefore, the lateral heat transfer of the top foil can be ignored. Assuming a uniform rotor temperature distribution, the rotor temperature and the top foil temperature are used as boundary conditions for the film temperature calculation.

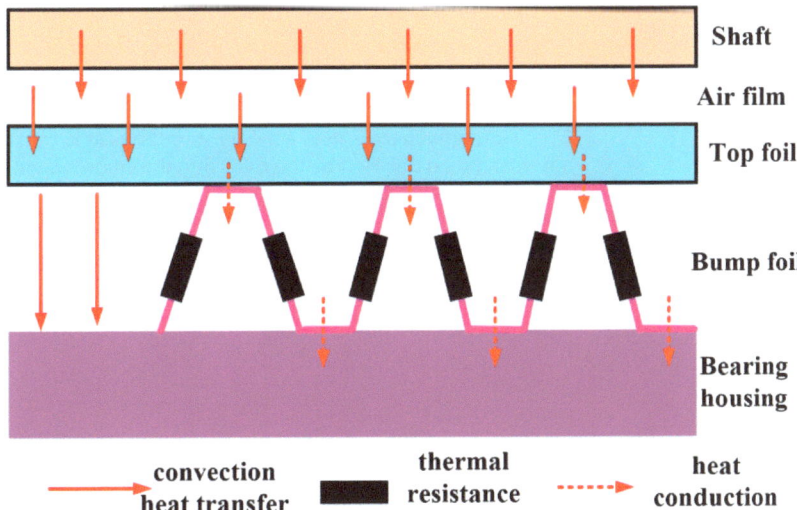

Figure 2. Schematic diagram of the heat transfer model of a BFJB.

As can be seen from Figure 2, the heat generated by the air film is divided into three parts. Part of the heat is transferred to the bump foil in the form of heat conduction through the top foil. Then, it is transferred to the air gap between the bump foil and the bearing seat in the form of thermal convection and, finally, it is diffused into the surrounding environment through the bearing seat. Another part of the heat is transferred directly through the top foil in the form of thermal convection into the air gap between the top foil and the bearing seat, and then the heat is dissipated into the environment through the bearing seat. The third part of the heat is transferred to the air gap between the top foil and the bearing seat in the form of thermal convection through the top foil, and then the heat is conducted to the bearing seat through the bump foil, and finally diffused into the environment through the bearing seat [46].

The thermal resistance of each part can be calculated using Equation (8).

$$R = \frac{h}{kA} \tag{8}$$

The thermal resistance of each part is calculated in Equation (8), where k (W/(m·K)) denotes the thermal conductivity, A (m^2) denotes the heat transfer area, R (K/W) denotes the thermal resistance, and h denotes the film thickness in the heat transfer direction.

The total thermal resistance can be calculated as [38]:

$$R_{tot} = R_T + \cfrac{1}{\cfrac{1}{R_{G1}} + \cfrac{1}{(R_{sec}+R_{G2})} + \cfrac{1}{R_B}} + R_H + R_{cf} \qquad (9)$$

where R_T denotes the thermal resistance of the top foil, R_G denotes the thermal resistance of the air gap of the bump foil, R_s is the thermal resistance of the bump foil, R_H is the thermal resistance of the bearing shell, R_{sec} is the thermal resistance of the top foil contact, and R_{cf} is the thermal resistance during natural convection heat dissipation.

When the heat transferred from the air film to the top foil and the heat transferred from the top foil are in dynamic equilibrium, the bearing temperature comes to a steady state. According to this equilibrium relationship, the heat balance equation can be presented as [43]:

$$-k_a A \frac{\partial T_F}{\partial z} = \frac{T_0 - T_F}{R_{tot}} \qquad (10)$$

After the dimensionless Formula (10), we can get:

$$\overline{T_F} + \gamma \frac{\partial \overline{T_F}}{\partial \overline{z}} = 0 \qquad (11)$$

where $\gamma = -\frac{k_a A R_{tot}}{h}$.

The supercharger uses a solid rotor; the heat is transferred from the gas film to the rotor, and then diffused by the rotor into the environment. Because of the high rotational speed of the rotor, the rotor temperature can be further assumed to be equal in the circumferential direction [24], and the temperature situation of the rotor can be simplified as a one-dimensional temperature model distributed along the axial direction [47,48].

$$Q_{conv}^j + \frac{k_R A_c}{\Delta y}\left(T_R^{j+1} + T_R^{j-1} - 2T_R^j\right) - \frac{T_R^j - T_0}{R_{in}} = 0 \qquad (12)$$

$$Q_{conv}^j = -k_a A_{R,i} \sum_i \frac{\partial T}{\partial z}\bigg|_{z=h} \qquad (13)$$

In the formula, k_R (W/(m·K)) is the thermal conductivity of the rotor; A_c (m^2) is the cross-sectional area of the rotor; T_R (K) is the rotor temperature; and R_{in} (K/W) is the thermal resistance in the thickness direction of the rotor shaft.

2.3. Calculation Process

In this paper, the Finite difference method and the Newton–Raphson method are combined to solve the model and the Taylor series expansion method is used to construct the difference scheme. In a homogeneous grid, Δx represents the space step in the x direction, and f represents an arbitrary function. In order to get the difference quotient of the partial derivatives of the function f, with respect to x, expand into a Taylor series in the neighborhood of x_i.

$$f_{i+1} = f_i + \frac{\partial f}{\partial x}\bigg|_i \Delta x + \frac{\partial^2 f}{\partial x^2}\bigg|_i \frac{\Delta x^2}{2!} + \frac{\partial^3 f}{\partial x^3}\bigg|_i \frac{\Delta x^3}{3!} + \frac{\partial^4 f}{\partial x^4}\bigg|_i \frac{\Delta x^4}{4!} + HOT \qquad (14)$$

$$f_{i-1} = f_i - \frac{\partial f}{\partial x}\bigg|_i \Delta x + \frac{\partial^2 f}{\partial x^2}\bigg|_i \frac{\Delta x^2}{2!} - \frac{\partial^3 f}{\partial x^3}\bigg|_i \frac{\Delta x^3}{3!} + \frac{\partial^4 f}{\partial x^4}\bigg|_i \frac{\Delta x^4}{4!} + HOT \qquad (15)$$

The aim of the finite difference method is essentially to approximate solutions to differential equations by replacing the derivatives in the differential equations with finite difference approximations. The grid of finite differences covers a continuous area, and the finite difference decomposition is defined at the nodes. Convert a continuous space to a discrete space grid with two sets of parallel lines containing $I \times J$ rectangular subdomains. The coordinates of the grid intersection can be presented as:

$$x = i\Delta x (i = 0, 1, \cdots, I), \; y = j\Delta y (j = 0, 1, \cdots, J) \tag{16}$$

$$\Delta x = X/I \tag{17}$$

$$\Delta y = Y/J \tag{18}$$

where Δx and Δy represent the space steps in the x and y directions, respectively, as shown in Figure 3.

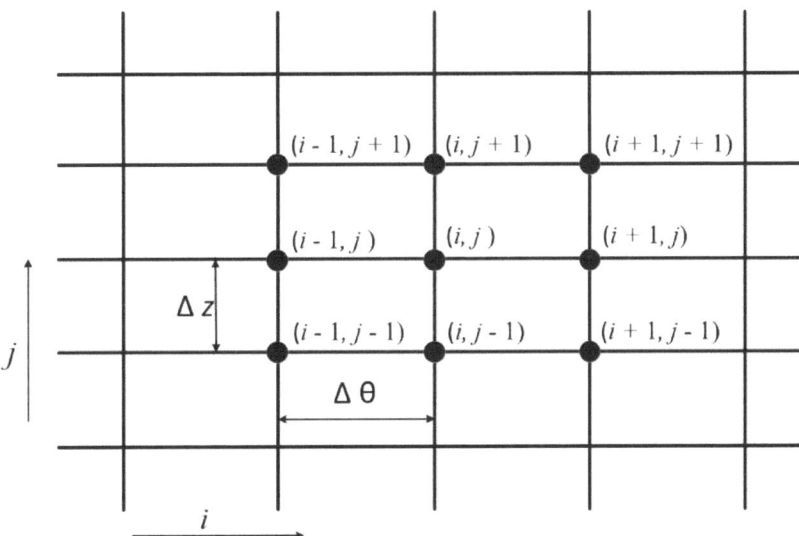

Figure 3. A portion of the computational grid of the film gap.

After finite difference processing, the Reynolds equation of the BFJB is transformed into a higher-order nonlinear equation system, which is difficult to solve directly. Therefore, the Newton–Raphson method is used to solve it iteratively. Newton–Raphson is a method of approximately solving equations in real and complex fields, and $(I - 2) \times (J - 2)$ equations can be obtained.

The calculation flow is shown in Figure 4.

In this paper, the BFJB is discretized into 125×125 points. This solves a system of linear equations, with 15,625 equations consuming a lot of time. The solution tool used in this research was the MATLAB commercial package, and the workstation was equipped with an Intel i7-1800H processor and 32 GB of memory, which manufactured by Dell Inc. in North America. Therefore, it could perform iterative calculations quickly and improve work efficiency.

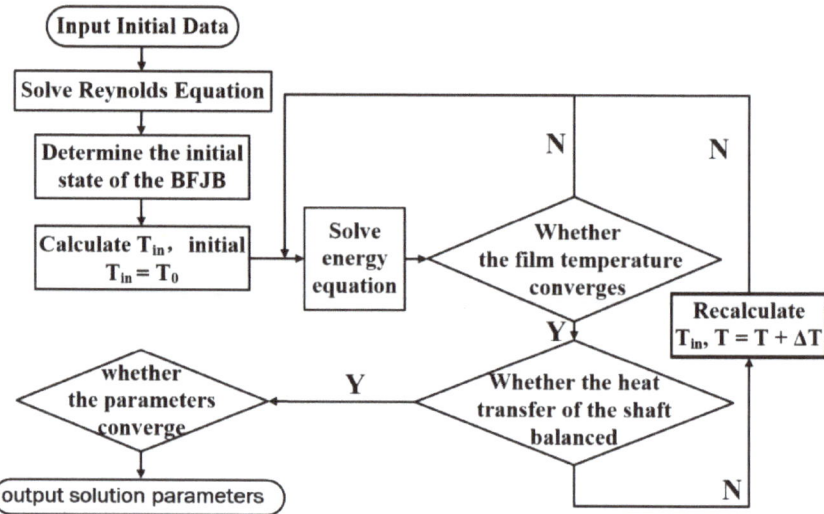

Figure 4. Flow chart of temperature calculation.

3. Experiments

3.1. Description of the Test BFJB

The test BFJB structure consisted of a top foil, bump foil and a bearing housing which was more stable than a traditional rigid bearing [49]. The support stiffness of the BFJB was determined by the structural parameters (as shown in Figure 5) [50]. When the shaft was stationary, the support stiffness of the BFJB was symmetrical. When the rotor turned from the free end of the foil to the fixed end, the top foil and the shaft formed a wedge-shaped area to generate aerodynamic pressure, and the support stiffness of the bump foil was no longer symmetrical.

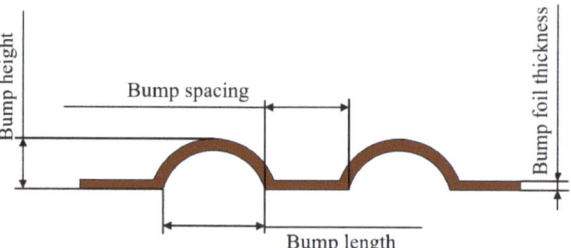

Figure 5. Configuration of bump foil.

Figure 6 demonstrates the production process of the experimental BFJB. The bump foil was stamped and formed by the bulging process, and the material was X750 stainless steel (Figure 6a). To obtain three different bump foils, firstly, stainless steel was cut into strips with a width of 6 mm and a length of 102 mm (Figure 6b). Then, corresponding bulging molds (Figure 6c) were made according to the structural parameters of each bump foil. Thirdly, the strips were placed in the bulging molds (Figure 6d). The semi-finished bump foil after bulging is shown in Figure 6e. Finally, the semi-finished BFJB was heated at 650 °C for 4 h and then cooled to room temperature in air (Figure 6f). The product after heat treatment is shown in Figure 6g.

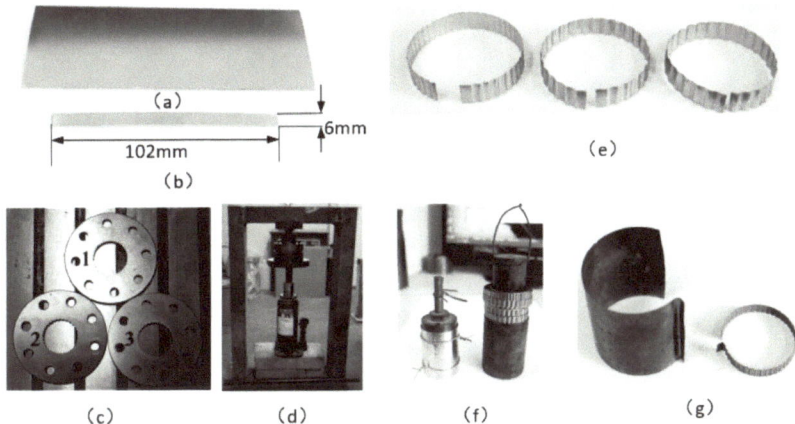

Figure 6. Test bearing preparation. (**a**) X750 Stainless steel; (**b**) Cut stainless steel bars; (**c**) Bump foil bulging model; (**d**) Bump foil bulging device; (**e**) Bulge-finished bump foil; (**f**) Foil heat treatment; (**g**) Finished bump foil.

The structure of the experimental BFJB is shown in Figure 7. The BFJB used in the test had two narrow grooves with a width of 0.2 mm in the bearing shell. The bump foil and the top foil were fixed on the grooves by an HT-CPS high-temperature structural adhesive, and the other end was free (Figure 7a). The HT-CPS high-temperature structural adhesive working temperature was −196~980 °C and the shear strength and the tensile strength after curing were 86 MPa and 38.5 MPa, respectively, which fully met the requirements of the BFJB. It should be noted that the distance between the bending point of the foil and the first bump must be consistent. This determines that all bumps are aligned in straight lines after the seven bump foils are installed. Symmetry is an important aspect of such structures, as shown in Figure 7b,c. The prepared BFJB object is shown in Figure 7d.

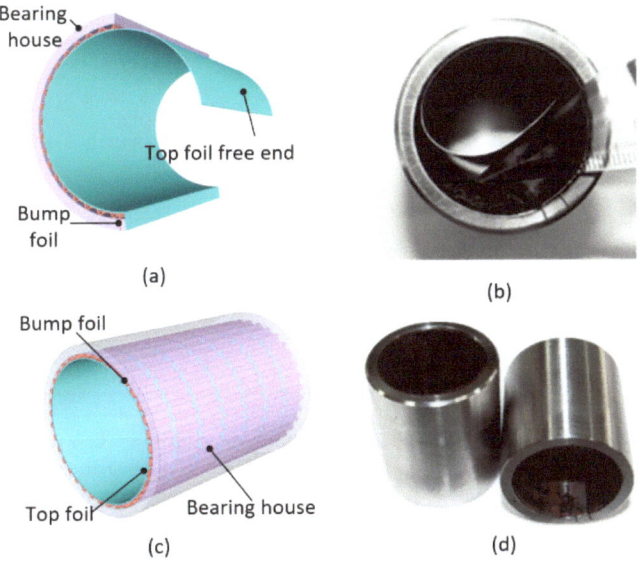

Figure 7. Installation structure of experimental bearing. (**a**) Assembly structure diagram; (**b**) Physical drawing of assembly structure; (**c**) Assembly structure perspective; (**d**) Test bearing.

The main parameters of the test bearings are shown in Table 1. No. 1, No. 2 and No. 3 test BFJBs had top foil with a thickness of 0.1 mm. No. 4, No. 5 and No. 6 test BFJBs had top foil with a thickness of 0.12 mm. The rest of the parameters were exactly the same.

Table 1. Main parameters of the bearings.

Number	B1	B2	B3	B4	B5	B6
Top foil thickness (mm)	0.1	0.1	0.1	0.12	0.12	0.12
Nominal clearance (μm)	76	76	76	56	56	56
Bump length (mm)	2	2.5	3	2	2.5	3
Bump foil thickness (mm)				0.1		
Bump spacing (mm)				0.7		
Bump height (mm)				0.5		
Bearing diameter (mm)				32		
Axial length (mm)				48		
Poissons ratio				0.29		
Elasticity modulus (Gpa)				213		

In the initial state, the BFJBs and connecting pieces weighed 1225 g. The different load tests carried out in this paper were based on the initial state.

3.2. Description of the Test Bench

This experiment adopts a self-designed test bench suitable for BFJB research. The components used in the test bench are shown in Table 2, and Figure 8 shows the mounting position of the relevant components.

Table 2. Test bench components.

Title	Performance	Function
Electric spindle	Max. speed 90,000 r/min	Drive shaft to rotate
SJH300 Inverter	Max. power 7.5 kW frequency 1500 Hz	Adjust the speed
HZ-8500 Eddy Current Displacement Sensor	range 1.5 mm, Nominal sensitivity frequency 4 kHz	Measure the displacement of the bearing during operation
Tension sensor and its accessories	range 20 kg precision 0.1% voltage signal 0–5 V	Measure the resistance torque of the bearing
Type K thermocouple	range 0~400 °C	Measure the bearing housing temperature
Electric Glow Plug		Different bearing temperatures available
data collection system	Max. frequency 4 kHz	Collect and record bearing temperature
work platform	800 mm × 800 mm thickened cast iron	Reduce the impact of vibration on EX results

The measurement principle of the BFJB test bench is shown in Figure 9. The motorized spindle was used to drive the rotor, and the bearing and the bearing seat were suspended on the shaft. The bearing resistance torque was converted into a tensile force measurement through a measuring rod fixed on the bearing seat, and the bearing load was increased by adding weights on the left side. An eddy current displacement sensor with a range of 0.5 mm was used to measure the rotational speed of the shaft, and two eddy current displacement sensors were used to measure the bearing displacement without contact, which were mounted symmetrically on both sides of the shaft [51]. The data from the two displacement sensors within 1 s were acquired as soon as the BFJB had fully lifted off and reached a steady state. These were transformed into horizontal and vertical coordinate dimensions. According to these coordinate dimensions, we drew the BFJB journal orbits and obtained the lift-off time.

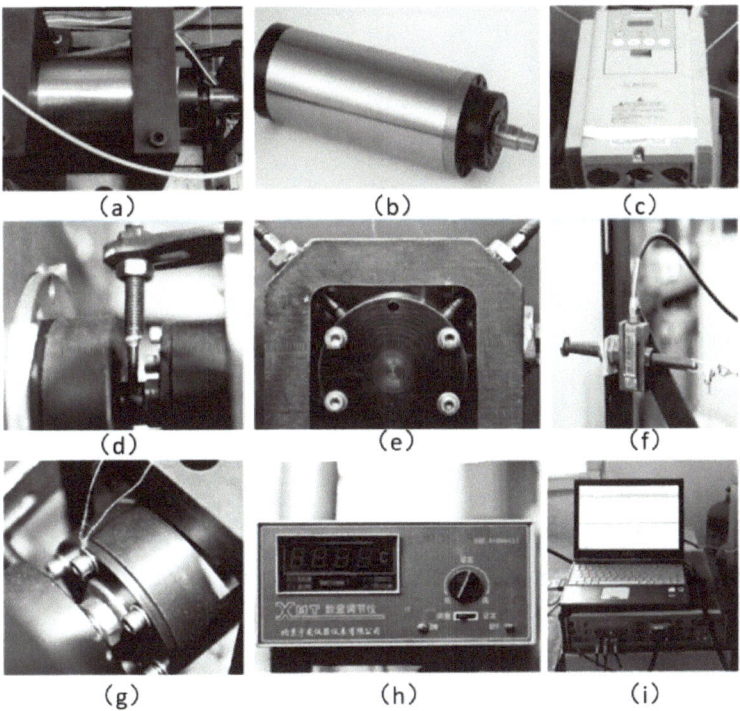

Figure 8. Configuration of the test bench. (**a**) Rotor system; (**b**) Electric spindle; (**c**) Inverter; (**d**) Eddy current speed sensor; (**e**) Eddy current displacement sensor; (**f**) Resistance torque sensor; (**g**) K-type thermocouple; (**h**) Temperature indicator; (**i**) Data collection system.

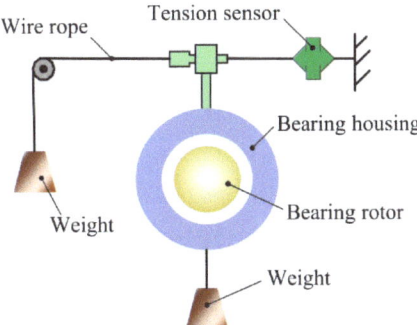

Figure 9. Schematic of the BFJB test bench.

In this test bench, the K-type thermocouple on the back of the bump foil and a digital display instrument were used to measure the temperature of the film. In this paper, the temperature collected by the thermocouple was approximated in regard to the temperature of the gas film [25]. Figure 10 shows the installation positions of 7 K-type thermocouples, which were symmetrically distributed along the middle of the BFJB to both ends.

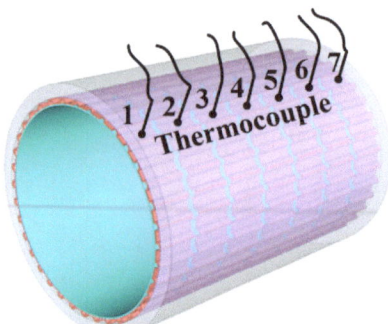

Figure 10. Location of the K-type thermocouple.

Figure 11 shows the BFJB test bench. All components of the test bench were arranged on an 800 mm × 800 mm working platform. To avoid the vibrations of the test bench affecting the test results, the working platform was made of thickened cast iron to minimize vibration.

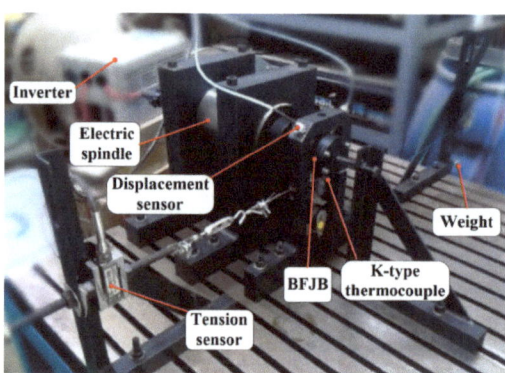

Figure 11. Bump foil radial air bearing test bench.

4. Results and Discussion

In order to verify the temperature characteristics of BFJB, the numerical results were compared with the experimental results. In this section, EX is the experimental result and NU is the numerical calculation result.

4.1. Influence of the Increase in Speed and Load on the Temperature Rise in the BFJB and the Shaft

Figure 12 shows the relationship between the BFJB temperature and the rotor speed when no load was applied, as measured by the first thermocouple in Figure 10. It can be seen that the rotational speed increased from 20,000 rpm to 90,000 rpm as the temperature of the gas film increased almost linearly. The results show that the rotation speed was lower than 65,000 rpm, and the experimental results were in good agreement with the numerical results. When the rotation speed exceeded 65,000 rpm, the EX temperature of the gas film fluctuated frequently. After analysis, it was concluded that this may have been related to the vibration of the rotor at the critical speed, which caused the rotor to make contact with the top foil, resulting in temperature fluctuations [52].

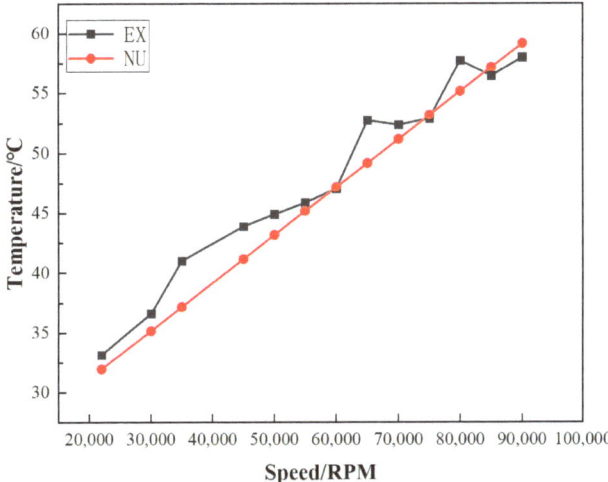

Figure 12. The effect of rotational speed on the maximum film temperature.

Figure 13 shows the variation of the maximum steady-state temperature of the test bearing with the load. A total of six load points of 16 kg were measured in the test. The load point of 0 kg does not mean that the load of the BFJB was 0, but means that the BFJB, bearing seat, wire rope and resistance torque measuring rod composed a total of 1225 g. The loads indicated in the figure are additional loads. The changes in the maximum bearing temperature with loads at three different speeds of 30,000 rpm, 50,000 rpm and 70,000 rpm were analyzed, and the applied load increased from 10 to 60 N. Compared to the NU and EX, the maximum temperature of the gas film gradually increased as the load increased from 10 to 60 N, but the increased range was not large. Compared with the changes in the film temperature caused by the rotation speed, the change in the film temperature caused by the load was relatively small. The temperature measured in the experiment was slightly higher than the temperature calculated by the numerical model.

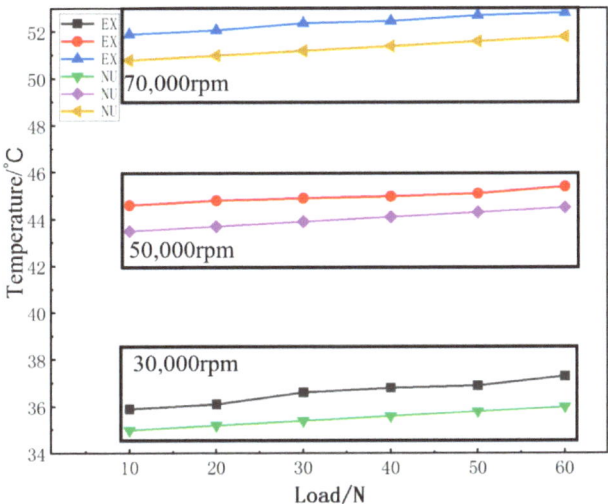

Figure 13. Effect of load on maximum bearing temperature.

As mentioned above, the supercharger adopted a solid rotor, and the heat was transferred to the rotor by the gas film, and then diffused by the rotor into the environment. The working speed of the rotor was relatively high, and it can be further assumed that the rotor temperature was equal in the circumferential direction. Thereby, the temperature of the rotor can be simplified as a one-dimensional temperature model distributed along the axial direction. The steady-state temperature measured by the thermocouples, as shown in Figure 10, represents the temperature distribution of the shaft along the axial direction. Figure 14 shows the change in the temperature of the shaft with the rotation speed, and seven temperature points in the axial direction were measured. It can be seen that the temperature in the middle of the rotor was the highest, which was then distributed symmetrically to both ends. After analysis, it was concluded that this was because the two ends of the rotating shaft exchanged heat with the environment, resulting in a lower temperature than the middle part of the rotor.

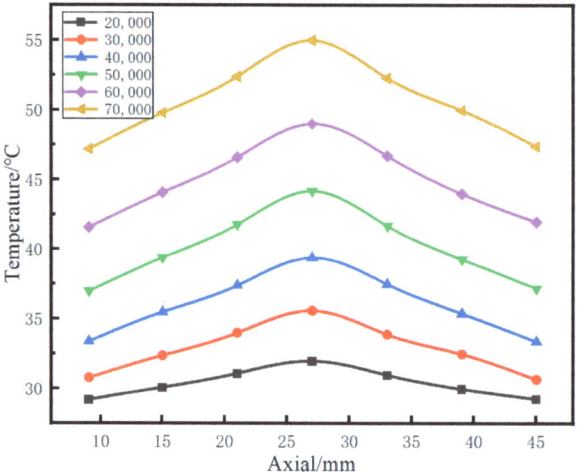

Figure 14. Axial temperature distribution of the rotor.

4.2. Influence of Structural Parameter Changes on the Performance of BFJB

Figure 15 shows the relationship between the rotational speed, the frictional resistance torque T, and the time measured in the experiment. It can be seen that the shaft started to rotate when the resistance torque of the bearing reached the maximum and, as the rotation speed of the shaft increased, the resistance torque of the BFJB gradually decreased. When the speed of the shaft reached a certain value, the resistance torque of the bearing tended to be stable; the resistance torque of the bearing did not change with rotating speed change, and the resistance torque of the bearing was very small at this time, with almost no resistance. The speed at which the BFJB resistance torque was in a steady state was defined as the lift-off speed of the BFJB. During the stopping process, when the speed dropped to the point where the bearing resistance torque was no longer stable, this speed of the BFJB was defined as the stop-contact speed. The lift-off speed and stop-contact speed measured by this method were slightly higher than other methods [53].

Figure 16 shows the effect of bearing structural parameters on lift-off speed. Taking No. 1–3 BFJBs as a group, and No. 4–6 BFJBs as the other group, it can be seen that the lift-off speed of the bearing under the same load increased with the increase in the width of the bump foil. As the width of the bump foil increased from 2 mm to 2.5 mm, and then increased from 2.5 mm to 3 mm, the lift-off speed of the BFJB increased by 5320 r/min and 13,417 r/min, respectively. Therefore, it is considered that the influence of the bump width on the support stiffness of the BFJB was nonlinear. Dividing B1 and B4 B2 and B5, B3 and B6 into three groups, it can be seen that the lift-off speed of the No. 1, 2, and 3 test BFJBs with 0.1 mm thickness top foil under each load was higher than that of No. 4, 5, and 6 test BFJBs with 0.12 mm thickness top foil. This result is consistent with the analysis of the influence of the width of the bump foil on the bearing lift-off speed. When the thickness of the top foil changed from 0.1 mm to 0.12 mm, the lift-off speed of the bearing decreased by an average of 5771 r/min. Comparing the influence of the thickness of the top foil on the bearing stiffnesses to the width of the bump foil on the bearing stiffnesses, it was found that the width of the bump foil had a significant effect on the bearing stiffness.

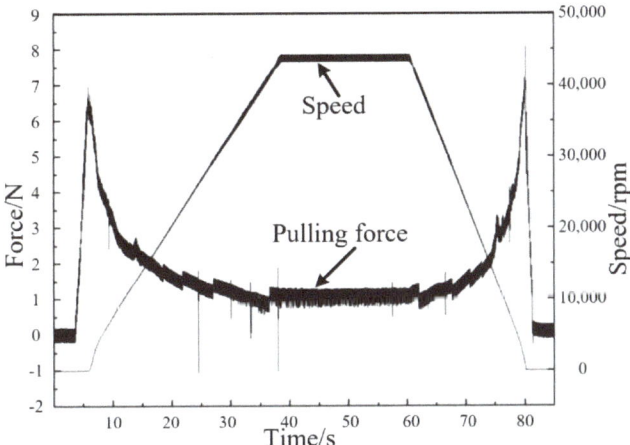

Figure 15. Resistance and speed of the BFJB during lift-off and stop-contact.

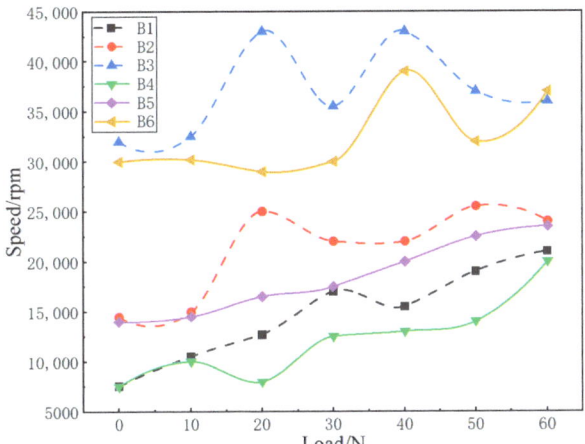

Figure 16. Influence of bearing structure parameters on lift-off speed.

Figure 17 shows the effect of the width of the bump foil on the maximum film temperature. From the test of No. 1–3 BFJBs, it can be seen that with the increase in the width of the bump foil, the maximum temperature of the air film gradually increases under the same load, but the increase was not large. From the EX of No. 4–6 BFJBs, the same rules were observed. Comparing No. 1–3 BFJBs and No. 4–6 BFJBs in Figure 16, it can be seen that with the increase in the thickness of the top foil, the gas film temperature decreased slightly. After analysis, it was concluded that this could be explained by the variation in the stiffness of the elastic support structure to the width of the bump foil or the thickness of the top foil.

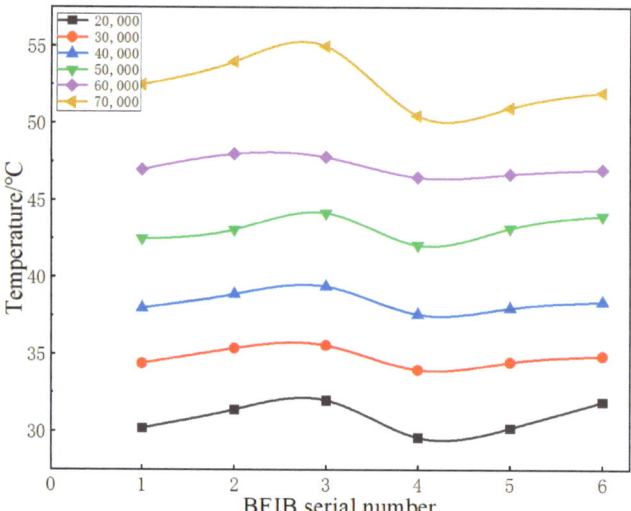

Figure 17. Influence of different structural parameters on gas film temperature.

Figure 18 shows that the No. 3 BFJB was accelerated to stability without additional load at room temperature, and then the bearing was heated to different specified temperatures with glow plugs. Subsequently, the speed of the electric spindle of the test bench gradually reduced until the measurement torque was unstable. Recording the speed at this time as the stop-contact speed at this temperature, it can be seen that the lift-off speed was slightly different for each measurement, which was caused by the different room temperatures and measurement deviation during the measurement. The stop-contact speed was always higher than the lift-off speed, and the stop-contact speed gradually increased with the temperature increases. The results can be explained by the thermal expansion of the bearing with the increase in temperature and the reduction in the bearing capacity with the decrease in the air gap and the degradation of the stiffness of the air film, resulting in the gradual increase in the stop-contact speed. Therefore, superchargers of APEs should be gradually cooled down during the stop to avoid the increased wear caused by excessive contact speed.

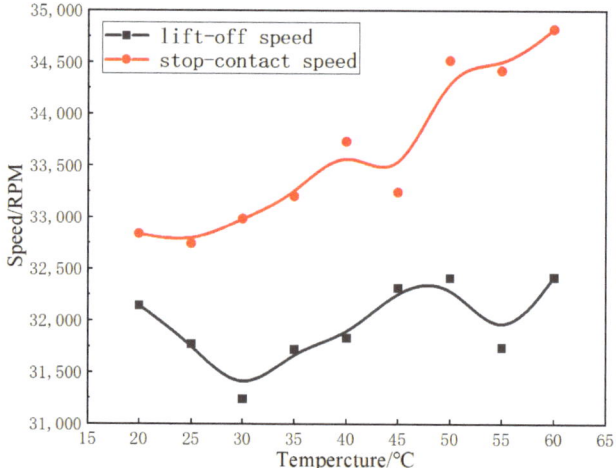

Figure 18. Variation of stop-contact speed with bearing temperature.

4.3. The Process of Stop-Contact under Different Load Conditions

Figure 19 shows the variations in the stop-contact speed and the spend time with the load measurements of the No. 3 bearing. Firstly, the rotor system was accelerated to 40,000 rpm at room temperature and the rotating shaft was in lift-off, at this time, to stop the rotor drive. It can be seen that the downtime was gradually shortened as the bearing load increased. When the speed was lower than the stop-contact speed, the rotor decelerated with the load increase. After analysis, it was found that the frictional resistance increased when the rotor was in sliding contact with the bearing as the load increased. The speed drop acceleration was not large before reaching the stop-contact speed, which can be explained by the film temperature rising with the load increase, resulting in the thermal expansion of the bearing and an increase in the stopping contact speed with increasing load.

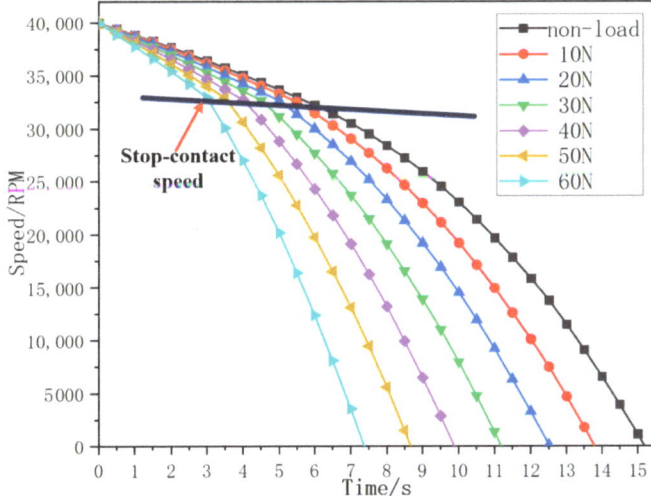

Figure 19. Time taken to stop under different load conditions.

5. Conclusions

In this paper, both experimental and theoretical methods were applied to investigate the thermal characteristics of BFJBs, and a method to evaluate the state of BFJBs by measuring the temperature of the gas film was proposed. Detailed conclusions are as follows:

(1) In our study, with the increased width of the bump foil, the support stiffness of the BFJB decreased, the lift-off speed of the BFJB accelerated and climbed, and the magnitude of the temperature rise increased. These results can be explained by the fact that the stiffness of the elastic support structure changed with the width of the bump foil or the thickness of the top foil, resulting in the lift-off speed change;

(2) In order to avoid damage to the bearing caused by excessive dry friction, which affects the safety of turbochargers, this paper proposed a method of monitoring temperature to predict BFJB performance. The fault of the supercharger can be diagnosed when the temperature suddenly changes, and the temperature of the supercharger can be managed according to temperature characteristics. During the stop process of an APE with a turbocharger, the power should be slowly reduced to cool the supercharger sufficiently. When the rotor system drops to the stop-contact speed, an additional load can be taken to accelerate the rotor system to stop. These measurements can avoid long-term dry friction and prolong bearing life;

(3) During the working of the engine, the temperature of the BFJB increases approximately linearly with the increase in the load and the rotational speed. However, compared

with the increase in the rotational speed, the bearing temperature rise caused by the increase in the load is smaller. Therefore, on the basis of the research into the rotor system conducted in this experiment, extended research into other rotor systems can be carried out, and a turbocharger suitable for higher-power engines can be developed.

Author Contributions: Methodology, Y.Z.; validation, L.S. and F.D.; formal analysis, Z.X.; investigation, S.Z.; resources, S.D.; writing—original draft preparation, K.Z. All authors have read and agreed to the published version of the manuscript.

Funding: This work was funded by the Basic Research Program of the National Nature Science Foundation of China, grant number 51775025 and 51775013, China Key Research and Development Plan (No. 2017YFB0102102, 2018YFB0104100).

Institutional Review Board Statement: Not applicable.

Informed Consent Statement: Not applicable.

Data Availability Statement: The data used to support the findings of this study are included within the article.

Conflicts of Interest: The authors declare no conflict of interest.

References

1. Andres, L.S.; Chirathadam, R.A. Measurements of Drag Torque, Lift-Off Journal Speed, and Temperature in a Metal Mesh Foil Bearing. *J. Eng. Gas Turbines Power* **2010**, *132*, 2201–2210.
2. Dellacorte, C.; Pinkus, O. Tribological Limitations in Gas Turbine Engines: A Workshop to Identify the Challenges and Set Future Directions. In Proceedings of the Tribological Limitations in Gas Turbine Engines, Albany, NY, USA, 1 May 2000.
3. Radil, K.; Zeszotek, M. An Experimental Investigation into the Temperature Profile of a Compliant Foil Air Bearing. *Tribol. Trans.* **2004**, *47*, 470–479. [CrossRef]
4. Agrawal, G.L. Foil Air/Gas Bearing Technology—An Overview. In Proceedings of the International Gas Turbine & Aeroengine Congress & Exhibition, Orlando, FL, USA, 2 June 1997.
5. Weibin, L. *Study of Bump-Type Foil Aero-Dynamic Bearing's Application on an Aero-Engine*; Civil Aviation University of China: Tianjin, China, 2017.
6. Rui, Y. *Study on Rotordynamic Characteristics and Experimental Verification of Rotor System of Oil-Free Turbocharger*; Hunan University: Changsha, China, 2017.
7. Walton, J.F.; Heshmat, H. Application of Foil Bearings to Turbomachinery Including Vertical Operation. *J. Eng. Gas Turbines Power* **2002**, *124*, 1032–1041. [CrossRef]
8. Shuiting, D.; Yue, S. Analysis on development trend and key technology of aircraft heavy fuel piston engine. *J. Aerosp. Power* **2021**, *36*, 1121–1136.
9. Yu, Z.; Tong, X. Digital-twin-driven geometric optimization of centrifugal impeller with free-form blades for five-axis flank milling. *J. Manuf. Syst.* **2020**, *58*, 22–35.
10. Weisong, X.; Xin, L. Review of Technique Application and performance Evaluation for Aerodynamic Elastic Foil Gas Bearing in Aero-engine. *Lubr. Eng.* **2018**, *43*, 136–147.
11. Yu, Z.; Yue, S. Parametric modeling method for integrated design and manufacturing of radial compressor impeller. *Int. J. Adv. Manuf. Technol.* **2020**, *112*, 1–15.
12. Qi, S.M.; Geng, H.P.; Yu, L. New Method for the Calculation of the Characteristics of Aerodynamic Bearings. *J. Mech. Strength* **2006**, *28*, 369–373.
13. San Andrés, L.; Ryu, K. Thermal management and rotordynamic performance of a hot rotor-gas foil bearings system. Part 1: Measurements. *ASME J. Eng. Gas Turbines Power* **2011**, *133*, 253–262. [CrossRef]
14. Rudloff, L. Experimental Analyses of a First Generation Foil Bearing: Start-Up Torque and Dynamic Coefficients. *J. Eng. Gas Turbines Power* **2011**, *133*, 92501. [CrossRef]
15. Radil, K.; Dellacorte, C. Foil Bearing Starting Considerations and Requirements for Rotorcraft Engine Applications. In Proceedings of the 65th Annual Forum Proceedings, Grapevine, TX, USA, 1 August 2009.
16. Rubio, D.; Andrés, L.S. Bump-Type Foil Bearing Structural Stiffness: Experiments and Predictions. *J. Eng. Gas Turbines Power* **2006**, *128*, 653–660. [CrossRef]
17. Xingjun, S.; Gang, X. Engineering Experimental Investigation on the Performance of Large Load Capacity Air Foil Bearing. *Jlubrication Eng.* **2017**, *42*, 125–131.
18. Dickman, J.R. *An Investigation of Gas Foil Thrust Bearing Performance and its Influencing Factors*; Case Western Reserve University: Cleveland, OH, USA, 2010.
19. Blok, H.; Rossum, J. The Foil Bearing—A New Departure in Hydrodynamic Lubrication. *Lubr. Eng.* **1953**, *9*, 316–320.

20. Hu, L.; Zhang, G. Performance analysis of multi-leaf oil lubricated foil bearing. *Proc. Inst. Mech. Eng. Part J J. Eng. Tribol.* **2013**, *227*, 962–979. [CrossRef]
21. Dellacorte, C. A New Foil Air Bearing Test Rig for Use to 700 °C and 70,000 rpm. *Tribol. Trans.* **2008**, *41*, 335–340. [CrossRef]
22. Braun, M.J.; Choy, F.K. Two-Dimensional Dynamic Simulation of a Continuous Foil Bearing. *Tribol. Int.* **1996**, *29*, 61–68. [CrossRef]
23. Feng, K.; Deng, Z.; Zhao, X. Test on static and temperature characteristics of gas foil bearing. *J. Aerosp. Power* **2017**, *32*, 1394–1399.
24. Xie, Y.Q. *Design and Experimental Study of an Oil Free Turbocharger*; Hunan University: Changsha, China, 2015; p. 81.
25. Pattnayak, M.R.; Pandey, R.K. Performance behaviours of a self-acting gas journal bearing with a new bore design. *Tribol. Int.* **2020**, *151*, 711–718. [CrossRef]
26. Ganai, P.; Pandey, R.K. Performance improvement of foil air journal bearing employing micro- pocket and textures on top compliant surface. *Surf. Topogr. Metrol. Prop.* **2021**, *9*, 025045. [CrossRef]
27. Samanta, P.; Murmu, N.C. The evolution of foil bearing technology. *Tribol. Int.* **2019**, *135*, 305–323. [CrossRef]
28. Liu, X.; Li, C.; Du, J.; Nan, G. Thermal Characteristics Study of the Bump Foil Thrust Gas Bearing. *Appl. Sci.* **2021**, *11*, 4311. [CrossRef]
29. Kumar, J.; Khamari, D.S. A review of thermohydrodynamic aspects of gas foil bearings. *Proc. Inst. Mech. Eng. Part J J. Eng. Tribol.* **2021**, 1591214344. [CrossRef]
30. Salehi, M.; Swanson, E. Thermal Features of Compliant Foil Bearings—Theory and Experiments. *J. Tribol.* **2001**, *123*, 566–571. [CrossRef]
31. Peng, Z.; Khonsari, M. A Thermohydrodynamic Analysis of Foil Journal Bearings. *J. Tribol.* **2006**, *128*, 534–541. [CrossRef]
32. Yu, Z.; Lifeng, H. Investigation on transient dynamics of rotor system in air turbine starter based on magnetic reduction gear. *J. Adv. Manuf. Sci. Technol.* **2021**, *1*, 2021009-1–2021009-9.
33. Feng, K.; Kaneko, S. A Thermohydrodynamic Sparse Mesh Model of Bump-Type Foil Bearings. *J. Eng. Gas Turbines Power* **2013**, *135*, 022501. [CrossRef]
34. Liu, L.j. *Theoretical Considerations of Hydrodynamic and Thermal Characteristics of Gas Foil Thrust Bearings*; Hunan University: Changsha, China, 2015.
35. Andrés, L.S.; Kim, T.H. Thermohydrodynamic Analysis of Bump Type Gas Foil Bearings: A Model Anchored to Test Data. *J. Eng. Gas Turbines Power Trans. ASME* **2010**, *132*, 042504. [CrossRef]
36. Andrés, L.S.; Kim, T.H. Thermal Management and Rotordynamic Performance of a Hot Rotor-Gas Foil Bearings System: Part 2—Predictions versus Test Data. In Proceedings of the ASME Turbo Expo 2010: Power for Land, Sea, and Air, American Society of Mechanical Engineers, Glasgow, UK, 14–18 June 2010; pp. 263–271.
37. Tianwei, L.; Yu, G. Numerical and Experimental Studies on Stability of Cryogenic Turbo-Expanderwith Protuberant Foil Gas Bearings. *Cryogenics* **2018**, *96*, 62–74.
38. Xu, C.R. Testing Analysis of the Bump Foil Bearing with MoS$_2$ Coating. *J. Tribol.* **1988**, *1*, 37–41.
39. Ren, S.Q.; Yang, Y.F. Effect of rotational error of precision centrifugal main axel on working radius. *J. Harbin Inst. Technol.* **2000**, *1*, 54–57.
40. Bonello, P.; Hassan, M. An experimental and theoretical analysis of a foil-air bearing rotor system. *J. Sound Vib.* **2018**, *413*, 395–420. [CrossRef]
41. Li, C.L.; Du, J.J. Temperature calculation and static and dynamic characteristics analysis of bump foil gas bearing. *J. Harbin Inst. Technol.* **2017**, *49*, 46–52.
42. Hassan, M.B.; Bonello, P. A new modal-based approach for modelling the bump foil structure in the simultaneous solution of foil-air bearing rotor dynamic problems. *J. Sound Vib.* **2017**, *396*, 255–273. [CrossRef]
43. Zhang, X.B.; Ding, S.T. Investigation on gas lubrication performance of porous gas bearing considering velocity slip boundary condition. *Friction* **2022**, *10*, 891–910. [CrossRef]
44. Feng, K.; Kaneko, S. A Study of Thermohydrodynamic Features of Multi Wound Foil Bearing Using Lobatto Point Quadrature. *ASME Turbo Expo Power Land Sea Air* **2009**, *43154*, 911–922.
45. Sim, K.; Kim, T.H. Thermohydrodynamic analysis of bump-type gas foil bearings using bump thermal contact and inlet flow mixing models. *Tribol. Int.* **2012**, *48*, 137–148. [CrossRef]
46. Kai, F. Thermal Characteristic Analysis of Novel Three-pad Radial Gas Foil Hydrodynamic Bearings. *J. Hunan Univ. (Nat. Sci.)* **2020**, *47*, 35–44.
47. Maraiy, S.Y.; Crosby, W.A. Thermohydrodynamic analysis of airfoil bearing based on bump foil structure. *Alex. Eng. J.* **2016**, *55*, 2473–2483. [CrossRef]
48. Lee, D.; Kim, D. Thermohydrodynamic Analyses of Bump Air Foil Bearings with Detailed Thermal Model of Foil Structures and Rotor. *J. Tribol.* **2010**, *132*, 16–22. [CrossRef]
49. Heshmat, H. Advancements in the Performance of Aerodynamic Foil Journal Bearings: High Speed and Load Capability. *J. Tribol.* **1994**, *116*, 287. [CrossRef]
50. Zhou, Y.; Shao, L.T. Numerical and experimental investigation on dynamic performance of bump foil journal bearing based on journal orbit. *Chin. J. Aeronaut.* **2021**, *02*, 586–600. [CrossRef]
51. Zhou, T.; Hu, M. Vibration features of rotor unbalance and rub-impact compound fault. *J. Adv. Manuf. Sci. Technol.* **2022**, *33*, 611–619. [CrossRef]

52. Guo, Z.Y. Effects of static and imbalance loads on nonlinear response of rigid rotor supported on gas foil bearings. *Mech. Syst. Signal Process.* **2018**, *133*, 106271. [CrossRef]
53. Liu, J.; Du, F.R. Experiment Study on Lift-Off Speed of Bump Foil Journal Bearings. *Bearing* **2013**, *8*, 27–29.

Article

MoHydroLib: An HMU Library for Gas Turbine Control System with Modelica

Yifu Long [1], Shubo Yang [1,*], Xi Wang [1], Zhen Jiang [1], Jiashuai Liu [1], Wenshuai Zhao [1], Meiyin Zhu [2], Huairong Chen [1], Keqiang Miao [1] and Yi Zhang [3]

1. School of Energy and Power Engineering, Beihang University, Beijing 100191, China; lyfleo@buaa.edu.cn (Y.L.); xwang@buaa.edu.cn (X.W.); zhenjiang@buaa.edu.cn (Z.J.); by1904028@buaa.edu.cn (J.L.); 18101379594@buaa.edu.cn (W.Z.); chenhuairong211@buaa.edu.cn (H.C.); kqmiao@buaa.edu.cn (K.M.)
2. Beihang Hangzhou Innovation Institute Yuhang, Hangzhou 310023, China; mecalzmy@buaa.edu.cn
3. Systems Engineering Research Institute, Beijing 100094, China; cuizhe@cssc-cmc.com
* Correspondence: yangshubo@buaa.edu.cn

Abstract: Modelica is an open-source, object-oriented equation-based modeling language. It is suitable for describing sophisticated dynamic systems (symmetry/asymmetry) as it uses mathematical acausal equations to express physical characteristics. The hydraulic mechanical units (HMU) of gas turbine engine control systems couple the contents of mechanical, hydraulic, symmetry, and other multidisciplinary fields. This paper focuses on the Modelica description method of those HMU models. The content of this work is threefold: firstly, the division form of basic elements in HMU is defined, and the method for describing these element models with Modelica is proposed; secondly, the organization of the element models is defined by using the inheritance characteristics of Modelica, and a lightweight (small code scale) component model is designed; and finally, the causal/acausal connections are designed according to bond graph theory, and the elements and components are integrated into a prototype modeling library. In this paper, the modeling library is verified by comparing simulation results of five typical HMU subsystem models with commercial modeling and simulation software.

Keywords: Modelica; hydraulic mechanical unit; modeling and simulation; HMU modeling library

1. Introduction

Hydraulic mechanical units (HMU) are widely used in aviation. For aeroengine control systems, HMUs are an important control actuator that contain lots of symmetry/asymmetry geometry, such as actuating pistons and special-shaped orifices. Its characteristics directly affect the design of control systems [1]. Therefore, it is necessary to carry out modeling and simulation research on HMUs.

The modeling and simulation of HMUs usually proceed in two ways: turning to mature commercial modeling and simulation software or studying and developing from the very beginning. Most of the underlying code of commercial software is invisible due to patents, copyright, and other problems, so it is difficult to understand the full process of modeling. In addition, commercial software is wide-ranging, and its organization is huge and complex, so its learning threshold is high. However, developing new HMU modeling tools expends a lot of time on learning, design, and testing. Therefore, the increasing demand for lightweight HMU modeling tool, which is easy to understand and use, is obvious.

In the field of modeling and simulation tools of HMUs, typical commercial software includes the simcenter AMESim of Siemens [2], Bathfp for hydraulic systems simulation developed by Bath University, UK [3], and the simulation software EASY5 for control systems developed by the MSC company (California, USA) [4]. In addition, there are some self-developed software tools. Min Xu et al. designed a set of HMU component model libraries in Matlab/Simulink and verified it on a certain aeroengine [5]. Marquis Favre W. et al. developed a planar mechanical modeling library based on AMESim and verified it on a planar seven-body mechanism test example [6,7]. Yuejun Xiao et al. developed an open-ended hydraulic module library based on the objected-oriented Visual Basic language [8]. W. Borutzky et al. introduced a library of hydraulic components, which was still under development [9]. There are also other authors devoting themselves to modeling and simulation of different fields [10–13]. However, both mature commercial software and self-developed modeling tools have some problems, such as being difficult to comprehend, requiring complex operations, having complicated functions, and being on a huge scale.

To solve the problems above, this paper proposed an HMU modeling library (MoHydroLib (Beijing, China)) based on Modelica. Modelica is an open-source dynamic modeling language suitable for large-scale, complex, and multidisciplinary physical systems. It describes the object through physical equations, and the model based on the equation descriptions is easy to understand in principle. Modelica uses equations to describe the working characteristics of different types of engineering components, and these components can be easily combined into subsystems, systems, and even architectural models [14]. This organization avoids code redundancy and can be used to design lightweight models. Modelica supports the solution of hybrid differential-algebraic equations, which has advantages for the solution of complex dynamic systems [15]. In addition, Modelica is an open-source language and supports the functional mock-up interface (FMI) protocol. The modeling library based on this language has a wider range of uses and can be applied to multi-platform and multi-domain co-simulation [16,17].

The main contributions of this paper are summarized as follows:

1. To lower the threshold for hydraulic system modeling, we developed a simple but powerful library with Modelica for gas turbine control systems, which is easy to access, easy to use, and easy to reshape;
2. We achieved strong compatibility by designing casual/acausal connections, where multi-platform and multi-domain co-simulation are supported;
3. We reduced code lines significantly through devising inheritance relationships and achieved this lightweight library.

The structure of the manuscript is as follows. Section 2 of this paper gives the basic design idea and connection organization process of the modeling library. The application examples of five typical HMU subsystems modeled by MoHydroLib are presented in Section 3, and their implementation functions and the verification of simulation results are discussed. Section 4 gives the summary of this paper.

2. Design of the Library

This section describes the design process of elements, components, and connections in MoHyroLib. Firstly, this section realizes the element design based on the physical equation description of HMU. Secondly, the component design is completed by utilizing the inheritance characteristics of Modelica, which can effectively reduce code redundancy. Finally, the acausal interface is defined by selecting acausal variable pairs in the HMU through applying bond graph theory, and the causal interface corresponding to the FMI interface protocol is used for exporting the packaging module or importing a C/C++ packaging module from other software to realize the co-simulation of multiple platforms.

2.1. Theory and Element Description

The core of HMUs is the control chamber and moving body [18]. The working theory of the control chamber is to control the pressure by changing the flow passing the inlet and outlet, and then determine the opening with the change of pressure. The pressure that acts on the moving body can alter its acceleration, speed, and displacement. The HMU can be divided into corresponding element types according to different functions, mainly including the orifice element, piston element, boundary element, and system element.

2.1.1. Orifice Element

The orifice element is one of the basic elements in HMUs. It can be used for throttling. It is a kind of hydraulic resistor, and its calculation principle is based on the Bernoulli equation [18]. The flow through the orifice is calculated from the pressure difference before and after it. The description equation is shown in Formula (1).

$$\begin{cases} \Delta P = P_1 - P_2 \\ Q_1 = sign(\Delta P) \cdot k_v \cdot \sqrt{2|\Delta P|/\rho} \cdot A \\ Q_1 = -Q_2 \end{cases} \quad (1)$$

where P_1 and P_2 are the pressures before and after the orifice, respectively, k_v represents the flow coefficient of the orifice, A is the flow area of the orifice, Q represents the volume flow rate through orifice and subscripts 1 and 2 indicate the inlet and outlet of this orifice.

Based on Formula (1), the displacement x_p of the moving body and the opening *underlap* of the orifice are introduced to obtain the orifice element with shielding. This element can be further subdivided into a circular orifice, a circular orifice with shielding and a special-shaped orifice with shielding.

2.1.2. Piston Element

The piston element can also be called a hydraulic cylinder; it is used for converting from hydraulic to mechanic energy. The main function of this element is to calculate the variables related to motion in the HMU. It calculates the acceleration, velocity, and displacement of the mass block according to the hydraulic pressure, spring force and friction. The equations it utilizes refer to Newton's second law. The positive and negative of the external force calculation are related to the action direction of the corresponding component force. In this element, the force consistent with the positive direction of the displacement of the moving body is defined as positive, and vice versa. The diagram view of this element in the library is shown in Figure 1, and the description equation of it can be seen in Formula (2).

$$m \cdot \ddot{x} = P_l \cdot A_l - P_r \cdot A_r - (k \cdot x + F_0) - c_f \cdot \dot{x} \quad (2)$$

where x, $\dot{x}$ and $\ddot{x}$ represent the displacement, velocity and acceleration of the piston, respectively, m is the mass of piston, P and A are the pressure and pressed area of the control chamber, the subscripts l and r represent the left and right control chamber, k is the stiffness of the spring, F_0 is the pre-tightening force of the spring, and c_f represents the coefficient of viscous friction.

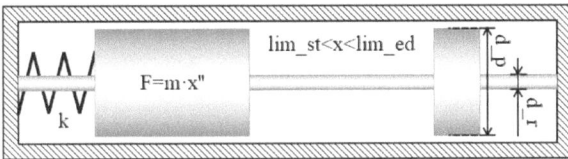

Figure 1. Diagram view of piston element.

2.1.3. Boundary Element

Boundary elements mainly refer to chamber elements and various pressure sources. The operating principle of chamber elements is depicted in Figure 2, and the equation they utilize is described by the pressure differential equation shown in Formula (3). The practical model designed in this paper considers the influence of volume change caused by the moving body as the volume flow rate $Q_v = A \times \dot{x}$.

$$\frac{dP_c}{dt} = \frac{B}{V_0}\sum Q \qquad (3)$$

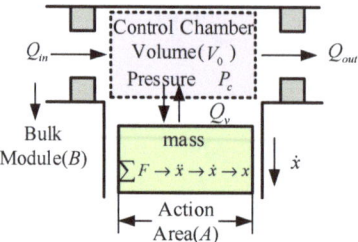

Figure 2. Operating principle of chamber.

As can be seen in Figure 2, P_c is the pressure of chamber, B is the bulk module of the working medium, V_0 is the volume of chamber, A is the pressed area, and Q is the volume flow rate. The boundary elements in the library also include various pressure sources. These pressure sources defined in this paper include the P source and pressure and the P-m source. The output pressure of the P source can be specified, and the output pressure and mass flow of the P-m source can be specified simultaneously. The function of the pressure source is to provide pressure to the HMU system, which is generally used as the input of other hydraulic components.

2.1.4. System Element

This element is used to define the type of medium working in the HMU, and it is described by some inherent characteristic parameters of the medium, such as the dynamic viscosity nu, bulk modulus B, and density ρ.

2.2. Component and Inheritance Relationship

In Section 2.1, several different basic elements are designed according to the working theory of HMU. In this section, the organization relationship between elements and the design of components in MoHydroLib is presented. The general organization form of the elements and components is shown in Figure 3.

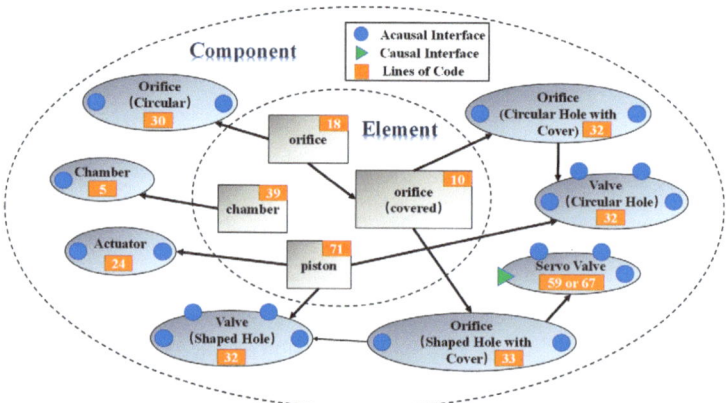

Figure 3. Relationship between element and component.

As can be seen in Figure 3, the library inherits or extends the element model by utilizing the inheritance characteristics of Modelica and adds interactive interfaces to form the component model. The total code of the model library applying this design method is only 452 lines, which is lightweight.

2.2.1. Orifice Component

On inheriting the orifice element, the orifice component introduces five parameters: the maximum flow coefficient *ckvmx*, the orifice diameter *d*, the number of orificse *n*, the hydrodynamic viscosity *nu* and the critical flow number *lamc*. These parameters are associated with the flow coefficient k_v and flow area A in the orifice element by an equation, as shown in Formula (4), and the interface is introduced as the input and output port to form a circular orifice component, as shown in Figure 4

$$\begin{cases} k_v = \frac{\tanh\sqrt{32|\Delta P|\cdot\rho\cdot d}}{nu\cdot lamc} \cdot ckvmx \\ A = \frac{\pi}{4}d^2 \times n \end{cases} \quad (4)$$

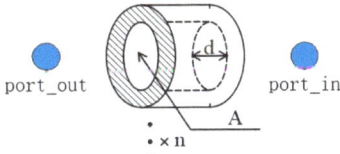

Figure 4. Diagram view of circle orifice.

The orifice component with a cover can be divided into a circular orifice with a cover and a special-shaped orifice with a cover, as shown in Figure 5. The core of these two kinds of orifices inherits the orifice element with a cover.

(**a**)   (**b**)

Figure 5. Diagram view of an orifice with a cover. (**a**) Circular orifice with cover, (**b**) special-shaped orifice with cover.

The major difference between the two components lies in the different ways of calculating the flow area according to the opening. The calculation of the circular hole is determined by the formula, while the special-shaped orifice is obtained by an a priori interpolation table from the opening to the flow area.

2.2.2. Valve Component

This component couples hydraulics and mechanics by inheriting the orifice element and the piston element. The variation in pressure in the control chamber (port_left and port_right) is converted to the variation in flow in the orifice. Valve components are divided into round hole valves and special-shaped hole valves according to their different orifice types, and the difference between these valves is only the inheritance type of the orifice component. The calculation of motion in valves is inherited from the piston element. The diagram view of a special-shaped hole valve is shown in Figure 6.

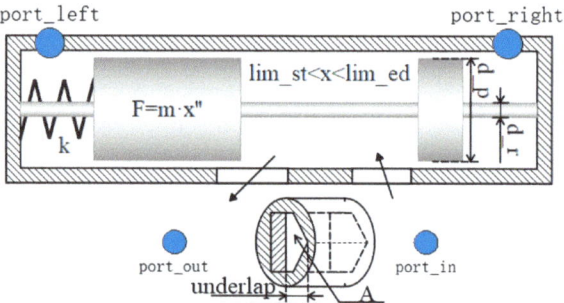

Figure 6. Diagram view of a special-shaped hole valve component.

2.2.3. Actuator Component

This component is used for implementing control commands by converting hydraulic flow to the actuating force. The actuator component inherits the codes of the piston element and introduces the external load force F_{psh} to participate in the calculation of displacement, velocity, and acceleration of the moving body. The diagram view of it is shown in Figure 7.

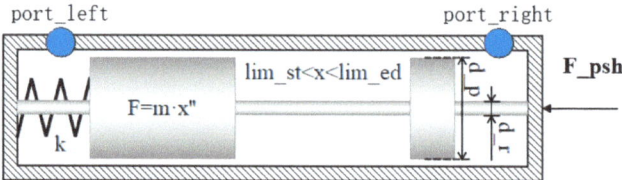

Figure 7. Diagram view of actuator component.

2.2.4. Servo Component

The servo component in MoHydroLib mainly refers to an electro-hydraulic servo valve, which regulates the opening according to the input current. This component realizes conversion from an electrical signal to hydraulic flow. Its motion calculation is different from the piston element, but the flow calculation can still inherit the special-shaped orifice component with a covering.

There are two position three-way valves (HSV23) and three position four-way valves (HSV34) in MoHydroLib. They have similar structures and calculation methods. The diagram views of HSV23 and HSV34 are shown in Figures 8 and 9, respectively.

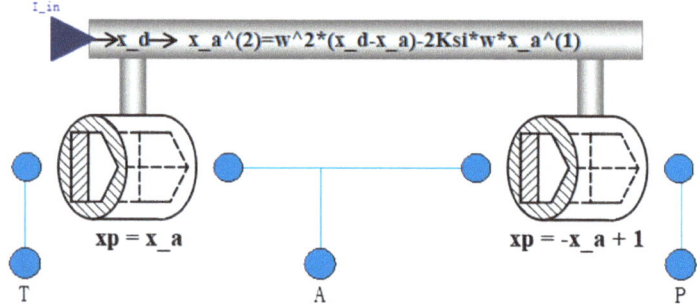

Figure 8. Diagram view of HSV23 component.

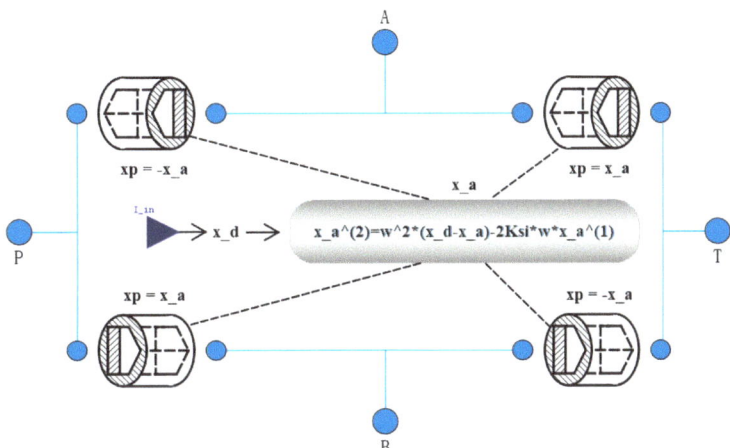

Figure 9. Diagram view of HSV34 component.

2.3. Connection Relationship

In order to facilitate the transmission of data and the exchange between designed elements and components, it is necessary to define the input and output interfaces for them. Modelica is an equation-oriented language, and MoHydroLib presented in this paper provides an object-oriented modeling method. This paper carries out interface design based on bond graph theory and combines these two different modeling concepts [19].

Bond graph theory describes the system through power flow and defines three different analogies: signals, components, and connections [20]. This section focuses on describing how to express the connection relationship in an integral system. In order to explain this expression, the concept of causality needs to be introduced: derivative operators need to know the future behavior, while this is numerically impossible [20]. Therefore, causality requires that the integration of variables is calculated by discrete addition, and the flow or unit power is not calculated by the differential form of variable integration; that is, the signal transmission in the system is directional.

According to whether the direction of signal transmission needs to be predetermined, the description of the connection relationship is divided into acausal connection and causal connection. In the MoHydroLib presented in this paper, acausal connection refers to the connection relationship between components and elements, and the power flow on these connections does not need to specify the direction in advance, while causal connection refers to the connection relationship between components or elements and the external environment, other objects, etc. The power flow on these connections needs to specify the direction, as shown in Figure 10.

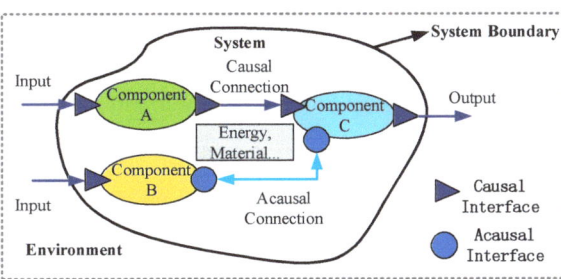

Figure 10. Connection relationship illustration.

2.3.1. Acausal Connections

The power flow on the acausal connection does not need to specify the direction. This paper describes the acausal connection by defining the form of the acausal interface. Bond graph theory usually employs a pair of variables to indicate power, such as unit mass work $e(t)$ and mass flow Q_m. The product of these two variables is power P_h. These variable pairs used in this paper are pressure P and volume flow rate Q. The product of the two variables is power, as shown in Formula (5).

$$P_h(t) = P(t) \cdot Q(t) \tag{5}$$

As these two variables can be used to describe power, they can serve as an intermediary for the interaction of information between elements and components. This intermediary is defined as an acausal interface. The connection between two acausal interfaces is an acausal connection. If multiple connections are connected to the same acausal interface, the acausal interface is called a junction. Based on bond graph theory [21], the junctions in the model are considered to have the following characteristics:

- The pressure on each bond/connection connected to the junction is equal:

$$P_1 = P_2 = \cdots = P_n; \tag{6}$$

- The algebraic sum of volume flow of all bonds/connections is zero:

$$Q_1 + Q_2 + \cdots + Q_n = 0. \tag{7}$$

For this kind of junction with equal pressure and a zero algebraic sum of the volume flow rate, the pressure needs to be given by other components other than the junction, that is, the boundary elements (chamber and pressure source) in the MoHydroLib. Therefore, the pressure and volume flow rate are used as an acausal interface for the interaction between elements and components, and the data transmission and interaction between components and the external environment are completed through the causal interface in the built-in library of Modelica. An example of causal/acausal connection is shown in Figure 11.

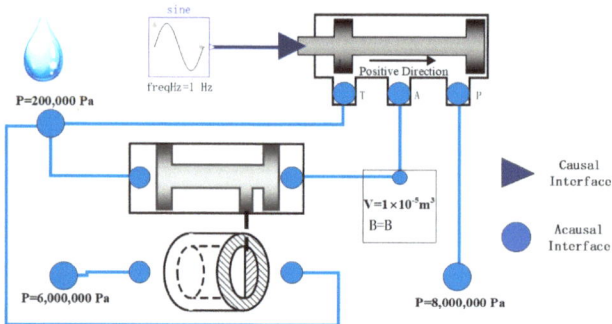

Figure 11. Causal/acausal connection example.

As shown in Figure 11, this example is the electro-hydraulic servo metering subsystem. The circular port includes pressure sources, which are acausal interfaces. The connections of two acausal interfaces are acausal connections. The formation of an acausal connection in the figure includes one-to-one and one-to-many. The data transmission between HMU components can be realized through these acausal interfaces. The triangular port in the figure is the causal interface, which is used to receive the external servo control current.

2.3.2. Causal Connections

Causal connections refers to those connections with clear directionality in MoHydroLib. In this paper, it is only used to represent the input and output interfaces between elements or components in the library and the external environment or other objects. Causal connection directly uses the input and output interface in the Modelica library. At present, the causal interface in the connection of elements and components in MoHydroLib only exists in the external command input of the specified pressure source in the boundary component and the control current input of the servo component. Causal connections are applicable to the type of connection (energy, information flow) that must specify the data direction. There is a clear sequence of calculations or solutions between any two elements or components connected.

In addition, causal connections also include connections with other software. The models constructed by elements or components in MoHydroLib can use the causal interface as the interaction interface with the model in other platforms. The model can also generate an FMU package file based on the FMI interface protocol and call it in other software. However, the FMU file generated in this way can only use the first-order Euler method by default, which is difficult to use to solve some complex dynamic systems. Therefore, packaging the models of other software and calling them under the platform supporting Modelica is an alternative way. For example, after generating the source code of a model, it can be packaged into a DLL file with a C/C++ compiler, and then the DLL file can be called through OpenModelica. In this way, the DAE solver provided by OpenModelica can be utilized.

3. Application Example

Aiming at the HMUs in the control systems of gas turbine engines, this chapter gives several model examples of typical HMU subsystems, including constant pressure valve subsystems, safety valve subsystems, differential pressure valve subsystems, servo metering subsystems and single-port chamber high-frequency dynamic simulation subsystems. The simulation results of these typical subsystem models are compared with the results of a model constructed by commercial HMU modeling software. Since these subsystems can hardly be tested independently, their functions are not verified experimentally. There are some experimental verifications carried out by others comparing simulation results (using the same commercial software) with the experimental data [22–24]. Therefore, the simulation of this manuscript is rational and reliable to some extent. Despite the simplicity (452 lines of code in total), the strong ability of the library is verified in this section, compared with the well-known commercial software.

3.1. Constant Pressure Valve Subsystem

The constant pressure valve subsystem is a hydraulic component used to maintain the constant control pressure. The structure of the corresponding subsystem is shown in Figure 12.

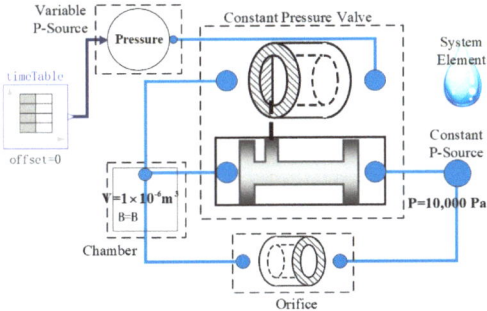

Figure 12. Constant pressure valve system.

The inlet pressure of the constant pressure valve is given by a variable pressure source element with a change trend as shown in Figure 13.

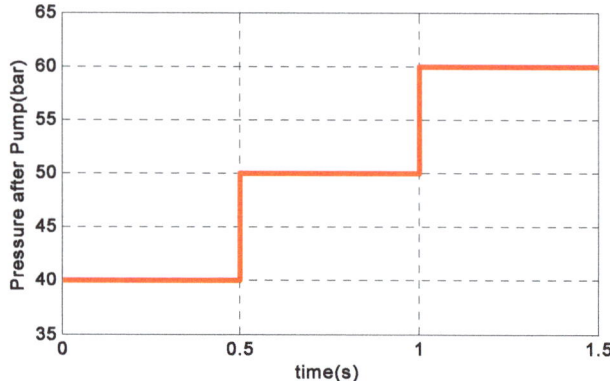

Figure 13. Input port pressure of a constant valve (pressure after pump).

The model calculates the outlet control pressure of the constant pressure valve and the displacement of the valve core under the above inlet pressure input conditions.

The simulation results are shown in Figure 14. The constant pressure valve subsystem established by MoHyroLib can realize the basic function, and its simulation results are basically consistent with those of commercial software.

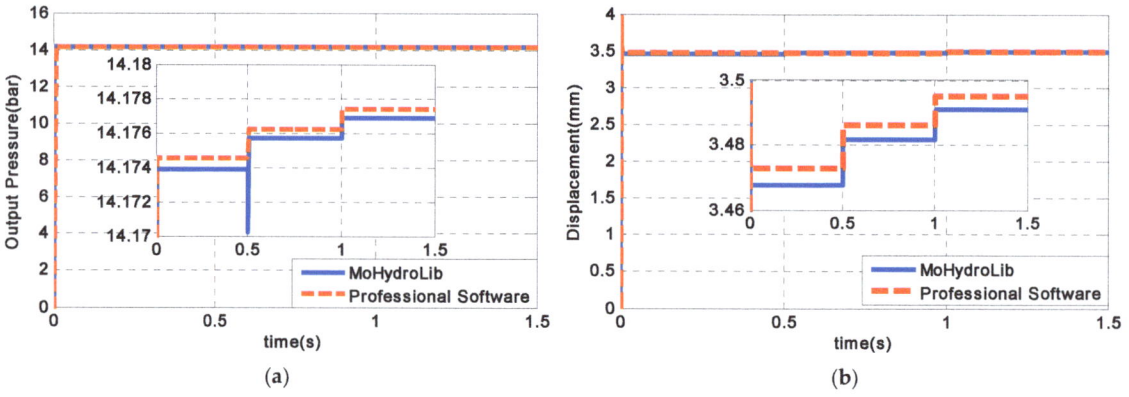

Figure 14. The simulation result of ouput pressure and displacement in different modeling tools. (**a**) Comparison of output port pressure in a constant valve calculated by two modeling tools; (**b**) comparison of displacement in a constant valve calculated by two modeling tools.

3.2. Safety Valve Subsystem

The safety valve subsystem is a hydraulic component to ensure that the pressure after the pump does not exceed the set safety value when the speed of the fuel pump increases continuously. The structure of the safety valve subsystem is shown in Figure 15.

The trend of the input speed of the pump is shown in Figure 16. The calculated pressure after the pump and the displacement of the safety valve under this input speed is shown in Figure 17.

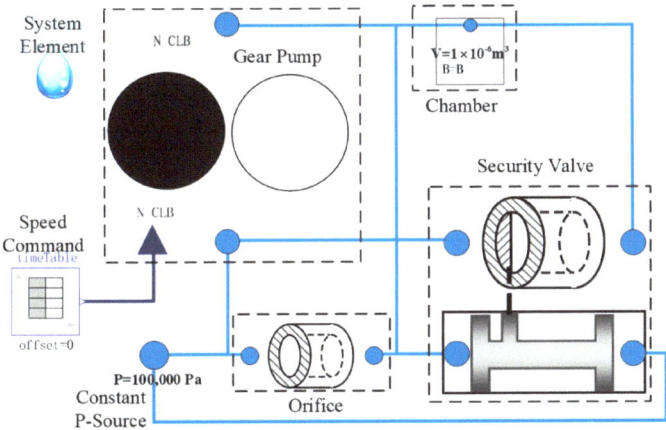

Figure 15. Safety valve system.

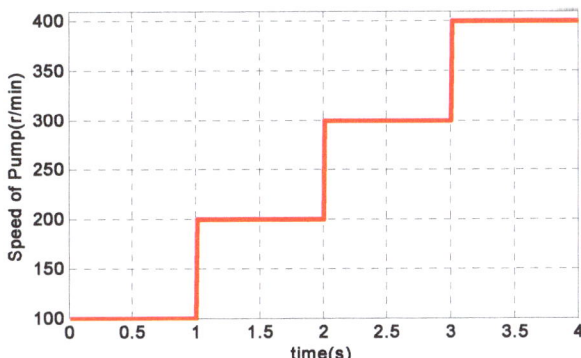

Figure 16. Input speed of the gear pump.

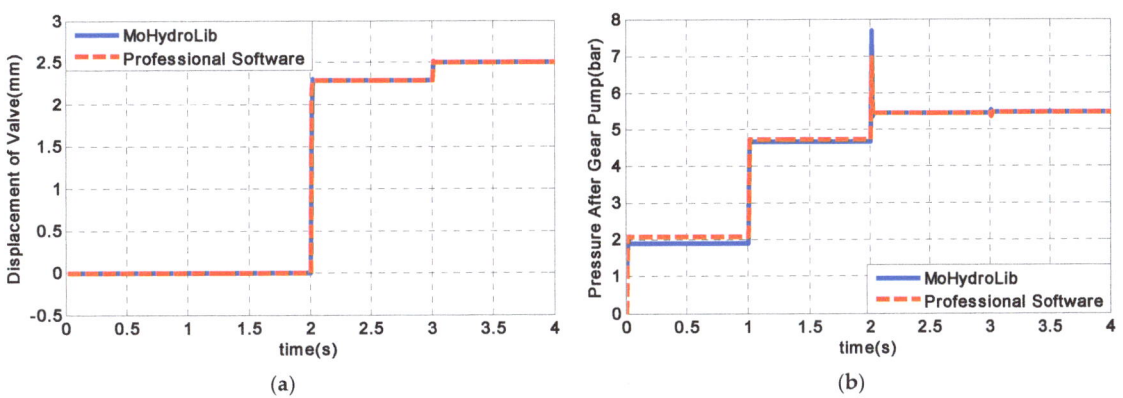

Figure 17. The results of the simulation of the displacement and pressure after pump by two modeling tools. (**a**) Comparison of displacement in he safety valve calculated by two modeling tools. (**b**) Comparison of pressure after pump calculated by two modeling tools.

Comparing the simulation results in Figure 17, the safety valve subsystem established by MoHydroLib can achieve the desired function, and the simulation results are basically consistent with the results of the commercial software.

3.3. Differential Pressure Valve Subsystem

The differential pressure valve subsystem is a hydraulic component that maintains the differential pressure at the inlet and outlet of the metering valve within a certain range. The model structure of the subsystem is shown in Figure 18. The change trend of the inlet pressure in the differential pressure valve subsystem is shown in Figure 19a. The model can calculate the outlet pressure (outlet pressure of the metering valve), pressure difference and spool displacement of the valve.

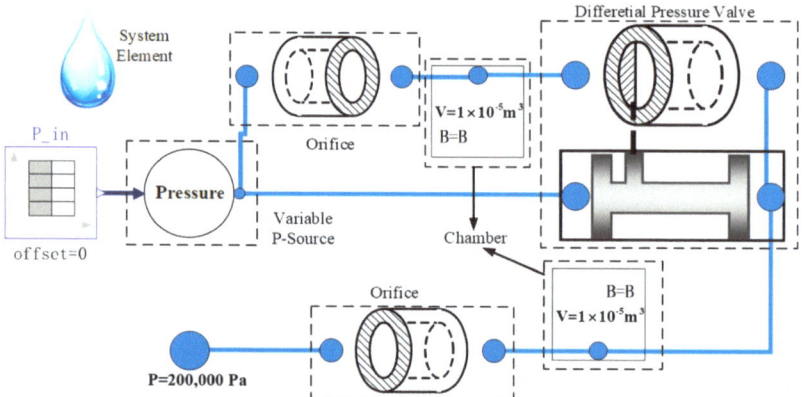

Figure 18. Differential pressure valve system.

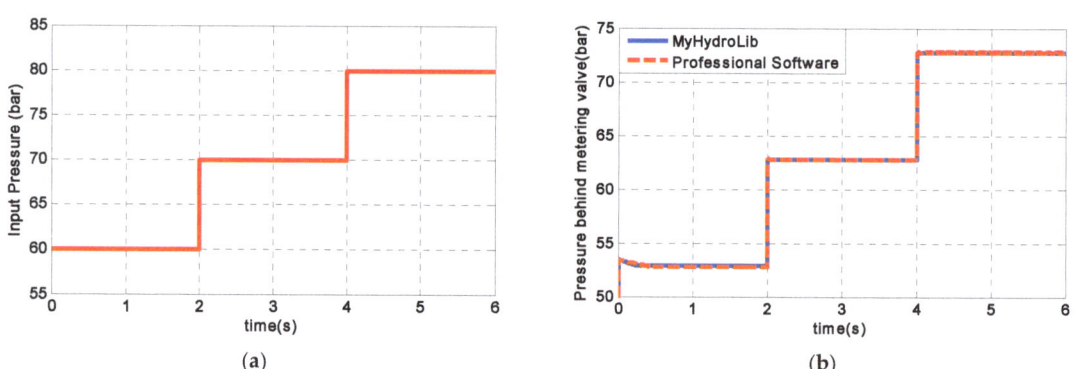

Figure 19. Input and outlet fuel pressure of the differential pressure valve. (**a**) Input fuel pressure of differential pressure valve given by P-source. (**b**) Comparison of outlet fuel pressure of the differential pressure valve calculated by two modeling tools.

Comparing the simulation results of these two models (Figures 19b and 20), the differential pressure valve subsystem established by MoHydroLib could realize its basic functions, and its simulation results are basically consistent with those of commercial software.

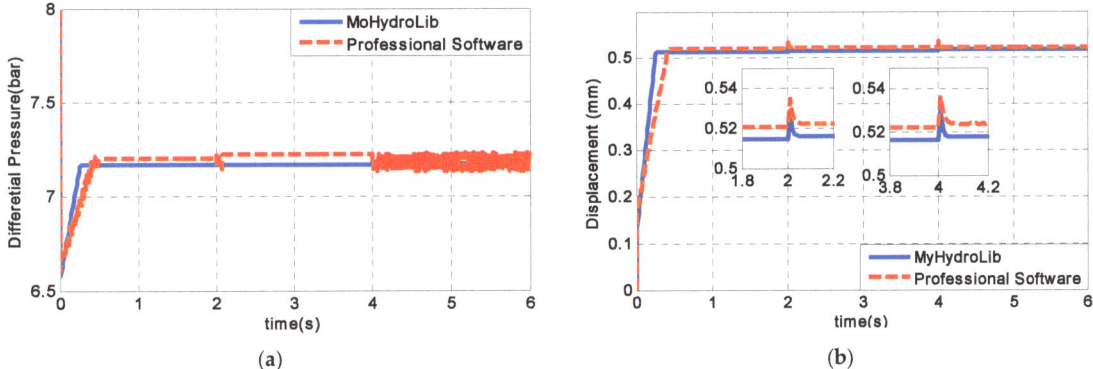

Figure 20. The results of the simulation of the differential pressure and displacement by two modeling tools. (**a**) Comparison of the differential pressure of the metering valve calculated by two modeling tools. (**b**) Comparison of the displacement in the differential pressure valve calculated by two modeling tools.

3.4. Servo Metering Subsystem

The servo metering subsystem is the basic hydraulic component of fuel supply system. It uses an electro-hydraulic servo valve to control the flow area of metering valve. Under the closed-loop feedback control, the command tracking of the metering valve displacement is realized. The corresponding subsystem structure is shown in Figure 21.

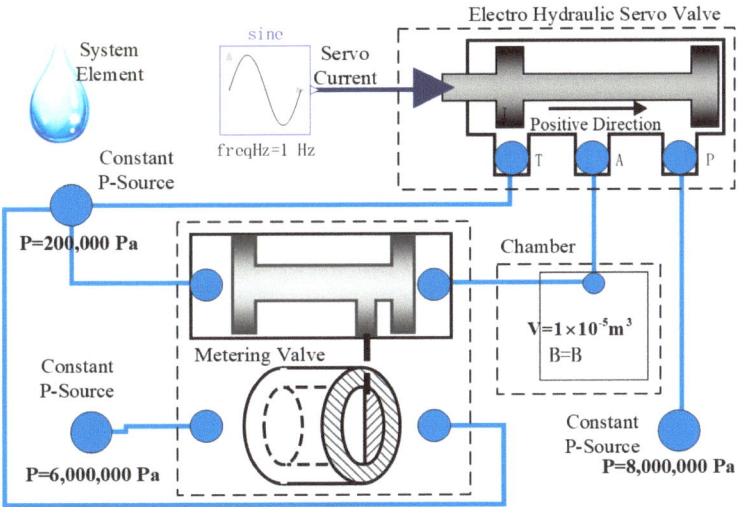

Figure 21. Servo valve metering system.

The change trend in current of servo valve is shown in Figure 22. Under this input condition, the displacement of the metering valve and the outlet metering flow of the subsystem can be calculated, and it is shown in Figure 23. Based on the model structure shown in Figure 21, the displacement closed-loop control of the metering valve is added, and the structure is shown in Figure 24.

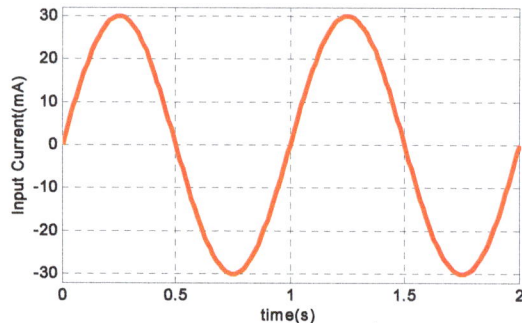

Figure 22. Input current of the electro hydraulic servo valve in the servo metering subsystem.

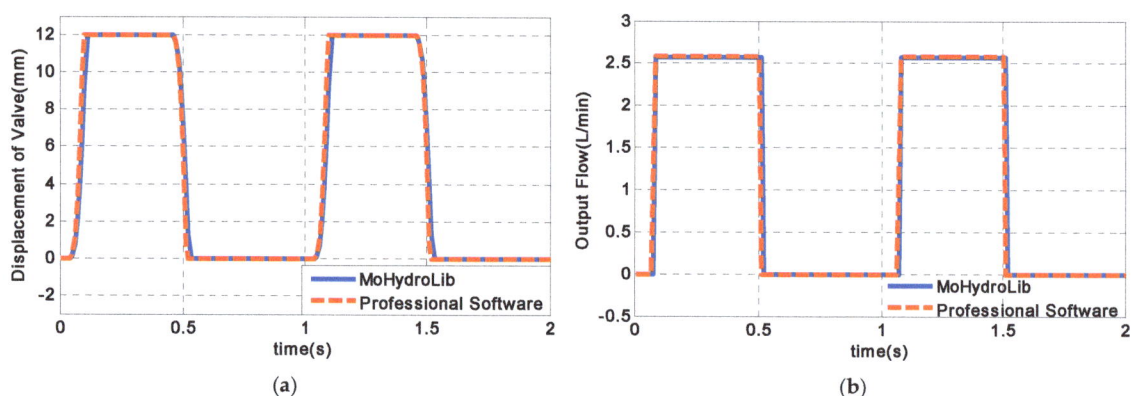

Figure 23. The results of the simulation of the displacement and ouput flow by two modeling tools. (**a**) Comparison of the displacement in the metering valve calculated by two modeling tools. (**b**) Comparison of the flow in the metering valve calculated by two modeling tools.

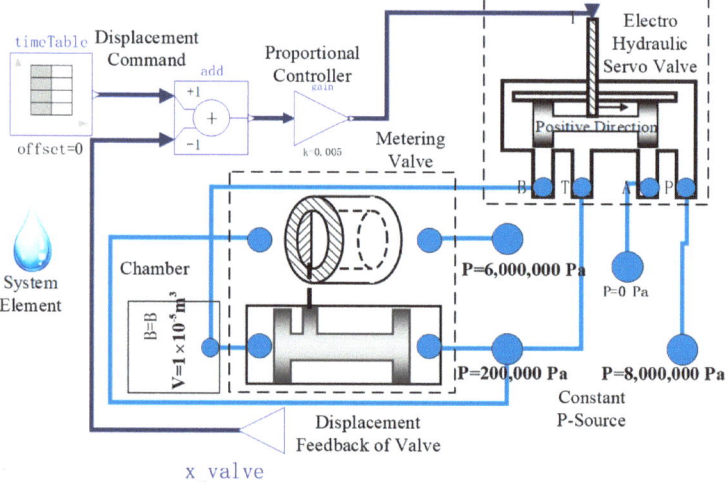

Figure 24. Servo metering subsystem with a close-loop control.

Given the displacement command of the closed-loop servo metering subsystem, as shown in Figure 25, the closed-loop displacement response and metering outlet flow of the subsystem can be obtained through model simulation, as can be seen in Figure 26.

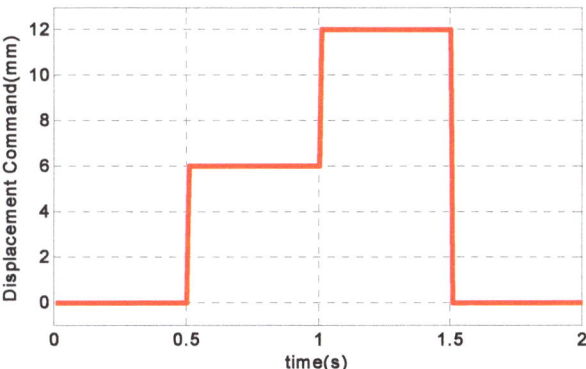

Figure 25. Displacement command signal of the metering valve.

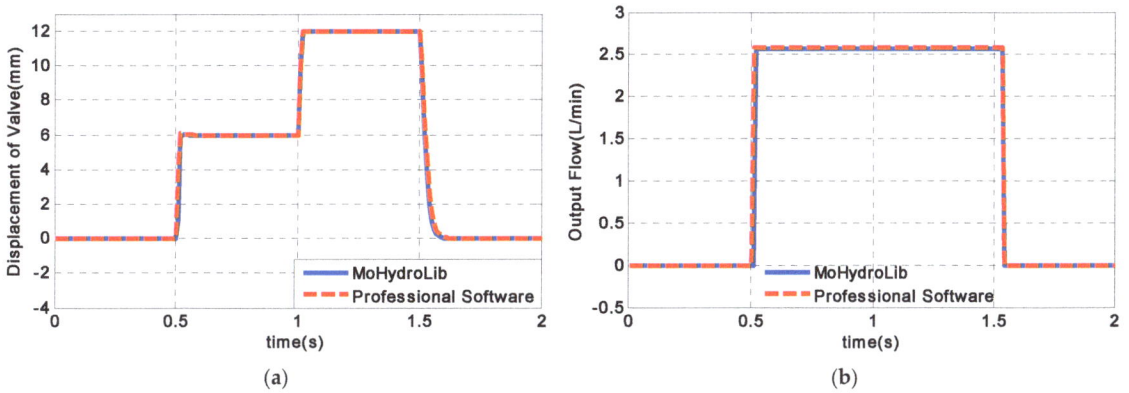

Figure 26. The results of the simulation of closed-loop displacement and ouput flow by two modeling tools. (**a**) Comparison of the displacement with a controller calculated by two modeling tools. (**b**) Comparison of the flow with a controller calculated by two modeling tools.

Comparing the simulation results of Figures 23 and 26, the servo metering subsystem established by MoHydroLib can realize the basic functions, and the simulation calculation results are basically consistent with the commercial software.

3.5. High Frequency Dynamic Solution Simulation

This section is intended to verify the calculation results of a subsystem shown in Figure 27. The fixed-opening valve component in this subsystem is the substitute for an orifice. According to the assumption of an incompressible working medium, since the volume of the chamber cannot be changed, when the inlet pressure of the orifice changes at high frequency, the inlet flow of the chamber should theoretically be zero. The simulation verifies whether the model can correctly calculate the inlet flow of the chamber under this condition. In this example, the calculation accuracy of different software is uniformly set with 1×10^{-4}. The change of the inlet pressure is shown in Figure 28.

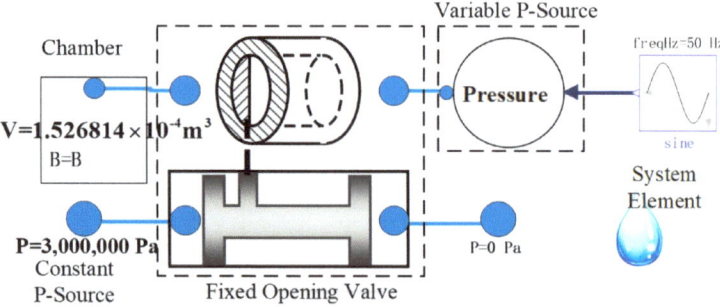

Figure 27. High-frequency dynamic solution subsystem.

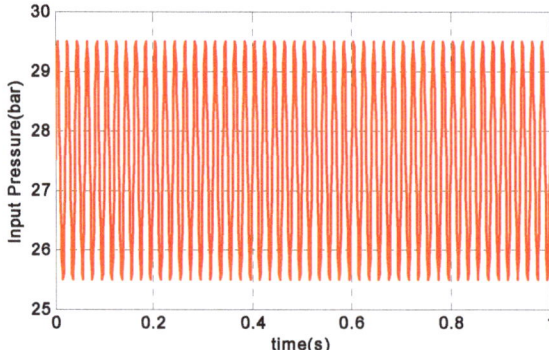

Figure 28. Input pressure of the orifice in the subsystem.

Under the input conditions, the calculation results are shown in Figure 29. The simulation of this example compares the calculation results of OpenModelica, a commercial HMU modeling and simulation software and a commercial mathematical modeling and simulation software. In addition, it compares the calculation results of different solvers (ODE and DAE) in a commercial mathematical modeling and simulation software.

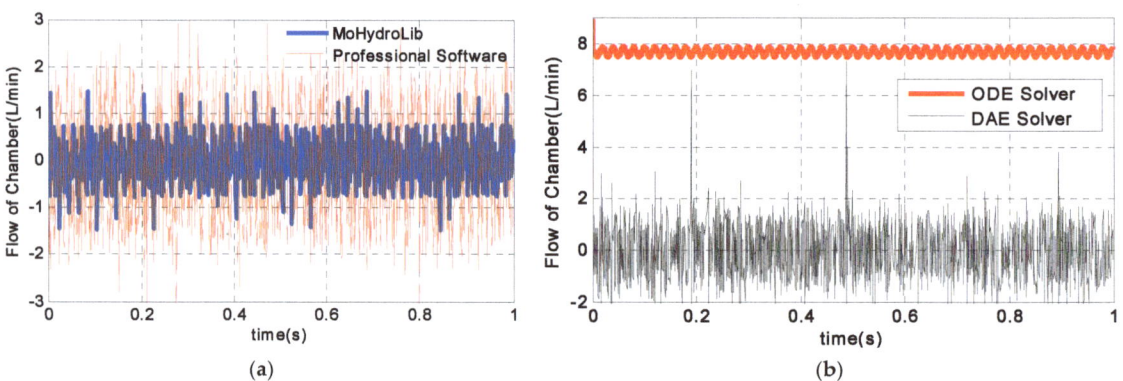

Figure 29. The simulation result of calculated flow by different modeling tools. (**a**) Comparison of the flow of the chamber calculated by two modeling tools. (**b**) Comparison of the flow of the chamber calculated by mathematical simulation software using two different solvers.

Figure 29 shows that the model flow calculation results based on MoHydroLib are satisfied under the given 1×10^{-4} calculation accuracy, which is close to the calculation results of a commercial HMU modeling and simulation software. When the commercial mathematical modeling and simulation software chooses the ode solver, due to the defects of its computer theory, it leads to incorrect calculation results, while using the DAE solver, the results are basically close to the other two models.

4. Discussion

This manuscript designs a lightweight HMU modeling library based on the Modelica language. The library is mainly applied for the modeling and simulation of HMUs in gas turbine engine control systems. This library completes the description of HMU elements, components and subsystems based on Modelica and devises acausal and causal interfaces to realize the interaction of information between various parts and also make multi-platform co-simulation possible. It has the following advantages:

1. It demonstrates a low threshold; the description of the basic elements in the library is based on the physical equations of the working theory of the HMU, such as Newton's second law in motion calculation and Bernoulli's principle in flow calculation, which is easy to understand in principle;
2. It is lightweightl the library adopts the Modelica language and uses the "inheritance" feature of the language. The total code of the modeling library is 452 lines, which is convenient for users to learn, modify and use;
3. It demonstrates a strong simulation capability. It is capable of calculating high-frequency dynamic responses of a single port chamber orifice system, and the simulation results are consistent with the results of a professional HMU modeling and simulation software;
4. It demonstrates good simulation compatibility. The library is implemented in the open-source Modelica language and supports the FMI protocol. It has good compatibility with other software and languages. Meanwhile, it supports the export of FMU format and the import of dynamic link library files for co-simulation calculations.

In the future, further developments based on the modeling library which are not limited to HMUs in gas turbine engine control systems could be considered. The modeling and simulation of sensors, controllers, and the engine itself could also be realized by the Modelica language.

Author Contributions: Conceptualization, Y.L. and S.Y.; methodology, Y.L. and Z.J.; software, Y.L., Z.J., J.L. and W.Z.; validation, X.W., W.Z. and M.Z.; investigation, H.C. and K.M.; resources, S.Y., X.W. and Y.Z.; data curation, H.C., K.M. and Y.Z.; writing—original draft preparation, Y.L.; writing—review and editing, S.Y. and M.Z.; visualization, S.Y.; supervision, X.W.; project administration, X.W.; funding acquisition, X.W. All authors have read and agreed to the published version of the manuscript.

Funding: This research is supported by National Science and Technology Major Project (2017-V-0015-0067) and AECC Sichuan Gas Turbine Establishment Stable Support Project (GJCZ-0011-19).

Institutional Review Board Statement: Not applicable.

Informed Consent Statement: Not applicable.

Data Availability Statement: Not applicable.

Conflicts of Interest: The authors declare no conflict of interest.

References

1. Padovani, D.; Rundo, M.; Altare, G. The Working Hydraulics of Valve-Controlled Mobile Machines. *J. Dyn. Syst. Meas. Control.* **2020**, *142*, 070801. [CrossRef]
2. Simcenter System Simulation. Available online: https://www.plm.automation.siemens.com/global/en/products/simcenter/simcenter-system-simulation.html (accessed on 1 March 2022).

3. Centre for Power Transmission & Motion Control Software. Available online: https://www.bath.ac.uk/case-studies/centre-for-power-transmission-motion-control-software (accessed on 1 March 2022).
4. Easy5: Advanced Control & Systems Simulation. Available online: https://www.mscsoftware.com/product/easy5 (accessed on 1 March 2022).
5. Xu, M.; Wang, X.; De-tang, Z. Simulation on hydromechanical controller of modern aero-engine. *J. Aerosp. Power* **2009**, *24*. [CrossRef]
6. Marquis-Favre, W.; Bideaux, E.; Scavarda, S. A planar mechanical library in the AMESim simulation software. Part I: Formulation of dynamics equations. *Simul. Model. Pract. Theory* **2006**, *14*, 25–46. [CrossRef]
7. Marquis-Favre, W.; Bideaux, E.; Scavarda, S. A planar mechanical library in the AMESim simulation software. Part II: Library composition and illustrative example. *Simul. Model. Pract. Theory* **2006**, *14*, 95–111. [CrossRef]
8. Xiao, Y.; Zhu, J.; Zhang, D. Study on visual modeling and simulation platform of hydraulic system based on bond graph. *Proc. SPIE—Int. Soc. Opt. Eng.* **2005**, *6041*, 440–444.
9. Borutzky, W.; Barnard, B.; Thoma, J.U. Describing bond graph models of hydraulic components in Modelica. *Math. Comput. Simul.* **2000**, *53*, 381–387. [CrossRef]
10. Zhu, M.; Wang, X. An Integral Type μ Synthesis Method for Temperature and Pressure Control of Flight Environment Simulation Volume. In Proceedings of the ASME Turbo Expo 2017: Turbomachinery Technical Conference and Exposition, Charlotte, NC, USA, 26–30 June 2017.
11. Graeber, M.; Kosowski, K.; Richter, C.; Tegethoff, W. Modelling of heat pumps with an object-oriented model library for thermodynamic systems. *Math. Comput. Model. Dyn. Syst.* **2010**, *16*, 195–209. [CrossRef]
12. Pei, X.; Liu, J.; Wang, X.; Zhu, M.; Zhang, L.; Dan, Z. Quasi-One-Dimensional Flow Modeling for Flight Environment Simulation System of Altitude Ground Test Facilities. *Processes* **2022**, *10*, 377. [CrossRef]
13. Jiang, Z.; Wang, X.; Chen, H.; Wang, Y. Design of Speed Closed-Loop Control with Variable Pressure Difference Valve for Aero Engine. In Proceedings of the ASME Turbo Expo 2020: Turbomachinery Technical Conference and Exposition, Virtual, Online, 21–25 September 2020.
14. Michael, M.T. Modelica by Example. Available online: https://mbe.modelica.university/front/intro/ (accessed on 2 March 2022).
15. Modelica Association. Modelica—A Unified Object-Oriented language for PHYSICAL Systems Modeling. Language Spec. v. 3.4. 2017. Available online: https://modelica.org/documents.html (accessed on 2 March 2022).
16. Pazold, M.; Burhenne, S.; Radon, J.; Herkel, S.; Antretter, F. Integration of Modelica models into an existing simulation software using FMI for Co-Simulation. In Proceedings of the 9th International Modelica Conference, Munich, Germany, 3–5 September 2012.
17. Blochwitz, T. The functional mockup interface for tool independent exchange of simulation models. In Proceedings of the 8th International Modelica Conference, Dresden, Germany, 20–22 March 2011.
18. Si-qi, F. *Aeroengine Control (Volume One)*, 1st ed.; Northwestern Polytechnic University Press: Xi'an, China, 2008; pp. 141–182.
19. Thoma, J.U.; Perelson, A.S. Introduction to Bond Graphs and Their Application. *IEEE Trans. Syst. Man Cybern.* **1976**, *6*, 797–798. [CrossRef]
20. Novák, P.; Sindelár, R. Component-Based Design of Simulation Models Utilizing Bond-Graph Theory. *IFAC Proc. Vol.* **2014**, *47*, 9229–9234. [CrossRef]
21. Gawthrop, P.J.; Bevan, G.P. Bond-graph modeling. *Control. Syst. IEEE* **2007**, *27*, 24–45.
22. Zhao, H.C.; Wang, B.; Ye, Z.F. Study on Co-Modeling Method for Digital Control Fuel Metering Unit. *J. Propuls. Technol.* **2016**, *37*, 1752–1758.
23. Zhang, D. Simulating and Experimental Study on Fuel Control System of a Certain Aeroengine. Master's Thesis, Nanjing University of Aeronautics and Astronautics, Nanjing, China, 1 March 2008.
24. Wang, B.; Zhao, H.; Yu, L.; Ye, Z. Study of Temperature Effect on Servovalve-Controlled Fuel Metering Unit. *ASME J. Eng. Gas Turbines Power* **2015**, *137*, 061503. [CrossRef]

Article

Transient Controller Design Based on Reinforcement Learning for a Turbofan Engine with Actuator Dynamics

Keqiang Miao [1], Xi Wang [1], Meiyin Zhu [2,*], Shubo Yang [1], Xitong Pei [3] and Zhen Jiang [1]

- [1] School of Energy and Power Engineering, Beihang University, Beijing 100191, China; kqmiao@buaa.edu.cn (K.M.); xwang@buaa.edu.cn (X.W.); yangshubo@buaa.edu.cn (S.Y.); zhenjiang@buaa.edu.cn (Z.J.)
- [2] Beihang Hangzhou Innovation Institute Yuhang, Hangzhou 310023, China
- [3] Research Institute of Aero-Engine, Beihang University, Beijing 100191, China; peixitong@buaa.edu.cn
- * Correspondence: mecalzmy@buaa.edu.cn

Abstract: To solve the problem of transient control design with uncertainties and degradation in the life cycle, a design method for a turbofan engine's transient controller based on reinforcement learning is proposed. The method adopts an actor–critic framework and deep deterministic policy gradient (DDPG) algorithm with the ability to train an agent with continuous action policy for the continuous and violent turbofan engine state change. Combined with a symmetrical acceleration and deceleration transient control plan, a reward function with the aim of servo tracking is proposed. Simulations under different conditions were carried out with a controller designed via the proposed method. The simulation results show that during the acceleration process of the engine from idle to an intermediate state, the controlled variables have no overshoot, and the settling time does not exceed 3.8 s. During the deceleration process of the engine from an intermediate state to idle, the corrected speed of high-pressure rotor has no overshoot, the corrected-speed overshoot of the low-pressure rotor does not exceed 1.5%, and the settling time does not exceed 3.3 s. A system with the designed transient controller can maintain the performance when uncertainties and degradation are considered.

Keywords: turbofan engine; transient control; reinforcement learning; deep deterministic policy gradient (DDPG)

1. Introduction

A turbo engine, which is a classical type of aero engine, is a sophisticated piece of thermal equipment with symmetrical geometry. In recent years, with the rapid development of the aerospace industry, the capacity for supersonic and hypersonic flight over a wider flight envelope has been demanded for aero engines [1]. To achieve these goals, more and more complex structures—such as variable cycle, adaptive cycle and turbine-based combined cycle (TBCC) systems—are applied to aero engines, making their modeling and control design more difficult [2]. The control design plays an essential role in the integral aero engine system for a controller, which is responsible for keeping the system asymptotically stable, minimizing the transient process time, and maintaining enough margin to keep the engine working in the event of a surge, extreme temperatures, or excess revolutions.

However, control design is becoming more challenging with modern control theory, making engines so sophisticated that they cannot be modeled accurately. One of the challenges is the existence of disturbances and uncertainties in the aero engine system, which can affect the performance and stability of the system. Therefore, rejection of disturbances and uncertainties has been a critical design objective, which is traditionally achieved by observer and robust control design methods. Observer control is widely used to reject the disturbance [3,4], while robust control can be applied with model uncertainties [5–7]. Both observer and robust control design depend on an accurate linear model, where the uncertainties are introduced into the system via linearization. When an accurate linear model is unable to be obtained, these methods design controllers with the idea of sacrificing some performance for robustness. Another challenge is that control parameters should be changed when the components degrade after operating under tough working conditions for a long period of time throughout the whole asymmetrical life cycle [8,9]. When performance degrades, a system with traditional controllers will be vulnerable [10].

Therefore, reinforcement learning is taken into consideration, because it combines the advantages of optimal control and adaptive control [11], meaning that the desired performance can be achieved and the parameters can be adjusted to obtain the required robustness. This is critical because the optimal controller can be obtained without knowing the full system dynamics. As a branch of artificial intelligence, the key to reinforcement learning in feedback control is training the agent with a policy deciding the action of the system. Many algorithms—such as policy iteration, value iteration, Q-learning, deep Q-networks (DQNs), and deep deterministic policy gradient (DDPG)—have been proposed to learn the parameters under different conditions. Policy iteration and value iteration are basic methods of finding optimal values and optimal policy by solving the Bellman equation where policy iteration converges to the optimal value in fewer steps, making value iteration easier to implement [12]. Q-learning methods are based on the Q-function, which is also called the action-value function [13]. The Q-function designs an adaptive control algorithm that converges online to the optimal control solution for complete unknown systems. The DQN is capable of solving problems of high-dimensional observation space, but it cannot be straightforwardly applied to continuous domains, since it relies on finding the action that maximizes the action-value function [14]. The development of DDPG, which is a method of learning continuous action policy, originates from the policy gradient (PG) method developed in 2000 [15]. In 2014, Silver presented deterministic policy gradient (DPG) [16]. More details about the development of DDPG from DPG were given by Lillicrap [13]. Studies using the reinforcement learning method described above have been carried out in the aeronautics and aerospace control industry for learning policies for autonomous planetary landing [17] and unmanned aerial vehicle control [18]. In [19], a deep reinforcement learning technique is applied to a conventional controller for spacecraft. The author of [20] demonstrated that deep reinforcement learning has a possibility to exceed the conventional model-based feedback control in the field of flow control. The DPG algorithm has been adopted to design coupled multivariable controllers for variable cycle engines at set points [21]. The DDPG algorithm has been used to adjust the engine pressure ratio control law online in order to decrease fuel consumption for an adaptive cycle engine [22]. Reinforcement learning is also applied to prediction of aero engines' gas path health state [23] and life-cycle maintenance [24]. However, few applications of reinforcement learning have been directly used for aero engine control design—especially the transient control design. Turbofan engines take action continuously, with states changing quickly in a wide range of working conditions with uncertainties and degradation. Reinforcement learning is likely to be an ideal way to design the controller for a turbofan engine, since its optimal design procedure is independent of the knowledge of full system dynamics. Therefore, a reinforcement-learning-based controller design method with the agent trained with the DDPG algorithm is proposed in this paper. This is achieved with the turbofan engine nonlinear model, which reduces the uncertainties introduced into the system via linearization. Moreover, this approach has the advantage of designing the set point controller and transient controller together with the same policy, which will restrain the jump from one to the other. A series of improvements are proposed to improve the stability of the closed-loop system, making the training process achievable with a nonlinear model by solving the problem of divergence. Symmetrical performance is achieved in the acceleration and deceleration process of the engine with the designed controller.

The rest of this paper is organized as follows: In Section 2, a nonlinear model of a dual-spool turbofan engine is built and linearized. In Section 3, a brief background of reinforcement learning and DDPG is given. In Section 4, the method of designing the controller for a turbofan engine with reinforcement learning is presented. In Section 5, a series of simulations are applied in different conditions. The simulation results are compared with a traditional gain-scheduled controller, which is designed with linear matrix inequality (LMI) based on an LPV model. Finally, conclusions are given in Section 6.

2. System Uncertainties Analysis

Consider the follow nonlinear system:

$$\begin{cases} \dot{x}(t) = f(x(t), u(t), d(t)) \\ y(t) = g(x(t), u(t), d(t)) \end{cases} \quad (1)$$

where $x(t) \in R^x$ is the state vector of the system, $u(t) \in R^u$ is the input vector of the system, $y(t) \in R^y$ is the output vector of the system, and $d(t) \in R^d$ is the disturbance vector.

Moreover, a linear system is proposed to approximate the dynamic of the nonlinear system for decreasing the nonlinear complexity and making it easier to design the controller $k(t)$, which regulates the $y(t)$ to the desired outputs $r(t)$, based on classic control theory or modern control theory. This can be described as follows:

$$\begin{cases} \delta \dot{x} = A(x - x_e) + B(u - u_e) + G(d - d_e) \\ \delta y = C(x - x_e) + D(u - u_e) + H(d - d_e) \end{cases} \quad (2)$$

where x_e is the steady-state vector of the system, u_e is the input vector that keeps the system working at x_e, y_e is the steady output vector of the system at state x_e with input u_e, and d_e is the disturbance vector at x_e. For a stable system, (x_e, u_e, d_e), which keeps the system working at steady state, always exists. The nonlinear system at steady state can be described as follows:

$$\begin{cases} 0 = f(x_e, u_e, d_e) \\ y_e = g(x_e, u_e, d_e) \end{cases} \quad (3)$$

Matrices of the linear system are usually obtained by linearizing at steady points. A is the state matrix, B is the input matrix, C is the output matrix, D is the feedforward matrix, G is a disturbance matrix, and H is a disturbance matrix. They are obtained as follows:

$$A = \left.\frac{\partial f}{\partial x}\right|_{(x_e, u_e, d_e)}, B = \left.\frac{\partial f}{\partial u}\right|_{(x_e, u_e, d_e)}, G = \left.\frac{\partial f}{\partial d}\right|_{(x_e, u_e, d_e)}, C = \left.\frac{\partial g}{\partial x}\right|_{(x_e, u_e, d_e)}, D = \left.\frac{\partial g}{\partial u}\right|_{(x_e, u_e, d_e)},$$
$$H = \left.\frac{\partial g}{\partial d}\right|_{(x_e, u_e, d_e)}.$$

As a result, uncertainties are introduced into the system. This is described as follows:

$$\begin{cases} \omega(\dot{x}) = f(x, u, d) - f(x_e, u_e, d_e) - A(x - x_e) - B(u - u_e) - G(d - d_e) \\ \omega(y) = g(x, u, d) - g(x_e, u_e, d_e) - C(x - x_e) - D(u - u_e) - H(d - d_e) \end{cases} \quad (4)$$

The uncertainties depend on the difference between the current state x and the linearized steady state x_e. In order to reduce the uncertainties of the linear system when it is used to approximate the nonlinear system over the whole working condition, the linear parameter-varying (LPV) model, which is widely used in modern control theory, is introduced.

For the nonlinear system shown in Equation (1), the LPV model can be obtained by linearizing at n different equilibrium points shown in Equation (3) and scheduling with a parameter. Then, the nonlinear system can be approximated with the LPV model as follows:

$$\begin{cases} \dot{x} = A(p)x + B(p)u \\ y = C(p)x + D(p)u \end{cases} \quad (5)$$

where $p = \begin{bmatrix} p_1 & p_2 & p_3 & \cdots & p_n \end{bmatrix}$ are the scheduled parameters and $\sum_{j=1}^{n} p_j = 1, p_j \geq 0$. System matrices are defined as follows:

$$\begin{cases} A(p) = \sum_{j=1}^{n} p_j A(j) & B(p) = \sum_{j=1}^{n} p_j B(j) \\ C(p) = \sum_{j=1}^{n} p_j C(j) & D(p) = \sum_{j=1}^{n} p_j D(j) \end{cases} \quad (6)$$

where $A(j)$, $B(j)$, $C(j)$, and $D(j)$ are the jth matrices obtained by linearizing the nonlinear system at equilibrium points $(x_e(j), u_e(j), d_e(j))$.

Although the LPV model is proposed with the idea of reducing the uncertainties by reducing the difference between x and x_e, the uncertainties introduced into the system in the linearization still exist, and are very hard to calculate.

Moreover, the dynamics and uncertainty of the actuators cannot be ignored in the transient process of an aero engine, and this is usually simplified as a first-order function, which is defined as follows:

$$u = \tau_a v \approx \tau_a v = \begin{bmatrix} \frac{k_1}{\tau_{a1}s+1} & 0 & \cdots & 0 \\ 0 & \frac{k_2}{\tau_{a2}s+1} & \cdots & 0 \\ \vdots & \vdots & \ddots & \vdots \\ 0 & 0 & \cdots & \frac{k_u}{\tau_{au}s+1} \end{bmatrix} v \quad (7)$$

where u represents the inputs of the engine and v represents the control signals given by the controller. As a result, uncertainties will be introduced into the system again.

The augmented nonlinear system can be described as follows:

$$\begin{cases} \dot{x}(t) = f(x, v, \tau_a, d) \\ y(t) = g(x, v, \tau_a, d) \end{cases} \quad (8)$$

The augmented LPV linear model is described as follows:

$$\begin{cases} A(p) = \sum_{j=1}^{n} p_j A(j) & B(p) = (\sum_{j=1}^{n} p_j B(j))\tau_a \\ C(p) = \sum_{j=1}^{n} p_j C(j) & D(p) = (\sum_{j=1}^{n} p_j D(j))\tau_a \end{cases} \quad (9)$$

Controller design with reinforcement learning is based on the augmented nonlinear model denoted in Equation (8), while controller design with modern control theory relies on the augmented LPV model with uncertainties denoted in Equation (9).

3. Reinforcement Learning Algorithm

3.1. Preliminaries

Definition 1. *The Markov decision process (MDP), which is one of the bases of reinforcement learning, is a memoryless stochastic process denoted with a tuple <S, A, P, R, γ>, where S is a finite set of state, A is the set of action, P is the state transition probability matrix, R is the reward function, and γ is the discounted factor.*

Definition 2. *Cumulative reward represents the sum of discounted future reward:*

$$r_t^{\gamma} = \sum_{i=t}^{\infty} \gamma^{(i-t)} r(s_i, a_i) \quad (10)$$

where discounted factor $\gamma \in [0,1]$, state $s_i \in S$, action $a_i \in A$, and reward function $r : S \times A \to \mathbb{R}$.

Theorem 1. *Supposing that the gradient of deterministic policy $\nabla_\theta \mu_\theta(s)$ and the gradient of action-value function $\nabla_a Q^\mu(s,a)$ exist, the parameter θ of the policy is adjusted in the direction of the performance gradient defined as follows:*

$$\begin{aligned}\nabla_\theta J(\mu_\theta) &= \int_S \rho^\mu(s) \nabla_\theta \mu_\theta(s) \nabla_a Q^\mu(s,a)|_{a=\mu_\theta(s)} ds \\ &= \mathbb{E}_{s \sim \rho^\mu}[\nabla_\theta \mu_\theta(s) \nabla_a Q^\mu(s,a)\big|_{a=\mu_\theta(s)}]\end{aligned} \quad (11)$$

where μ_θ is a deterministic policy with parameter θ, ρ^μ is the discounted state distribution with policy μ, Q^μ is the action-value function with policy μ, and J is the performance objective $s \in S$, $a \in A$ [15].

3.2. Framework of Reinforcement Learning

The structure of reinforcement learning consists of an agent and an environment. At time t, the agent observes the state s_t and reward r_t of the environment, and executes action a_t following the internal policy π_θ. Then, the environment outputs the next state s_{t+1} and reward r_{t+1}. This is shown in Figure 1.

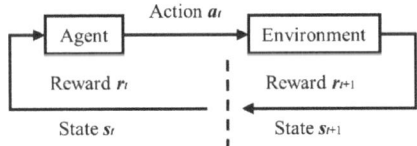

Figure 1. The structure of reinforcement learning.

The purpose of reinforcement learning is to obtain the agent that contains a policy π_θ maximizing the cumulative reward defined in Equation (10) by interacting with the environment. The procedure of training the agent with DDPG is shown in Figure 2.

In this procedure, the DDPG algorithm trains the agent with the actor–critic framework shown in Figure 3. Both critic and actor combine two neural networks: the estimation network, and the target network. The four neural networks are denoted as $\mu(s|\theta^\mu)$, $Q(s,a|\theta^Q)$, $\mu'(s|\theta^{\mu'})$, and $Q'(s,\mu'(s|\theta^{\mu'})|\theta^{Q'})$, with weights θ^μ, θ^Q, $\theta^{\mu'}$, and $\theta^{Q'}$, respectively.

(1) The actor network $\mu(s|\theta^\mu)$ represents the optimal action policy. It is responsible for iteratively updating the network weights θ^μ, choosing the current action a_i based on the current state s_i, and obtaining the next state s_{i+1} and the reward r_i;

(2) The critic network $Q(s,a|\theta^Q)$ represents the Q-value obtained after taking action following the policy defined with the actor network at every state s. It is used to update the network weights θ^Q and calculate the current $Q(s_i,a_i|\theta^Q)$;

(3) The target actor network $\mu'(s|\theta^{\mu'})$ is a copy of actor network $\mu(s|\theta^\mu)$. The weights of the target actor network are updated with the following soft update algorithm:

$$\theta^{\mu'} \leftarrow \tau\theta^\mu + (1-\tau)\theta^{\mu'} \quad (12)$$

where τ is the updating factor;

(4) The target critic network $Q'(s,\mu'(s|\theta^{\mu'})|\theta^{Q'})$ is a copy of the critic network, and is used to calculate y_i. Similarly, the weights are updated with the following soft update algorithm:

$$\theta^{Q'} \leftarrow \tau\theta^Q + (1-\tau)\theta^{Q'} \quad (13)$$

where τ is the updating factor.

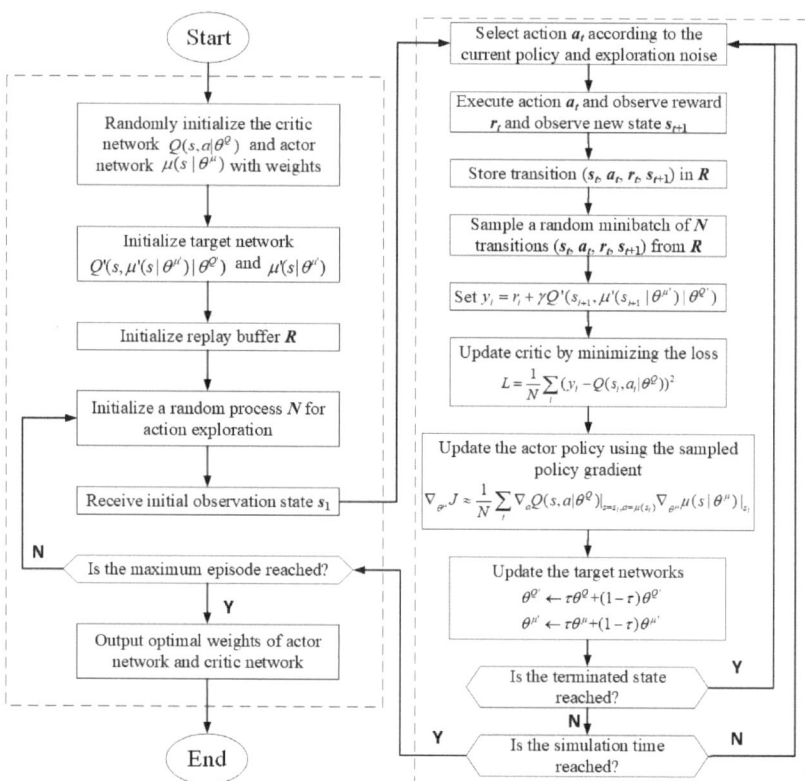

Figure 2. The training process with DDPG.

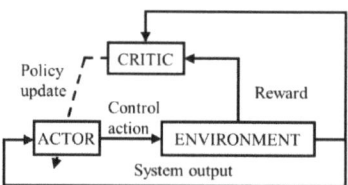

Figure 3. Actor–critic reinforcement learning framework.

With a relatively small τ, the weight of the target network will change slowly, increasing the stability of the system and making the training process easier to converge. The relationships of the four networks are shown in Figure 4. The purpose of the training process is to find the optimal weights of the networks.

Finally, the optimal weights of the actor network and critic network are obtained. Therefore, the actor and critic networks will work together to achieve the user-prescribed goal.

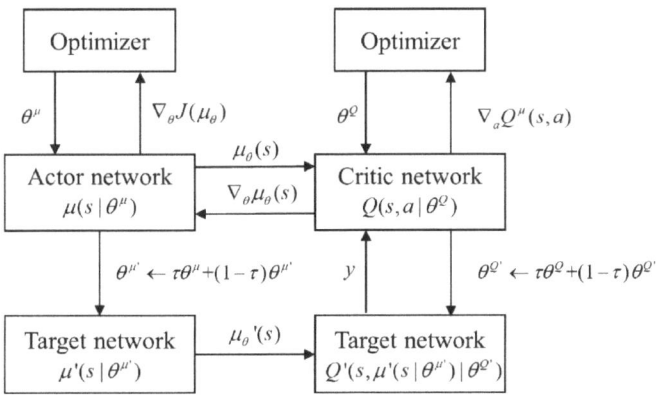

Figure 4. Relationships of the neural networks in the actor–critic framework.

4. Reinforcement Learning Controller Design Procedure for Turbofan Engines

In this section, the procedure of how to apply reinforcement learning to turbofan engines' transient control design is provided.

4.1. Framework Definition

As an example of the nonlinear system described in Equation (1), the dynamic of a two-spool turbofan engine can be modeled as follows [24]:

$$\begin{cases} \dot{n}_1 = f_1(n_1, n_2, v, d) \\ \dot{n}_2 = f_2(n_1, n_2, v, d) \end{cases} \quad (14)$$

where n_1 is the low-pressure rotor speed, n_2 is the high-pressure rotor speed, v is the control variable vector, and d is the disturbance vector.

Because the corrected rotor represents the characteristic of the turbofan better, the outputs of the nonlinear system can be denoted as $y = \begin{bmatrix} n_{1cor} & n_{2cor} \end{bmatrix}^T$, and the states of the nonlinear system can be denoted as $x = \begin{bmatrix} n_{1cor} & n_{2cor} \end{bmatrix}^T$.

$$\begin{cases} n_{1cor} = n_1 \sqrt{\frac{288.15}{T_1}} \\ n_{2cor} = n_2 \sqrt{\frac{288.15}{T_2}} \end{cases} \quad (15)$$

where T_1 is the temperature before the fan, and T_2 is the temperature before the compressor.

Moreover, the control value could be defined as $v(t) = \begin{bmatrix} A_8(t) & W_f(t) \end{bmatrix}^T$ in general, where A_8 is the throat area of the nozzle, and W_f is the fuel flow in the burner.

Ways to implement the desired objectives during the transient process could include open-loop fuel–air ratio control or closed-loop $\dot{n}$ control. For the fuel–air ratio command to be transformed into rotor speed, the trajectory of n_1 and n_2 was selected as the command in this paper. This is defined as $r(t) = \begin{bmatrix} n_{1cor}(t) & n_{2cor}(t) \end{bmatrix}^T$.

The simplest structure of a closed-loop output feedback control system is shown in Figure 5. In order to implement transient control with a reinforcement learning method with the set command and control values above, the controller module is replaced with observations, reward, and agent modules.

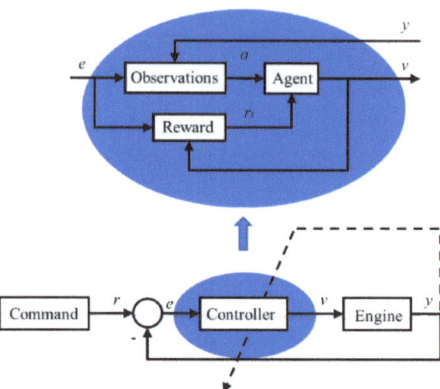

Figure 5. Framework of a traditional feedback control system and a feedback control system based on reinforcement learning.

The inputs of observation are system output y and the error between reference command r and system outputs $e = r - y$. The output of observation is the observed signal $o = [\int e \, dt \quad e \quad y]^T$. The inputs of reward are e and u, which are related to the accuracy and energy of the control. The output of reward is the cumulative reward r_t. The input of the agent is o and the output of the agent is v.

4.2. DDPG Agent Creation

The agent is trained with the aforementioned DDPG algorithm and actor–critic structure. Value functions are approximated with neural networks. This process is called representation. The number of layers, the number of neurons, and the connection of each layer should be defined based on the complexity of the problem. Moreover, both critic and actor representation options consist of learning rate and gradient threshold. The DDPG agent options include sample time, target smooth factor, discount factor, mini-batch size, experience buffer length, noise options variance, and noise options' variance decay rate. Then, the DDPG agent can be created with the specified critic representation, actor representation, and agent options.

4.3. Reward Function

In order to track the trajectory that contains the prescribed performance of the transient process of the engine, the reward function should reflect the error between the command and the output of the engine. Moreover, the error in the past should be taken into consideration, because the transient process is a continuous process where the state changes rapidly. Finally, a positive constant value should be given to keep the training process working from start to finish, because the agent will have the tendency to stop early with the purpose of not counting the cumulative error. Therefore, the reward function is set as follows:

$$r_t = 1 - |e_t| - 0.1|e_{t-1}| \tag{16}$$

4.4. Problems and Solutions

After applying the aforementioned settings, the reinforcement learning method can be preliminarily introduced into the controller design process for a turbofan engine. However, some problems still need to be solved. One of the most important problems is how to keep the environment convergent when the agent is trained. The turbofan engine model used in this paper in the training process was solved with the Newton–Raphson method, which means that the initial conditions cannot be far away from reasonable values. Otherwise, the engine model will be divergent and the reward will be uncontrolled and unbelievable.

Moreover, the training results will be invalid, and a lot of time will be wasted calculating the meaningless results. Therefore, some improvements must be added to the design process.

First of all, the initial parameters of the engine must be scheduled with the state n_{2cor}. When the reinforcement learning explores the performance of the environment, n_{2cor} represents the state of the engine. n_{1cor} should match n_{2cor}, as well as other coupled parameters, such as temperature and internal pressure. The initial condition of the nonlinear model is defined with the following parameters: mass flow of air, bypass ratio, pressure ratio of the compressor, pressure of the fan, and pressure ratio of the turbine.

Secondly, the control structure should be added into the closed-loop system between the agent and the engine. If the controller is designed with the traditional reinforcement learning method, meaning that the outputs of the agent are control actions, it will be very hard for the turbofan engine to be convergent. Therefore, a more efficient control structure should be introduced. With the new structure shown in Figure 6, the outputs become the parameters of the controller rather than the control value. For example, if the controller adopts the PI control law, the outputs of the agent are parameters K_p and K_i, which are scheduled with neural networks within the agent.

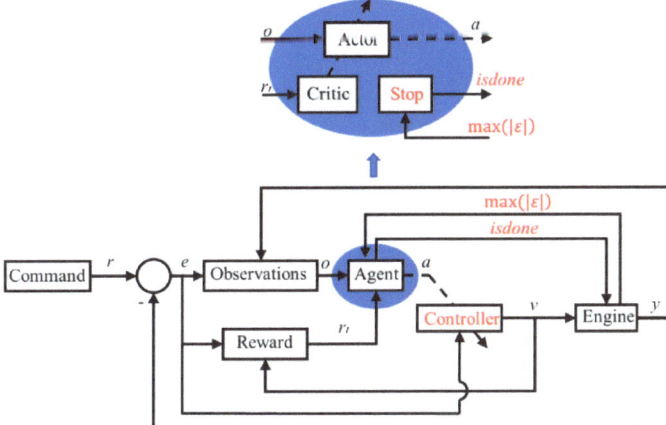

Figure 6. Improved framework of a feedback control system based on reinforcement learning with a DDPG algorithm.

Thirdly, the stop simulation module should be adopted. The convergent condition is defined as $|\varepsilon| < 10^{-6}$ in the engine model. The termination signal is configured as follows:

$$isdone = \begin{cases} 0 & \max(|\varepsilon|) \geq 1 \\ 1 & else \end{cases} \quad (17)$$

where ε is the iteration error vector of the engine model, which implies the astringency of the model.

With the termination signal, the training during an episode will stop in a timely manner when the model becomes divergent and the outputs become invalid.

4.5. Training Options

Training options specify the parameters in the process of agent training. These include the maximum number of episodes, the maximum steps per episode, the score-averaging window length, and the stop training value.

In conclusion, the process of designing a controller for a turbofan engine can be listed as follows:

Step 1: Set up the training framework by setting the inputs and outputs of the observation, reward, and agent;

Step 2: Create the DDPG agent by specifying critic representation, actor representation, and agent options;

Step 3: Set the reference signal that represents the desired performance for the turbofan engine, and define the reward function according to the purpose of the training process;

Step 4: Modify the structure with the aim of improving the convergence of the system;

Step 5: Train the agent with the DDPG algorithm.

5. Simulation and Verification

In this section, an example of designing a controller for a dual-spool turbofan engine using reinforcement learning is given, and it is compared with the gain-scheduled proportion integral (GSPI) controller designed with LMI in reference [25]. In order to validate the effectiveness of the reinforcement learning design method, the chosen control structure is also PI, which is widely used in turbofan engine control. The structure of the reinforcement learning proportion integral (RLPI) controller is shown in Figure 6, where a = [K_p K_i].

5.1. Options Specification

Training options are set with the parameters in Table 1. Training scope includes conditions from the idle to intermediate states, where n_{1cor} ranges from 7733 r/min to 10,065 r/min, and n_{2cor} ranges from 9904 r/min to 10,588 r/min.

Table 1. Training options.

Function	Description	Value
Critic Representation Options	Learn Rate	0.001
	Gradient Threshold	1
Actor Representation Option	Learn Rate	0.0001
	Gradient Threshold	1
DDPG Agent Options	Sample time	0.01
	Target Smooth Factor	0.003
	Discount Factor	1
	Mini-Batch Size	64
	Experience Buffer Length	1,000,000
	Noise Options Variance	0.3
	Noise Options' Variance Decay Rate	0.00001
Training Options	Sample time	0.01
	Maximum Episodes	20,000
	Maximum Steps per Episode	1000
	Score-Averaging Window Length	2
	Stop Training Value	996

5.2. Simulation Results in Ideal Conditions and with Uncertainties

The performance of the system with the RLPI controller is validated with the step response of acceleration from idle to intermediate and deceleration from intermediate to idle. Meanwhile, the reinforcement learning design method is only applied to the fuel flow control loop, with the purpose of minimizing the disturbance. Parameters of the nozzle area are shown in Figure 7. As mentioned in Section 2, the uncertainties of the actuators result from the order uncertainty or the dynamic uncertainty. In this section, the time constant of the actuator is changed from 0.1 to 0.2. Simulation results are shown in Figure 8 and Table 2. It is illustrated in Figure 8 that the controllers have the ability to track the command with a settling time of no more than 3.8 s and overshooting by no more than 1.5% when $\tau_a = 0.1$, which are defined as ideal conditions. The performances are very close to one another when the time constant of the actuator changes. This means that when the uncertainties of the actuator exist, the RLPI controller still maintains the performance of the closed-loop system. It is illustrated in Table 3 that the settling time changes by no more than 0.7 s and the overshooting increases by 0.73% when the time constant of the actuator increases.

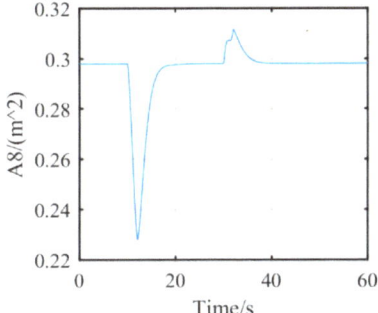

Figure 7. Nozzle area command.

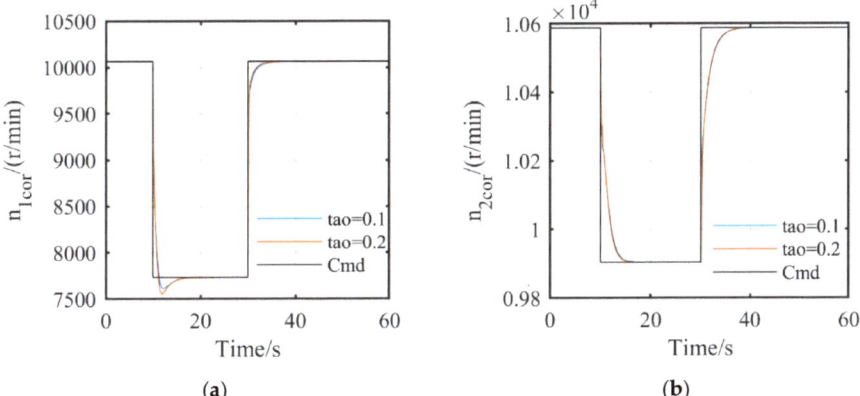

Figure 8. Response of corrected rotor speed with RLPI controllers with different time constants: (a) Comparison of n_{1cor} response. (b) Comparison of n_{2cor} response.

Table 2. Performance of the system with an RLPI controller in different conditions.

τ_a	Speed	T_s/s	σ%	State
0.1	n_{1cor}	1.43	0	Acceleration
		2.23	1.50	Deceleration
	n_{2cor}	3.84	0	Acceleration
		3.33	0	Deceleration
0.2	n_{1cor}	1.18	0	Acceleration
		2.90	2.23	Deceleration
	n_{2cor}	3.71	0	Acceleration
		3.15	0	Deceleration

Table 3. Performance of the systems with GSPI and RLPI controllers when degradation takes place.

Controller	Speed	T_s/s	σ%	State
GSPI	n_{1cor}	2.67	0	Acceleration
		3.78	0.39	Deceleration
	n_{2cor}	5.31	0	Acceleration
		3.17	0	Deceleration
RLPI	n_{1cor}	2.53	0	Acceleration
		4.40	0.75	Deceleration
	n_{2cor}	5.27	0	Acceleration
		4.50	0	Deceleration

5.3. Simulation Results with Degradation

Degradation occurs inevitably with the life cycle of the turbofan engine, meaning the system cannot work as effectively as before. Traditionally, more fuel will be consumed to obtain the desired performance.

In this section, the degradation is simulated by reducing the compressor efficiency by 5%. Simulation results are shown in Figures 9–11 and Table 3. It can be concluded that the transient performance of the system decreases when the efficiency of the compressor decreases for both deceleration and acceleration processes compared with ideal conditions. It is shown in Figure 10 that the maximum difference in n_{1cor} is about 300 r/min and the maximum difference in n_{2cor} is about 100 r/min. The n_{1cor} change with GSPI is smaller than that with RLPI, and the n_{2or} change with RLPI is smaller than that with GSPI. Moreover, it is shown in Figure 9 that the response of the corrected rotor speed is smoother with RLPI, due to the nonlinearity of RLPI. The transient process with RLPI also has faster transient response and smaller transient error.

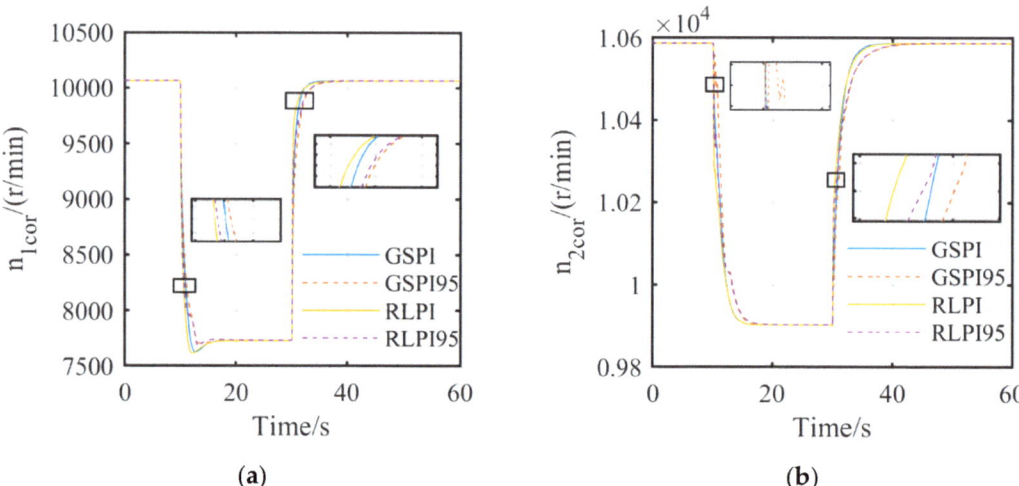

Figure 9. Response of corrected rotor speed with GSPI and RLPI controllers before and after degradation: (a) Comparison of n_{1cor} response. (b) Comparison of n_{2cor} response.

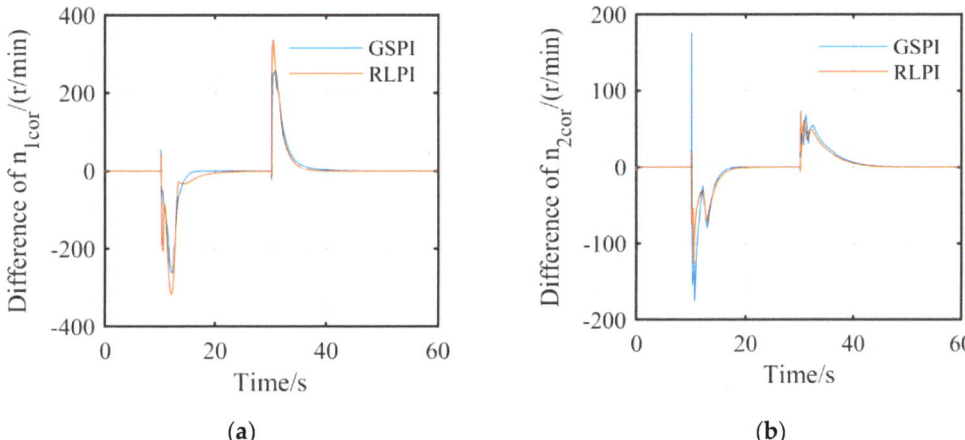

Figure 10. Difference in corrected rotor speed control performance with GSPI and RLPI controllers before and after degradation: (**a**) Difference in n_{1cor} control performance before and after degradation. (**b**) Difference in n_{2cor} control performance before and after degradation.

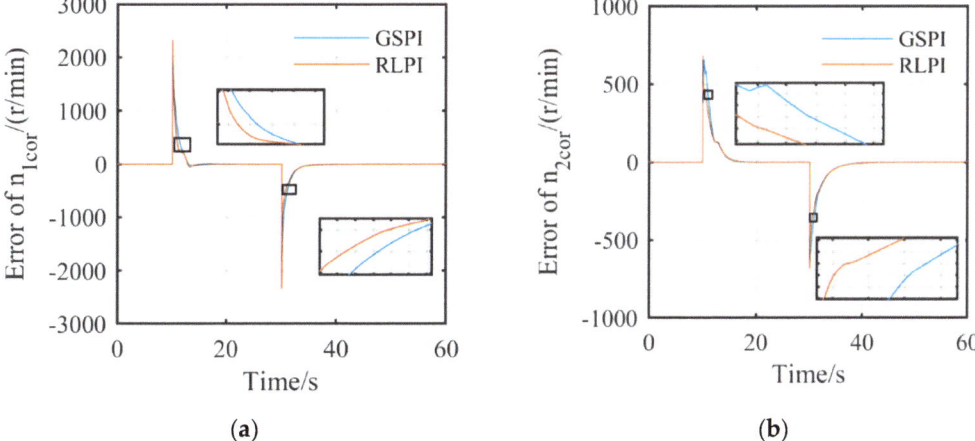

Figure 11. Corrected rotor speed error with GSPI and RLPI controllers when degradation takes place: (**a**) Comparison of n_{1cor} response error. (**b**) Comparison of n_{2cor} response error.

Cumulative error is used as a criterion of servo tracking, which represents the performance of transient control. The result is shown in Figure 12, where cumulative error (CE) is defined as follows:

$$CE = (n - n_{cmd})/n \qquad (18)$$

where n_{cmd} is the command. The results show that the cumulative error of both GSPI and RLPI increases, but the cumulative error of RLPI is much better. For n_{1cor}, the performance of the turbofan controlled with the RLPI controller after degradation is almost equal to the performance of the turbofan controlled with the GSPI controller before degradation.

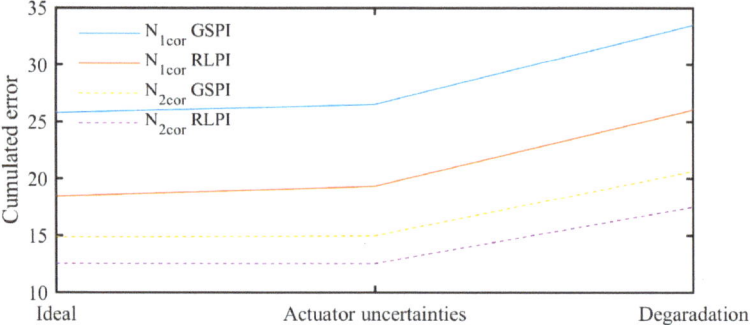

Figure 12. Cumulative error in the transient process under different conditions.

The results in Figure 13 show that when the components' efficiency decreases, more fuel is needed to keep the engine working at the desired speed. It is also illustrated in Figure 13 that the system with RLPI reduces fuel flow faster during deceleration, and adds fuel faster during acceleration. This explains the results, where RLPI tracked the command better in all conditions. It should also be noted that the surge margin (SM) [26] reduces from 17.17 to 12.19 when the degradation takes place.

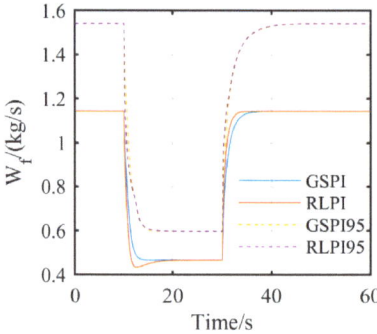

Figure 13. Comparison of W_f with the GSPI and RLPI controllers under different conditions.

The comparison of settling time under different conditions is shown in Figure 14, where the solid lines represent the performance with GSPI and the dotted lines represent the performance with RLPI. In the case of settling time, the RLPI has similar performance to GSPI. However, as noted above, the GSPI has better ability in terms of command tracking.

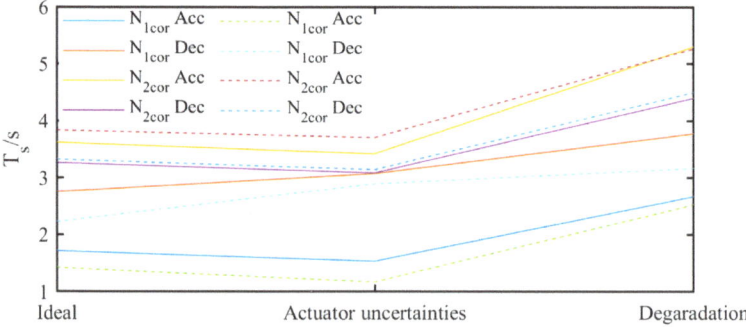

Figure 14. Comparison of settling time under different conditions.

6. Conclusions

This paper presents a method of designing a reinforcement learning controller for turbofan engines. A nonlinear model of the engine was developed as the environment for the reinforcement training process. A DDPG algorithm with actor–critic architecture and a traditional control structure—which was PI in this paper—is presented. The performance of the designed RLPI controller was verified under the chosen conditions and compared with GSPI controllers. The simulation results show that the closed-loop system with the RLPI controller has the desired performance in the transient process. Additionally, the RLPI controller's ability to deal with large uncertainties and degradation is proven by comparing the simulation results of actuators' uncertainties, and by compressor efficiency decreasing with the ideal condition.

Studies that should be carried out in the future are as follows: Firstly, the fuel flow and nozzle area need to be trained together in order to achieve better performance, but the training process will be harder to converge. Secondly, different working conditions in the flight envelope should be considered with all-round simulation. Thirdly, control structures other than PI should also be validated in the future. Finally, it is expected that experiments could be carried out with real engines in the future.

Author Contributions: Conceptualization, K.M. and X.W.; methodology, K.M. and X.W., software, K.M., M.Z. and S.Y.; validation, K.M., M.Z., S.Y. and Z.J.; writing—original draft preparation, K.M.; writing—review and editing, K.M., M.Z. and X.P.; visualization, K.M.; supervision, M.Z.; project administration, X.W. All authors have read and agreed to the published version of the manuscript.

Funding: This research was funded by AECC Sichuan Gas Turbine Establishment Stable Support Project, grant number GJCZ-0011-19, and by the National Science and Technology Major Project, grant number 2017-V-0015-0067.

Institutional Review Board Statement: Not applicable.

Informed Consent Statement: Not applicable.

Data Availability Statement: Not applicable.

Conflicts of Interest: The authors declare no conflict of interest.

References

1. Zhu, M.; Wang, X.; Dan, Z.; Zhang, S.; Pei, X. Two freedom linear parameter varying μ synthesis control for flight environment testbed. *Chin. J. Aeronaut.* **2019**, *32*, 1204–1214. [CrossRef]
2. Zhu, M.; Wang, X.; Miao, K.; Pei, X.; Liu, J. Two Degree-of-freedom μ Synthesis Control for Turbofan Engine with Slow Actuator Dynamics and Uncertainties. *J. Phys. Conf. Ser.* **2021**, *1828*, 012144. [CrossRef]
3. Gu, N.N.; Wang, X.; Lin, F.Q. Design of Disturbance Extended State Observer (D-ESO)-Based Constrained Full-State Model Predictive Controller for the Integrated Turbo-Shaft Engine/Rotor System. *Energies* **2019**, *12*, 4496. [CrossRef]
4. Dan, Z.H.; Zhang, S.; Bai, K.Q.; Qian, Q.M.; Pei, X.T.; Wang, X. Air Intake Environment Simulation of Altitude Test Facility Control Based on Extended State Observer. *J. Propuls. Technol.* **2020**, *in press*.
5. Zhu, M.; Wang, X.; Pei, X.; Zhang, S.; Liu, J. Modified robust optimal adaptive control for flight environment simulation system with heat transfer uncertainty. *Chin. J. Aeronaut.* **2021**, *34*, 420. [CrossRef]
6. Miao, K.Q.; Wang, X.; Zhu, M.Y. Full Flight Envelope Transient Main Control Loop Design Based on LMI Optimization. In Proceedings of the ASME Turbo Expo 2020, Virtual Online, 21–25 September 2020.
7. Gu, B.B. *Robust Fuzzy Control for Aeroengines*; Nanjing University of Aeronautics and Astronautics: Nanjing, China, 2018.
8. Amgad, M.; Shakirah, M.T.; Suliman, M.F.; Hitham, A. Deep-Learning Based Prognosis Approach for Remaining Useful Life Prediction of Turbofan Engine. *Symmetry* **2021**, *13*, 1861. [CrossRef]
9. Zhang, X.H.; Liu, J.X.; Li, M.; Gen, J.; Song, Z.P. Fusion Control of Two Kinds of Control Schedules in Aeroengine Acceleration Process. *J. Propuls. Technol.* **2021**, *in press*.
10. Yin, X.; Shi, G.; Peng, S.; Zhang, Y.; Zhang, B.; Su, W. Health State Prediction of Aero-Engine Gas Path System Considering Multiple Working Conditions Based on Time Domain Analysis and Belief Rule Base. *Symmetry* **2022**, *14*, 26. [CrossRef]
11. Frank, L.L.; Draguna, V.; Kyriakos, G.V. Reinforcement learning and feedback control. *IEEE Control Syst. Mag.* **2012**, *32*, 76–105.
12. Sutton, R.S.; Barto, A.G. *Reinforcement Learning: An Introduction*; MIT Press: Cambridge, MA, USA, 1998.
13. Lillicrap, T.P.; Hunt, J.J.; Pritzel, A.; Heess, N.; Erez, T.; Tassa, Y.; Silver, D.; Wierstra, D. Continuous control with deep reinforcement learning. *arXiv* **2016**, arXiv:1509.02971.

14. Richard, S.S.; David, M.; Satinder, S.; Yishay, M. Policy Gradient Methods for Reinforcement Learning with Function Approximation. *Adv. Neural Inf. Process. Syst.* **2000**, *12*, 1057–1063.
15. Silver, D.; Lever, G.; Heess, N.; Degris, T.; Wierstra, D.; Riedmiller, M. Deterministic Policy Gradient Algorithms. In Proceedings of the International Conference on Machine Learning, Bejing, China, 22–24 June 2014; pp. 387–395.
16. Giulia, C.; Shreyansh, D.; Roverto, C. Learning Transferable Policies for Autonomous Planetary Landing via Deep Reinforcement Learning. In Proceedings of the ASCEND, Las Vegas, NV, USA, 15–17 November 2021.
17. Sun, D.; Gao, D.; Zheng, J.H.; Han, P. Reinforcement learning with demonstrations for UAV control. *J. Beijing Univ. Aeronaut. Astronaut.* 2021, in press.
18. Kirk, H.; Steve, U. On Deep Reinforcement Learning for Spacecraft Guidance. In Proceedings of the AIAA SciTech Forum, Orlando, FL, USA, 6–10 January 2020.
19. Hiroshi, K.; Seiji, T.; Eiji, S. Feedback Control of Karman Vortex Shedding from a Cylinder using Deep Reinforcement Learning. In Proceedings of the AIAA AVIATION Forum, Atlanta, GA, USA, 25–29 June 2018.
20. Hu, X. *Design of Intelligent Controller for Variable Cycle Engine*; Dalian University of Technology: Dalian, China, 2020.
21. Li, Y.; Nie, L.C.; Mu, C.H.; Song, Z.P. Online Intelligent Optimization Algorithm for Adaptive Cycle Engine Performance. *J. Propuls. Technol.* **2021**, *42*, 1716–1724.
22. Wang, F. *Research on Prediction of Civil Aero-Engine Gas Path Health State And Modeling Method of Spare Engine Allocation*; Harbin Institute of Technology: Harbin, China, 2020.
23. Li, Z. *Research on Life-Cycle Maintenance Strategy Optimization of Civil Aeroengine Fleet*; Harbin Institute of Technology: Harbin, China, 2019.
24. Richter, H. *Advanced Control of Turbofan Engines*; National Defense Industry Press: Beijing, China, 2013; p. 16.
25. Miao, K.Q.; Wang, X.; Zhu, M.Y. Dynamic Main Close-loop Control Optimal Design Based on LMI Method. *J. Beijing Univ. Aeronaut. Astronaut.* 2021; in press.
26. Zeyan, P.; Gang, L.; Xingmin, G.; Yong, H. *Principle of Aviation Gas Turbine*; National Defense Industry Press: Beijing, China, 2008; p. 111.

Article

Quasi-Analytical Solution of Optimum and Maximum Depth of Transverse V-Groove for Drag Reduction at Different Reynolds Numbers

Zhiping Li [1,2,*], Long He [1,2] and Yixuan Zheng [1,2]

1. Research Institute of Aero-Engine, Beihang University, Beijing 100191, China; longhe@buaa.edu.cn (L.H.); zhengyixuan@buaa.edu.cn (Y.Z.)
2. National Key Laboratory of Science and Technology on Aero-Engine Thermodynamics, Beihang University, Beijing 100191, China
* Correspondence: leezip@buaa.edu.cn

Abstract: Reducing the skin-friction drag of a vehicle is an important way to reduce carbon emissions. Previous studies have investigated the drag reduction mechanisms of transverse grooves. However, it is more practical to investigate which groove geometry is optimal at different inflow conditions for engineering. The purpose of this paper is to establish the physical model describing the relationship between the dimensionless depth ($H^+ = Hu_\tau/\nu$) of the transverse groove, the dimensionless inflow velocity ($U_\infty^+ = U_\infty/u_\tau$), and the drag reduction rate (η) to quasi-analytically solve the optimal and maximum transverse groove depth according to the Reynolds numbers. Firstly, we use the LES with the dynamic subgrid model to investigate the drag reduction characteristics of transverse V-grooves with different depths (h = 0.05~0.9 mm) at different Reynolds numbers ($1.09 \times 10^4 \sim 5.44 \times 10^5$) and find that H^+ and U_∞^+ affect the magnitude of slip velocity (U_s^+), thus driving the variation of the viscous drag reduction rate (η_v) and the increased rate of pressure drag (η_p). Moreover, the relationship between U_s^+, η_v, and η_p is established based on the slip theory and the law of pressure distribution. Finally, the quasi-analytical solutions for the optimal and maximum depths are solved by adjusting U_s^+ to balance the cost (η_p) and benefit (η_v). This solution is in good agreement with the present numerical simulations and previous experimental results.

Keywords: drag reduction; transverse groove; optimal depth; maximum depth

1. Introduction

In recent years, due to the increase in energy consumption and the strict requirements of fuel efficiency, the technologies of reducing carbon emissions associated with skin-friction drag reduction have drawn much attention [1]. For Lufthansa Cargo's Boeing 777F freighters, reducing the skin-friction drag by 1% means annual savings of around 3700 tons of kerosene and just under 11,700 tons of CO_2 emissions [2]. Compared with traditional drag reduction methods, bionic microstructures have a better potential for engineering applications because of their remarkable drag reduction properties and good applicability [3–5]. Previous studies have shown that there are two types of microstructure, one is the riblets imitating shark shin [6,7] and the other is the transverse grooves imitating dolphin skin [8–11], which are parallel and perpendicular to the flow direction, respectively. It has been reported that longitudinal riblets are capable of delivering a reduction of surface friction drag around 10% [12]. The drag reduction mechanism of longitudinal riblets is attributed to the damping of crossflow fluctuations or the uplift of turbulent streamwise vortices above the riblet valley [13–15].

The better drag reduction properties of transverse grooves have been proved by the latest research. Lee et al. [16] observed that the maximum measured drag reduction of 40% was achieved by a nanoporous transverse grooved plate. Liu et al. [17]

used large eddy simulation (LES) technology to analyze entropy generation in the flow over a transverse grooved plate. The results showed that the total entropy generation in the near-wall region decreased by approximately 25%. Wang et al. [18] reported that an 18.76% net drag reduction was achieved by transverse grooves. These studies indicate that transverse grooves exhibit more significant potential for engineering applications than streamwise riblets.

At present, most studies focus on the drag reduction mechanisms of transverse grooves. Some studies suggested that the vortices formed within the grooves weaken the turbulence structure in the boundary layer near the wall [19,20]. The more popular perspective indicated that the vortices within the transverse grooves change the sliding friction into rolling friction at the solid–liquid interface, which is also named the "micro air-bearing phenomenon" [21–25]. Several studies focused on the Cassie–Baxter state to reduce the solid–liquid contact area and create a superhydrophobic effect over the transverse grooves [26–28]. The studies of Seo et al. [29], Lang et al. [30], and Mariotti et al. [31] showed that the vortices formed in the transverse grooves increase the momentum in the boundary layer near the wall, thus effectively controlling the flow separation.

All the above studies are of great significance to qualitatively explain the drag reduction mechanism of the transverse groove. However, it is of more practical interest to understand which groove geometry is optimal at different inflow conditions for engineering. To the best of the authors' knowledge, compared with a large number of studies in the context of streamwise riblets, less is known regarding the parameter studies on transverse grooves. The drag reduction characteristics of transverse grooves are mainly determined by the shape, AR (ratio of groove width to depth), and depth. Cui et al. [22] conducted a numerical simulation on the pressure drop in microchannel flow over different transverse-grooved surfaces and found that the drag-reduction rate of V-shaped transverse grooves is better than that of rectangular transverse grooves. The experimental results of Liu et al. [32] demonstrated that the drag reduction performance was best when the AR is 2 at a Reynolds number of 50,000. The purpose of this paper is to establish the physical model describing the relationship between the dimensionless depth ($H^+ = Hu_\tau/v$) of the transverse groove, the dimensionless inflow velocity ($U_\infty^+ = U_\infty/u_\tau$), and the drag reduction rate (η), so as to quasi-analytically solve the optimal and maximum transverse groove depth according to the different Reynolds numbers. The optimum depth of the transverse groove corresponds to the maximum drag reduction rate, and the maximum depth corresponds to the drag reduction rate of zero, which is the limit for the allowable machining error for engineering applications.

This paper is organized as follows. First, the numerical methodology is formulated in Section 2. Secondly, the drag reduction characteristics of transverse V-grooves with different depths at different Reynolds numbers are investigated by LES in Section 3. Then, in Section 4, the theoretical model for the optimal and maximum depth of the transverse groove is established. Based on this model, the quasi-analytical solution of the groove depth has been solved and several grooved plates with different groove depths have been designed to verify the drag reduction characteristics in Section 5. Finally, the conclusions are presented in Section 6.

2. Numerical Methodology
2.1. Solving Methods

The large eddy simulation (LES) method is used in the commercial software FLUENT 18.0 to obtain the induced drag reduction and flow characteristics [33,34]. A dynamic subgrid-scale (SGS) is chosen to model the unresolved small flow field scale motions [35,36]. The discretized continuity equation is solved using the Rhie and Chow method [37] to compute the mass flux at each face. The diffusion terms and the advection terms in the discretized momentum equations are solved using a second-order-accurate central-differencing discretization scheme and a second-order upwind scheme, respectively. Moreover, a second-order implicit time-stepping approach is used for temporal discretization.

Thus, the space and time resolution of the numerical method is of second-order accuracy. For the pressure–velocity coupling, SIMPLE (Semi-Implicit method for pressure linked equations) is used to enforce the mass conservation and obtain the pressure fields. For the evaluation of gradients and derivatives, the least-squares cell-based gradient method is employed. The dimensionless physical timestep $\Delta t U/H \approx 0.02$ [38] is used, where U denotes the uniform velocity at the inlet and H represents the depth of grooves. The dimensionless time of statistical averaging is above $TU/H \approx 400$ [34] to obtain the time-averaged results.

2.2. Computational Domain and Boundary Conditions

The 3D computational domain and the boundary conditions are shown schematically in Figure 1 and Table 1. The height and spanwise length of the computational domain are 25 mm ($L_y^+ = L_z^+ = 103$) [34], which covers at least the thickness of the boundary layer under all the flow conditions. The total length of the computational domain is 200 mm ($L_x^+ = 833$). A smooth wall with a length of 160 mm is placed upstream of a grooved wall for the development of turbulence from laminar flow. Another smooth wall with a length of 20 mm is located downstream of the grooved wall to prevent the propagations of pressure perturbations at the outlet. The simulated grooved wall is about 20 mm long, consisting of different symmetric V-grooves profiles—whose AR is 2 (aspect ratios, S/H, where S represents the width of an individual groove) and depths are 0.05–0.9 mm (a baseline plate without grooves is also simulated in the same position). Reynolds numbers range from 1.09×10^4 to 5.44×10^5, which is based on the length of the flat wall placed upstream of a grooved wall (160 mm), the freestream velocity, and the viscosity of the fluid. At all the solid walls, the no-slip condition is specified. At the inlet of the computational domain, an ideal gas flows with uniform velocity and a freestream turbulence intensity of 1%, and the pressure condition is set at the outlet.

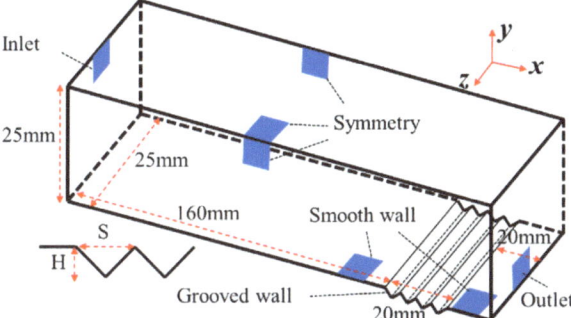

Figure 1. Computational domain and boundary conditions.

Table 1. Size of the computational domain and the corresponding number of grid nodes.

			Dimensionless Parameters	Nodes
	L_x^+		667 + 83 + 83	100 + 1200 + 25
	L_y^+		103	50
	L_z^+		103	25
Δx^+		Groove	0.05	1200
		Other	<10	100 + 25
	Δy^+		0.02~10	50

In order to verify that the normal height (L_y^+) and spanwise length (L_z^+) of the computational domain meet the requirements of LES, two normal heights ($L_y^+ = 103$ and 123.6) and two spanwise lengths ($L_z^+ = 103$ and 123.6) are chosen to investigate the effect on outcomes. Table 2 shows the simulation results of the total drag of the grooved plate. The drag hardly changes when the normal height and the spanwise length are greater than 103, which means that the results conducted with $L_x^+ = 103$ and $L_z^+ = 103$ in the present are adequate for the requirements of LES.

Table 2. Domain independence test (h = 0.1 mm, Re = 1.09×10^4).

L_x^+	L_z^+	Drag (N)
103	103	0.00272
	123.6	0.00273
123.6	103	0.00273
	123.6	0.00276

2.3. Grid Independence Study

The structured mesh is generated by commercial software, ICEM, as shown in Figure 2. The grid resolution and the number of grid nodes are shown in Table 1. The grids are clustered near the wall surface and the normal distance from the wall surface to the nearest grid points Y^+ is 0.04. The maximum normal grid resolution Δy_{max}^+ is less than 10. The streamwise grid resolution Δx^+ is 0.07 within the grooves, and Δx_{max}^+ is less than 10 in other streamwise positions. The spanwise grid resolution Δz^+ is 4.1 [34].

Figure 2. Mesh distribution around transverse V-grooves.

In order to verify that the grid resolution meets the requirements of the large eddy simulation, two streamwise grid resolutions within the groove Δx^+ (0.07 and 0.32) and two spanwise grid resolutions Δz^+ (4.1 and 10) are chosen to investigate the effect on the outcomes. Table 3 shows the simulation results for the drag of the grooved plate and the streamline inside the groove. The resistances of a grooved plate hardly change when $\Delta x^+ = 0.07$ (groove) and $\Delta z^+ = 4.1$, which are selected for deriving all the other results.

Table 3. Verification of grid resolution (h = 0.1 mm, Re = 1.09×10^4).

Δx^+ (Groove)	Δz^+	Drag (N)	Streamline
0.07	4.1	0.00274	→ 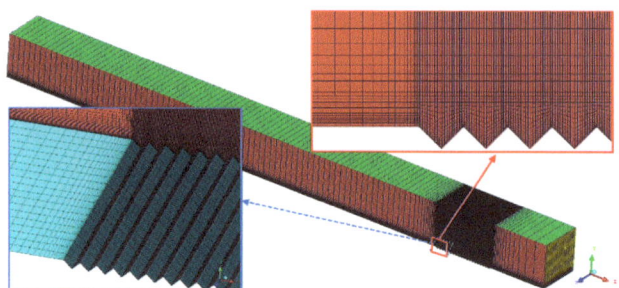
	10	0.00273	
0.32	4.1	0.00273	
	10	0.00275	→

Figure 3 shows the relative error of the total resistance of a grooved plate (with symmetric V-grooves whose AR is 2) compared with that of a smooth plate (based on the total drag in the case of 2,403,465 grids) at five different grid-refinement levels. The resistances of a grooved plate and the baseline plate no longer change when the number of grid cells is greater than 1,603,491, and the relative difference is 0.0465%, which is used as the grid resolution in deriving all the other results.

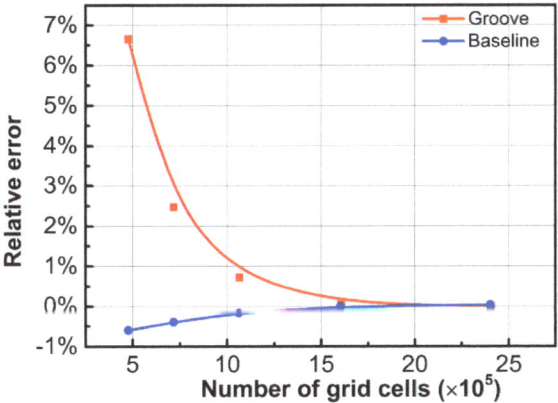

Figure 3. Verification of grid independence.

2.4. Experimental Validation

In order to check the accuracy of the obtained numerical results, the results of the numerical simulation are compared with the relevant experimental data reported previously. The numerical results of velocity profiles over the grooved plates with a groove depth of 1.62 mm and groove width of 3.57 mm are compared with the experimental results obtained by Ahmadi-Baloutaki et al. [39] in Figure 4a,b (the Reynolds number based on the length of the grooved plate is 1.85×10^5, and the turbulence intensity is 0.4% and 4.4%, respectively). The simulated velocity profiles agree well with the previously published experimental results, suggesting that the present numerical approach has sufficient accuracy for predicting the drag reduction of transverse grooves (the relative error is found to be less than 3%). Moreover, Figure 4c,d show the experimental results [40] and numerical results of the velocity vector over the grooved plate, respectively. It is apparent that the simulation results show good predictions of the location and structure of the vortex formed within the groove, which indicates that the CFD method can accurately simulate the flow details over the grooved plate.

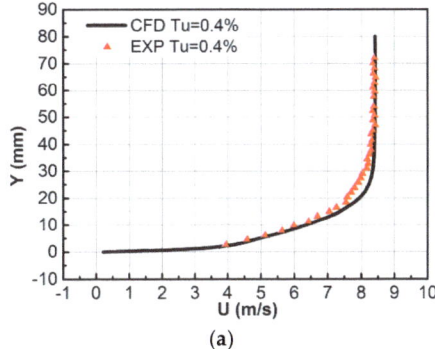
(a)

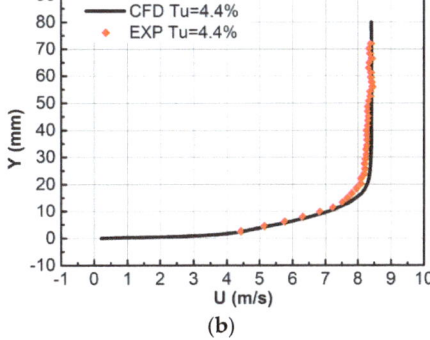
(b)

Figure 4. *Cont.*

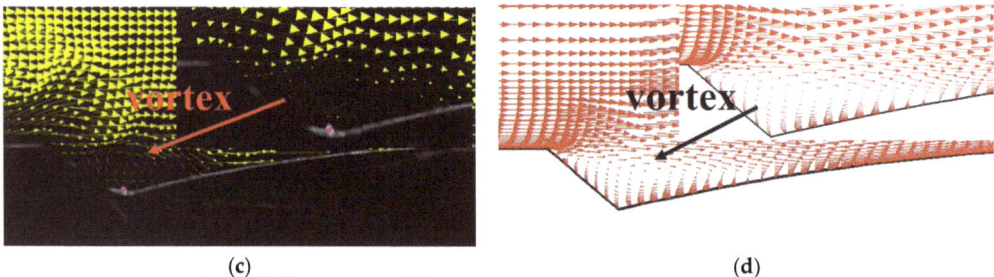

Figure 4. The comparison of numerical and experimental results. Mean velocity profile over the grooved plate when the turbulence intensity is 0.4% (**a**) and 4.4% (**b**); The instantaneous velocity vector over the grooved plate of experimental data [40] (**c**) and numerical result (**d**).

3. Characteristics of Drag Reduction Induced by Transverse V-Grooves with Different Depths at Different Reynolds Numbers

Before establishing the prediction model of the optimal and maximum depth, it is necessary to analyze the key physical factors affecting the drag reduction characteristics. In this section, the amounts of drag reduction induced by grooves with different depths at different Reynolds numbers are computed and the mechanisms of drag reduction are analyzed.

Figure 5 illustrates the variation in drag-reduction rate with depth at different Reynolds numbers. The drag-reduction rate is defined as,

$$\eta = \frac{F_G - F_R}{F_R} \qquad (1)$$

in which F_G and F_R represent the resistance of the grooved plate and the baseline plate, respectively. The results show that the drag-reduction rate induced by the grooved plate first increases and then decreases with the increase of depth. At each Reynolds number, there is an optimal groove depth for maximum drag reduction and a maximum groove depth corresponding to the zero-drag reduction rate, which is qualitatively consistent with the results of the previous study in the context of streamwise riblets [12]. Interestingly, the optimal and maximum depth decrease with the increase in Reynolds number, as shown in Table 4. It is worth noting that because the groove depth in the simulation cases is discrete, the optimal groove depth chosen from Table 4 actually corresponds to the 'near-maximum' drag reduction rate, and the maximum depth actually corresponds to the 'near-zero' drag reduction rate.

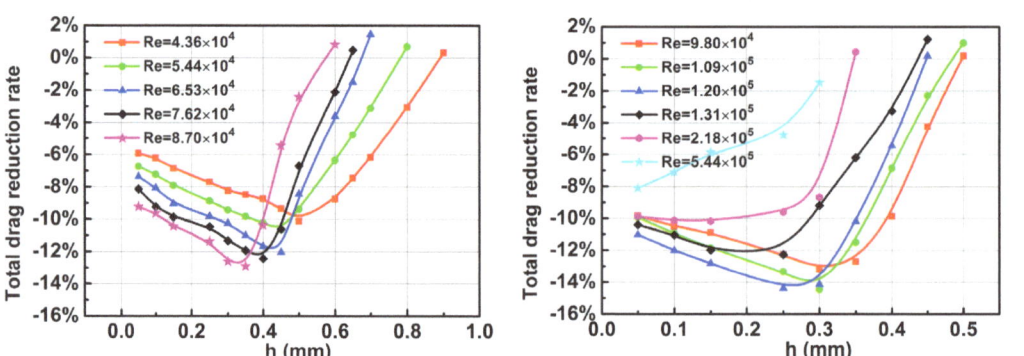

Figure 5. Total drag-reduction rate versus depth (AR = 2).

Table 4. Drag reduction rate of transverse grooves with different depths at different Reynolds numbers.

Re\h(mm)	0.05	0.1	0.15	0.25	0.3	0.35	0.4	0.45	0.5	0.6	0.65	0.7	0.8	0.9
	−1.4%	−1.7%	−2.1%	−2.5%	−2.6%	−2.8%	−3.3%	−4.0%	−4.6%					
	−4.2%	−4.8%	−5.2%	−5.6%	−5.8%	−6.0%	−6.4%	−7.0%	−7.9%					
	−5.0%	−5.5%	−6.1%	−6.8%	−7.1%	−7.3%	−7.6%	−8.2%	−8.8%					
	−5.9%	−6.2%	−6.8%	−7.7%	−8.2%	−8.5%	−8.7%	−9.3%	−10.1%	−8.7%	−7.4%	−6.1%	−3.0%	0.3%
	−6.7%	−7.2%	−7.9%	−8.8%	−9.4%	−9.8%	−10.2%	−10.7%	−9.4%	−6.3%	−4.7%	−3.1%	0.7%	
	−7.3%	−8.1%	−9.0%	−9.8%	−10.2%	−11.0%	−11.7%	−12.0%	−8.4%	−3.6%	−1.5%	1.5%		
	−8.1%	−9.2%	−9.9%	−10.5%	−11.3%	−11.9%	−12.4%	−10.6%	−6.7%	−2.1%	0.5%			
	−9.2%	−9.6%	−10.4%	−11.4%	−12.6%	−12.9%	−10.4%	−5.4%	−2.4%	0.8%				
	−9.8%	−10.5%	−10.9%	−12.2%	−13.2%	−12.7%	−9.8%	−4.2%	0.2%					
	−9.9%	−10.9%	−11.8%	−13.3%	−14.4%	−11.5%	−6.8%	−2.3%	1.0%					
	−11.0%	−12.0%	−12.8%	−14.4%	−14.1%	−10.2%	−5.4%	0.2%						
	−10.4%	−11.0%	−12.0%	−12.3%	−9.2%	−6.2%	−3.3%	1.2%						
	−9.9%	−10.1%	−10.2%	−9.6%	−8.7%	0.4%								
	−8.1%	−7.1%	−5.9%	−4.8%	−1.5%									

Optimal depth
Maximum depth

Figures 6 and 7 show the flow details at different depths (h = 0.3 mm and h = 0.4 mm) and Reynolds numbers (to facilitate model building in the next section, we convert the Reynolds number into the dimensionless velocity, U_∞^+, according to Equation (2) [41]) to analyze the physical mechanism driving the variation in total drag.

$$U_\infty^+ = \frac{U_\infty}{u_\tau} = \frac{1}{\sqrt{0.029 Re^{-0.2}}} \quad (2)$$

here, u_τ represents the local friction velocity.

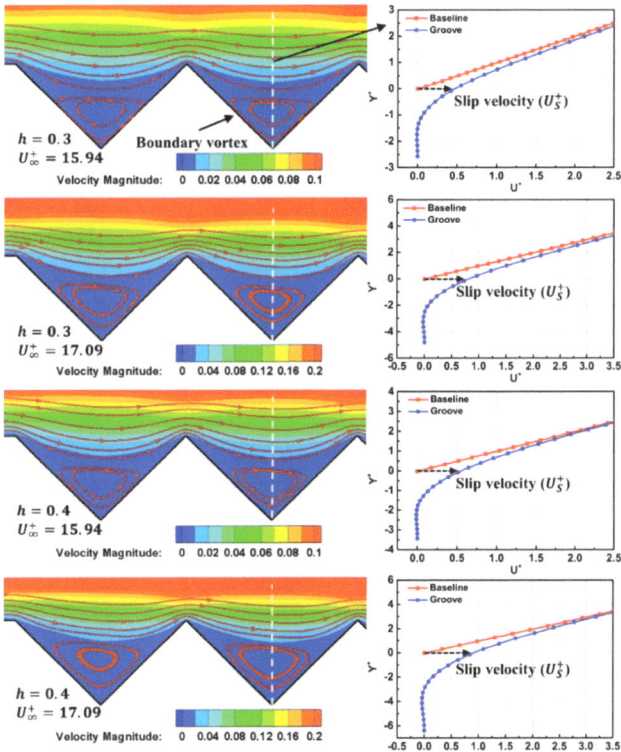

Figure 6. Velocity distribution in grooves.

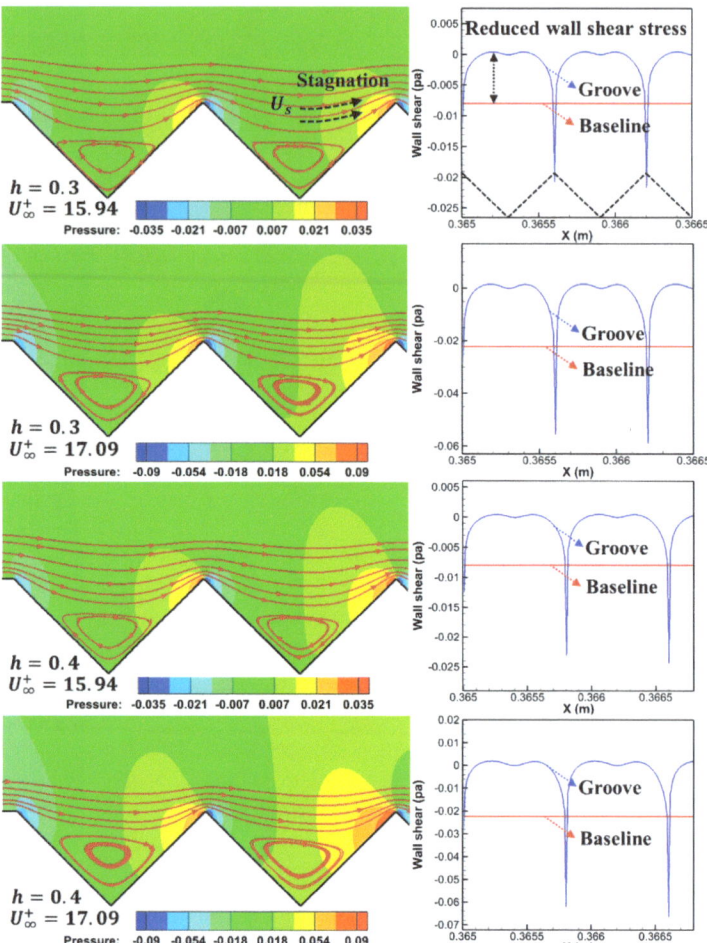

Figure 7. Pressure (**left**) and wall shear stress (**right**) distribution in grooves.

The streamline patterns and velocity that contour over the grooved plate are shown in Figure 6. Some vortices formed in the grooves can be distinctly observed as perpendicular to the flow direction. These boundary vortices act as "air bearings", which separate the boundary layer from the solid wall, resulting in fluid sliding over the grooved plate. In order to measure the sliding degree, the dimensionless velocity profiles ($Y^+ = yu_\tau/\nu$, and $U^+ = U/u_\tau$) at the centerline of the grooved plate and the baseline plate at the corresponding position are compared (Figure 6, right). The results show that the velocity gradient over the grooved plate is less than that on the baseline plate, and an induced slip velocity (U_s^+) on the horizontal line can be selected as the quantitative parameter to describe the slip phenomenon. Moreover, the comparison results of the different numerical cases indicate that the slip velocity is affected by the groove depth (h) and inflow velocity (U_∞^+).

The wall shear distribution and pressure contour shown in Figure 7 further illustrate that the magnitude of slip velocities drives the variation in the total viscous drag and pressure drag. On the one hand, these slip velocities reduce the velocity gradient over the groove plate, thus reducing the total viscous resistance. This point is further proved by the shear stress distribution diagram, which shows that the shear stress of the grooved wall is significantly less than that of the baseline plate. By comparing the numerical cases of grooved plates with the same depth at different inflow velocities, it can be seen that the larger the slip velocity, the smaller the velocity gradient. As well, the corresponding shear stress decreases the most, so the total viscous resistance decreases the most in turn. On the other hand, the "slip fluids" induced by the vortices separate on the leeward side of the groove and stagnation occurs on the windward side, resulting in additional pressure drag compared to the baseline plate—which increases the total resistance of the grooved plate. The larger the slip velocity, the greater the stagnation pressure on the windward side, resulting in greater additional pressure drag. In summary, the grooved plate reduces the viscous drag (benefits) and increases the pressure drag (costs), and the optimal drag reduction is the result of balancing the benefits and costs.

From the above analysis, the total drag of the grooved plate consists of the viscous drag (F_{GV}) and pressure drag (F_{GP}), which are expressed by Equations (3) and (4), respectively. F_{GV} and F_{GP} are determined by calculating the corresponding local stress—namely, shear ($\overline{\tau}$) and pressure (p) at the wall—and integrating the projected stress in the drag direction ($\overline{e}_x$, that is the unit vectors in the x direction shown in Figure 1 along the wetted wall (l_s).

$$F_{GV} = \int_0^{l_s} \overline{\tau} \cdot \overline{e}_x dl \qquad (3)$$

$$F_{GP} = \int_0^{l_s} (p - p_\infty) \overline{n} \cdot \overline{e}_x dl \qquad (4)$$

here, p_∞ represents the ambient pressure, l is the unit area along the groove wall, and $\overline{n}$ denotes the normal vector to the wall. Therefore, Equation (1) is transformed into Equation (5):

$$\eta = \frac{F_{GV} - F_R}{F_R} + \frac{F_{GP}}{F_R} = \eta_v + \eta_p \qquad (5)$$

here, $\eta_v = (F_{GV} - F_R)/F_R$ denotes the reduction rate for viscous drag, and "$\eta_p = F_{GP}/F_R$" denotes the increased rate of pressure drag.

Figure 8 shows the variation of η_v and η_p with U_s, u_t, and U_s^+ (in order to compare the model results with the experimental results in the following sections, four different grooves with h of 0.05, 0.1, 0.15, and 0.3 mm are used for analysis). It can be seen that the absolute values of η_v and η_p increase with the increase of U_s^+ ($U_s^+ = U_s/u_t$, where $u_t = \sqrt{\tau_w/\rho}$ is only a unit commonly used for dimensionless values. Therefore, there is no physical relationship between u_t and the drag reduction rate), which further indicates that the slip velocity drives the variation in the total viscous drag and pressure drag. In conclusion, the essence of balancing the benefit (η_v) and cost (η_p) to obtain the optimal drag reduction is how to obtain the optimal slip velocity by matching the depth of the groove and the inflow velocity.

Based on the above analysis, we propose a model to match the relationship between inflow velocity (U_∞^+) and groove depth (h) by determining the optimal slip velocity (U_s^+). The establishment process of this model is shown in Figure 9.

Step 1. Construct the relationship between inflow velocity (U_∞^+), depth ($H^+ = hu_\tau/\nu$), and slip velocity (U_s^+).
Step 2. Construct the physical relation between slip velocity (U_s^+) and viscous drag-reduction rate (η_v).
Step 3. Construct the physical relation between slip velocity (U_s^+) and pressure drag-increase rate (η_p).
Step 4. Balance η_v and η_p to design the groove depth (H^+) in different inflow velocities (U_∞^+).

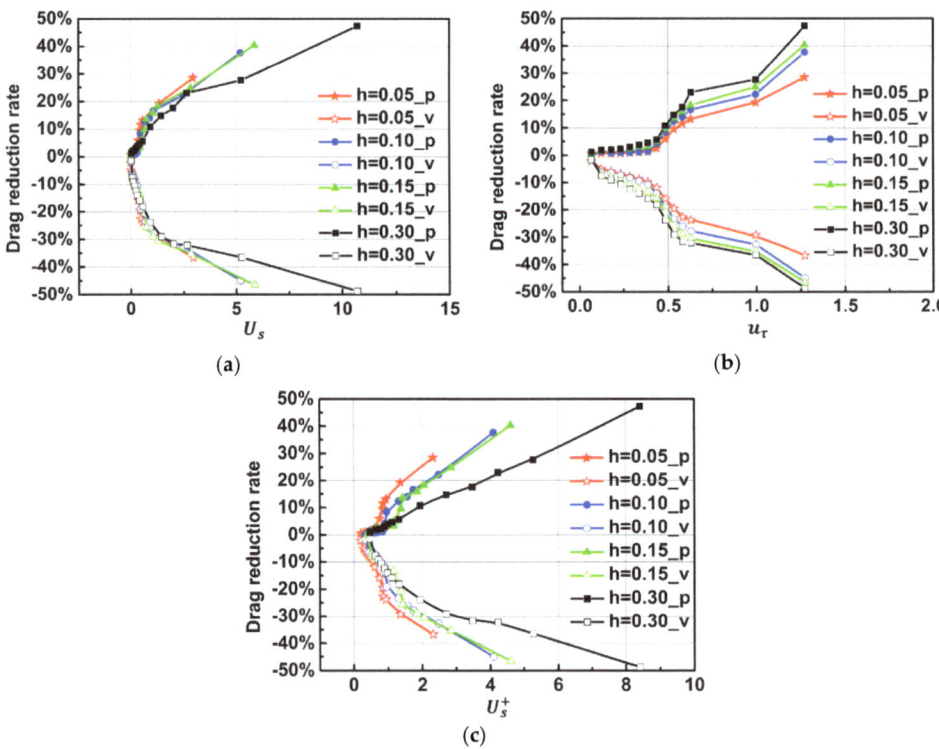

Figure 8. Drag reduction rate versus (**a**) slip velocity; (**b**) local friction velocity; (**c**) dimensionless slip velocity.

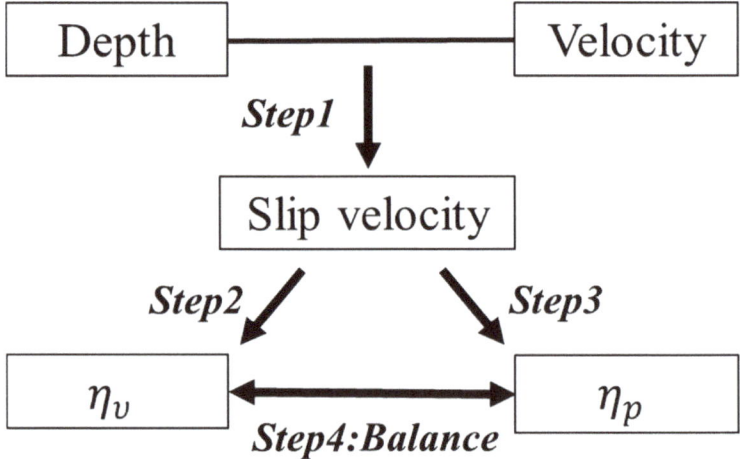

Figure 9. Process of model construction.

4. Construction the Theoretical Model Describing the Relationship between the Dimensionless Depth (H^+), the Dimensionless Inflow Velocity (U_∞^+), and Drag Reduction Rate (η)

4.1. The Relationship between Slip Velocity (U_s^+), Depth (H^+), and Inflow Velocity (U_∞^+)

The streamline patterns shown in Figure 7 reveal that the boundary vortices are not full inside the grooves. Therefore, we assume that the distances from the vortex center to the groove bottom and the slip surface are Kh and L_s (slip length, $L_s = (1-K)h$), respectively—as shown in Figure 10. Since the slip surface shown in Figure 10 (i.e., the position of the baseline plate) is at the viscous bottom layer, the total viscous stress can be expressed as

$$\tau|_{slip} = \mu \frac{\partial u}{\partial y} = \tau_\omega \qquad (6)$$

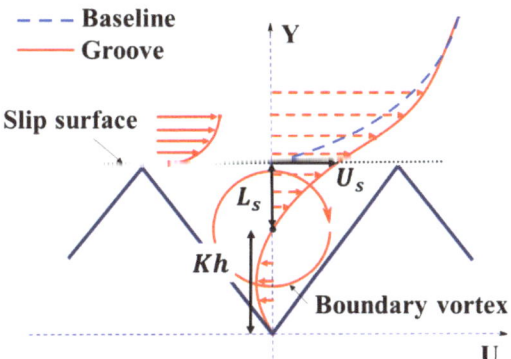

Figure 10. Schematic diagram of velocity profiles within the groove.

Over the grooved plate, the velocity at the vortex core and the slip surface increases from 0 to U_s, and both U_s and L_s are small quantities, so the velocity gradient of U in the Y direction at the midpoint of the slip surface can be approximated as

$$\frac{\partial u}{\partial y} = \frac{\tau_\omega}{\mu} \approx \frac{U_s}{L_s} \qquad (7)$$

in which $\tau_\omega = \rho(u_\tau)^2$, then Equation (7) can be transformed into Equation (8).

$$\frac{U_s}{u_\tau} = \frac{\rho u_\tau L_s}{\mu} \qquad (8)$$

therefore, substituting "$L_s = (1-K)h$" into Equation (8) yields the relation between the dimensionless slip velocity and dimensionless depth, which is described as

$$U_s^+ = (1-K)H^+ \qquad (9)$$

here, U_s^+ is equal to U_s/u_τ, H^+ is equal to $\rho u_\tau h/\mu$, and K represents the fullness of the vortex within the groove, which varies with different inflow velocities and groove depths, as shown in the streamline patterns of Figure 11.

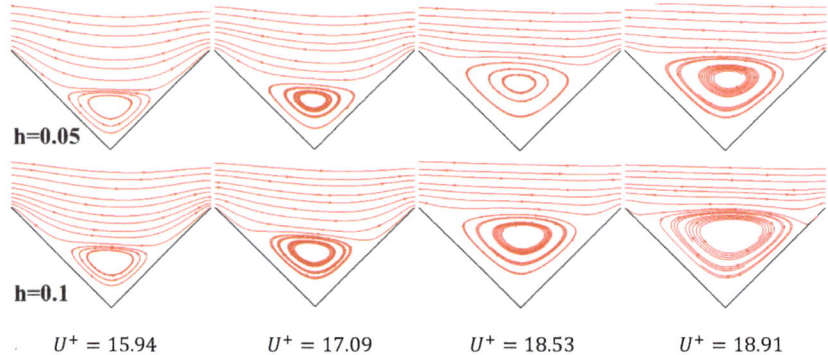

Figure 11. The position of boundary vortex varies with dimensionless velocity.

To quantitatively estimate the variation of K, we use the Boltzmann function to fit the data points (see Figure 12), which can be expressed as Equation (10). The resulting fits are accurate to better than 2% at a 95% confidence interval.

$$K = A_2 + \left(\frac{A_1 - A_2}{1 + e^{((U_\infty^+ - U_0)/du)}} \right) \tag{10}$$

where A_1, A_2, u_0, and du are the control parameters of the Boltzmann function, which can be described by Equation (11) (h is in millimeters).

$$\begin{cases} A_1 = 0.45809 - 0.32244h + 0.79245h^2 \\ A_2 = 0.74836 + 0.44569h - 1.12162h^2 \\ U_0 = 18.5287 - 9.83878h + 6.70225h^2 \\ du = 0.33396 + 4.79502h - 10.6357h^2 \end{cases} \tag{11}$$

therefore, substituting Equations (10) and (11) into Equation (9) yields Equation (12), which can predict the slip velocity over the groove plate with depth H^+ at the inflow velocity U_∞^+.

$$U_s^+ = f_1(H^+, U_\infty^+) \tag{12}$$

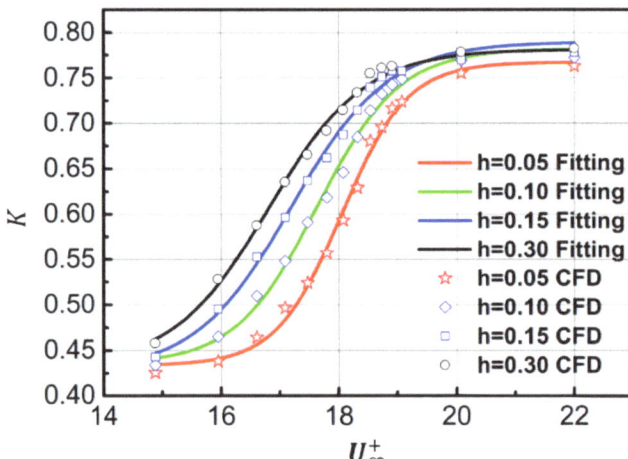

Figure 12. Boltzmann fitting of K value.

Figure 13 compares the CFD data with the results of Equation (12). The results show that the prediction value of Equation (12) are qualitatively in agreement with the numerical simulations. In detail, when $h \leq 0.1$, the prediction accuracy decreases with the increase of U_∞^+, so Equation (12) is more adequate for the $14.88 < U_\infty^+ < 18.53$. However, when $h > 0.1$, Equation (12) has a greater prediction accuracy when the dimensionless inflow velocity is between 14.9 and 22.

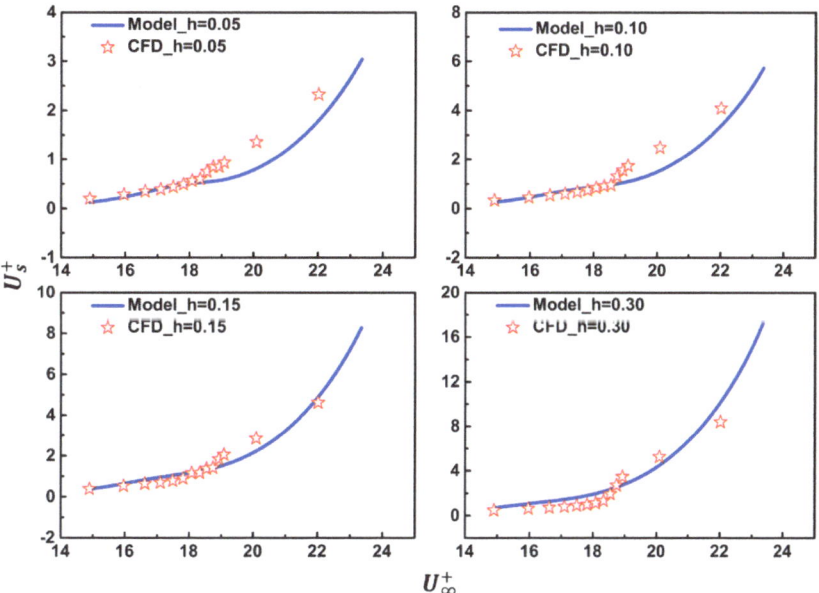

Figure 13. Comparison of the results of Equation (12) with CFD data.

4.2. The Relationship between Slip Velocity (U_s^+) and Viscous Drag-Reduction Rate (η_v)

Fukagata and Kasagi [42] proposed a theoretical prediction model for the drag reduction rate achieved by superhydrophobic surfaces in a turbulent channel flow. By comparing the bulk mean velocity of the baseline plate and the hydrophobic surface, they successfully established the relationship between the streamwise and spanwise slip length and the drag reduction rate. Different from their work, our work aims to establish the relationship between slip velocity, U_s^+, and viscous drag reduction rate, η_v, over the grooved plate.

According to Equation (5), the viscous drag reduction rate η_v is expressed by $(F_{GV} - F_R)/F_R$, where $F_{GV} \propto u_{\tau g}$ and $F_R \propto u_{\tau b}$, so η_v can be expressed as Equation (13).

$$\eta_v = \frac{F_{GV} - F_R}{F_R} = \left(\frac{u_{\tau g}}{u_{\tau b}}\right)^2 - 1 \quad (13)$$

here, $u_{\tau g}$ and $u_{\tau b}$ represent the local friction velocities of the grooved plate and baseline plate, respectively. Therefore, η_v can be estimated if the relationship between U_s^+ and $u_{\tau g}/u_{\tau b}$ is established.

$$U_g = U_{g-s} + U_s \quad (14)$$

The velocity profile over the grooved plate, as shown in Figure 14, can be regarded as the superposition of slip velocity, U_s, and the no-slip flow velocity profile (U_{g-s}), which can be described as Equation (14). Figure 15 shows the dimensionless velocity profiles of U_g and U_{g-s} ($U_\infty^+ = 22, h = 0.1$). The results show that if the local friction velocities of the grooved plate and baseline plate ($u_{\tau g}$ and $u_{\tau b}$) are used as the dimensionless units of U_s and U_{g-s}, respectively, then both $U_g/u_{\tau g}$ and $U_{g-s}/u_{\tau b}$ satisfy the logarithmic law

($U^+ = \frac{1}{\kappa} \ln Y^+ + B$, $\kappa = 0.41$, $B = 5.0$) [43] in the logarithmic region ($50 < Y^+ < 1000$), which was also demonstrated by the work of Min and Kim [44]. Therefore, U_g and U_{g-s} at $Y = K_\delta \delta$ can be described as Equations (15) and (16), respectively, where δ stands for the boundary layer thickness and K_δ is an adjustable constant ($K_\delta = 0.1 \sim 0.2$ in the logarithmic region [41]). For the predicted equation, $u_{\tau g}$ over the grooved plate is unknown before numerical calculation or experimental measurement, so all $u_{\tau g}$ in Equation (15) is converted to $\frac{u_{\tau g}}{u_{\tau b}} u_{\tau b}$.

$$U_g = \left(\frac{1}{\kappa} \ln \frac{K_\delta \delta u_{\tau g}}{\nu} + B \right) u_{\tau g} = \left[\frac{1}{\kappa} \ln \left(\frac{K_\delta \delta u_{\tau b}}{\nu} \frac{u_{\tau g}}{u_{\tau b}} \right) + B \right] \frac{u_{\tau g}}{u_{\tau b}} u_{\tau b} \quad (15)$$

$$U_{g-s} = \left(\frac{1}{\kappa} \ln \frac{K_\delta \delta u_{\tau b}}{\nu} + B \right) u_{\tau b} \quad (16)$$

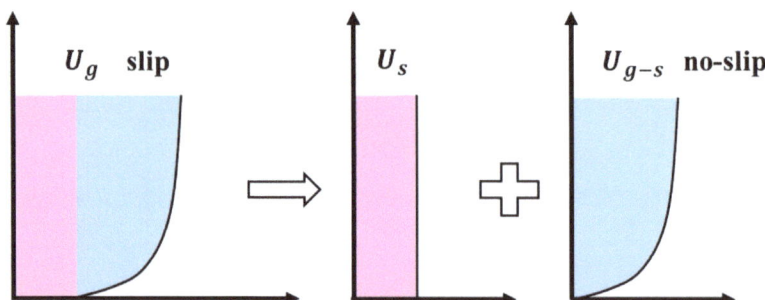

Figure 14. The velocity profile over the grooved plate is the superposition of slip velocity and no-slip flow velocity profile.

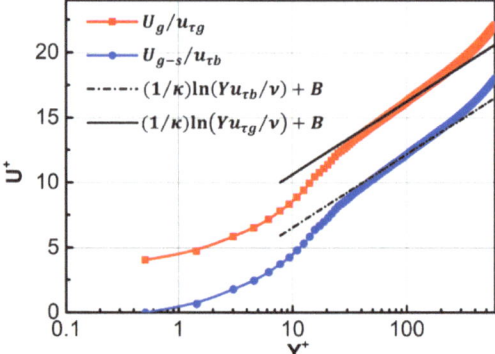

Figure 15. The velocity profile of U_g and U_{g-s} ($U_\infty^+ = 22$, $h = 0.1$).

Similarly, U_s in Equation (14) can be expressed as Equation (17) with $u_{\tau b}$.

$$U_s = U_s^+ u_{\tau g} = U_s^+ \frac{u_{\tau g}}{u_{\tau b}} u_{\tau b} \quad (17)$$

Therefore, substituting Equations (13) and (15)–(17) into Equation (14) yields the relationship between η_v and U_s^+.

$$U_s^+ = \frac{\frac{1}{\kappa} \ln \frac{K_\delta \delta u_{\tau b}}{\nu} + B}{\sqrt{1 + \eta_v}} - \frac{1}{\kappa} \ln \frac{K_\delta \delta u_{\tau b}}{\nu} - B - \frac{1}{\kappa} \ln \sqrt{1 + \eta_v} \quad (18)$$

With different U_s^+ in Figure 13 and the corresponding $U_\infty^+ \in [14.9, 22.0]$ as an input. Implicit Equation (18) is solved by the dichotomy to obtain η_v. In Figure 16, we compare the prediction of Equation (18) with CFD data in the present, The CFD data reported by Wu et al. (2019) [5], and the experimental data reported by Liu et al. (2020) [32]. The reason why the relationship between η_v and U_∞^+ is used, $\eta_v U_s^+$, is shown in Figure 16 and is mainly to facilitate the comparison between the results predicted by Equation (18) and the data obtained by previous studies. It is worth noting that the Reynolds number in all the U_∞^+ according to Equation (2), and the groove depths in our simulation are consistent with those in the references (h is 0.05, 0.1, and 0.3, respectively). $14.9 < U_\infty^+ < 20.35$, the predicted drag reduction rates are in good agreement with those from the present numerical simulations and the previous experimental data $U_\infty^+ > 20.35$, there is a significant error between the prediction of the model and the numerical results of Wu et al. [5], which indicates that this model is not applicable at a high Reynolds number.

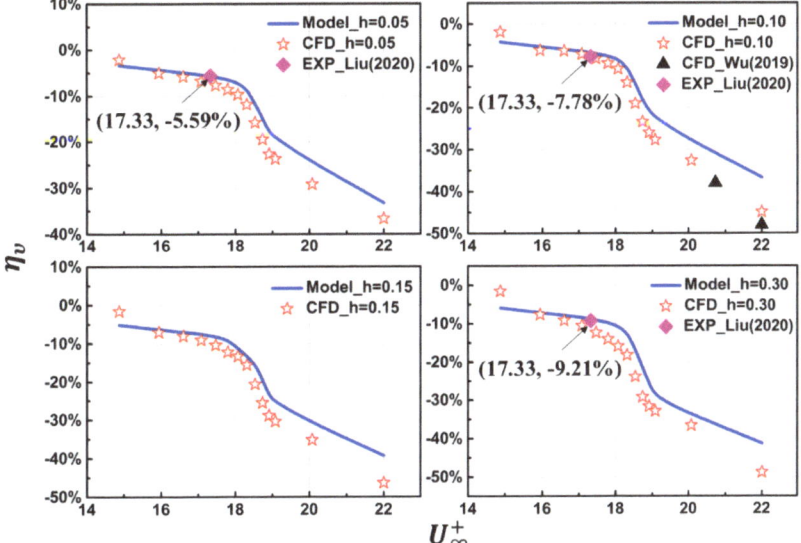

Figure 16. Comparison of the results of Equation (18) with CFD data and experimental data.

4.3. The Relationship between Slip Velocity (U_s^+) and Pressure Drag-Increase Rate (η_p)

Predicting the viscous drag reduction rate (η_v) of the grooved plate according to the corresponding slip velocity (U_s^+) has been conducted in Section 4.2. In this section, the relationship between U_s^+ and the pressure drag-increase rate (η_p) will be established by analyzing the effect of the slip velocity on the pressure distribution of the groove wall.

According to Equation (5), the pressure drag-increase rate η_p is expressed by Equation (19).

$$\eta_p = \frac{F_{GP}}{F_R} \tag{19}$$

here, the drag of the baseline plate can be estimated as

$$F_R = N \int_0^{S^+} \tau_\omega dx = NS^+ \rho \mu_{\tau b}^2 \tag{20}$$

where N denotes the number of transverse grooves in the corresponding length groove plate, and S^+ represents the dimensionless width of a groove.

Figure 17 shows the gauge pressure distribution on the groove wall ($h = 0.05$) at different inflow velocities. The results show that the high-pressure region is formed with the stagnation of slip velocity induced by the boundary vortex on the groove windward side and the corresponding low-pressure region is formed on the groove leeward side due to local flow separation. Therefore, the additional pressure drag (F_{GP}) of the grooved plate is equal to the integral of the difference between the high and low pressure in each groove. For a groove, taking the axis with zero-gauge pressure as the abscissa axis and the centerline of the groove as the ordinate axis, the pressure distribution is a function of x, as shown in Figure 18. Therefore, the total pressure drag of the grooved plate is expressed as

$$F_{GP} = 2N \int_0^{S^+/2} p(x) \sin\theta dx \tan\theta \tag{21}$$

here θ is 45 degrees (AR = 2).

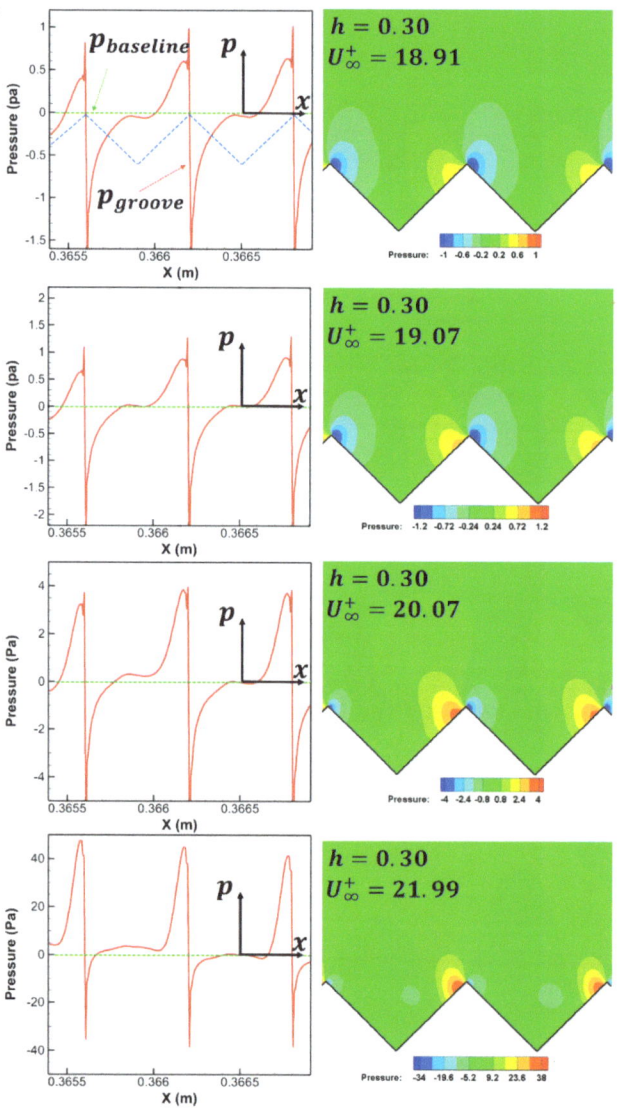

Figure 17. Gauge pressure distribution (the atmospheric pressure is 101,325 pa).

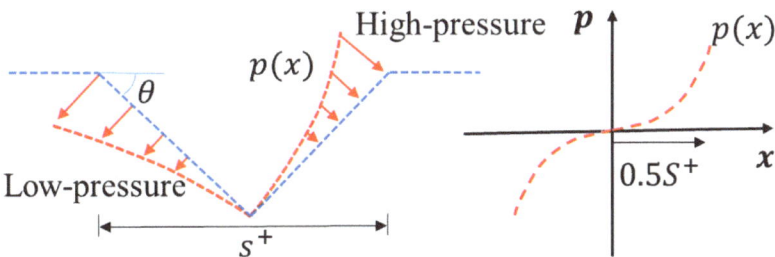

Figure 18. Schematic diagram of $p(x)$.

For the model description of $p(x)$, the experimental results of Feng et al. [45] show that the static pressure distribution in the groove can be described by an exponent or polynomial. On the basis of their work, combined with the pressure distribution in Figures 17 and 18, we assume that $p(x)$ can be described by Equation (22).

$$p(x) = C_1 x^3 + C_2 x^2 + C_3 x \tag{22}$$

where $C_3 = 0$, and C_1 and C_2 are adjustable variables that aim to obtain high-precision model results. Then, Equation (21) is transformed into Equation (23).

$$F_{GP} = N\sqrt{2}\frac{S^+}{8}\left[\frac{C_1}{8}(S^+)^3 + \frac{C_2}{3}(S^+)^2\right] = N\frac{\sqrt{2}p\left(\frac{S^+}{2}\right)}{8}S^+ + N\frac{\sqrt{2}C_2}{96}(S^+)^3 \tag{23}$$

where $p\left(\frac{S^+}{2}\right)$ represents the maximum pressure on the windward side, which is caused by the stagnation of the slip velocity, so $p\left(\frac{S^+}{2}\right)$ can be estimated as

$$p\left(\frac{S^+}{2}\right) = K_p 0.5\rho(U_s)^2 \tag{24}$$

in which K_p is an adjustable variable and "$0.5\rho(U_s)^2$" represents the dynamic pressure of the fluid velocity from U_s stagnation to 0. Substituting Equations (20) and (23) into Equation (19) yields η_p as

$$\eta_p = K_1\left(U_s^+\right)^2 + K_2\frac{(H^+)^2}{\mu_{\tau b}^2} \tag{25}$$

where $K_1 = \frac{K_p\sqrt{2}}{16}$ and $K_2 = \frac{\sqrt{2}C_2}{24\rho}$. Equation (25) shows that η_p is related to the slip velocity and the depth of the groove.

By adjusting the appropriate K_1 and K_2 values, the η_p predicted by Equation (25) can be consistent with the numerical results. Figure 19 shows the prediction of Equation (25) with different K_1 and K_2. Taking the fitting degree R^2 as the objective function, K_1 and K_2—which have the largest fitting degree—are selected as the model parameters, where $K_1 = 6.93 \times 10^{-6} h^{-1.428}$ and $K_2 = 0.00955 h^{-0.31646}$.

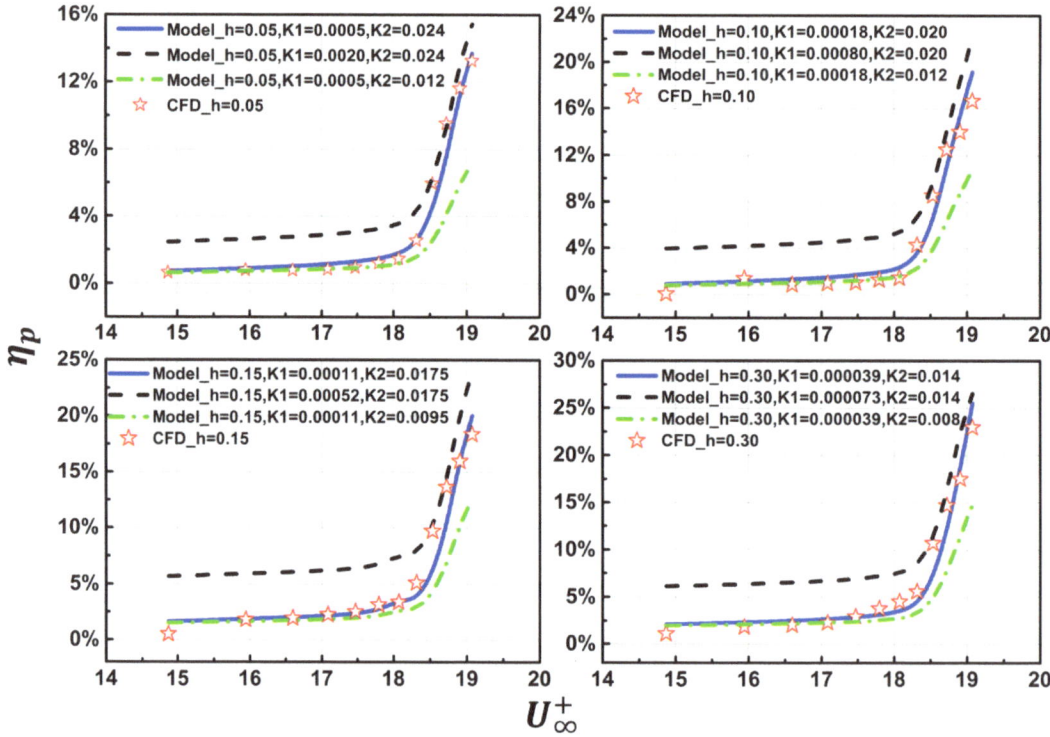

Figure 19. Comparison of the results of Equation (25) with CFD data.

5. Solution and Verification of the Model

5.1. Balance η_v and η_p to Solve the Quasi-Analytical Solution of Groove Depth (H^+) in Different Inflow Velocities (U_∞^+)

Simultaneous Equations (12), (18) and (25) yield Equation (26), which can predict the optimal depth and maximum depth of the transverse groove according to different inflow velocities. The steps for solving this equation are shown in Figure 20.

$$\begin{cases} U_s^+ = f_1(H^+, U_\infty^+) \\ U_s^+ = \dfrac{\frac{1}{\kappa}\ln\frac{K_\delta \delta u_{\tau b}}{\nu}+B}{\sqrt{1+\eta_v}} - \frac{1}{\kappa}\ln\frac{K_\delta \delta u_{\tau b}}{\nu} - B - \frac{1}{\kappa}\ln\sqrt{1+\eta_v} = f_2(U_\infty^+, \eta_v) \\ \eta_p = K_1(U_s^+)^2 + K_2\dfrac{(H^+)^2}{u_{\tau b}^2} = f_3(H^+, U_s^+) \end{cases} \quad (26)$$

Step 1: We input different H^+ and U_∞^+ to get different slip velocities U_s^+ of the transverse grooves.

Step 2: Input U_s^+ and U_∞^+ into implicit Equation (18) to solve η_v, and input U_s^+ and H^+ into Equation (25) to get η_p.

Step 3: Taking the sum of η_v and η_p as the objective value. When this value is the maximum, the relationship between optimal depth H_{opt}^+ and U_∞^+ is obtained, and when this value is equal to 0, the relationship between maximum depth H_{max}^+ and U_∞^+ is obtained. Figure 21 shows the solution when the step of U is 1 ($U \in [1, 70]$, the corresponding U_∞^+ is calculated by Equation (2)) and the step of h is 0.0006 ($h \in [0.01, 1.2]$, H^+ is calculated by corresponding h).

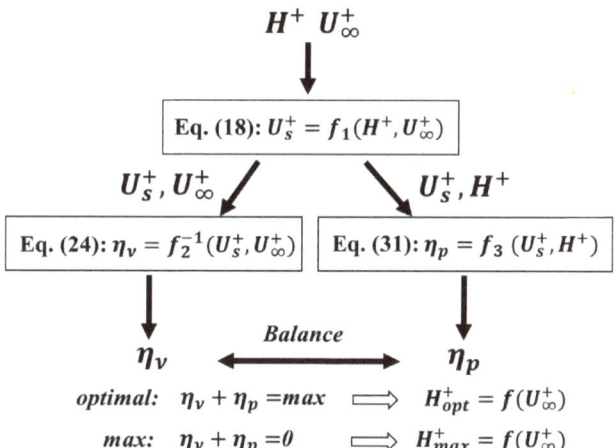

Figure 20. Steps for solving Equation (26).

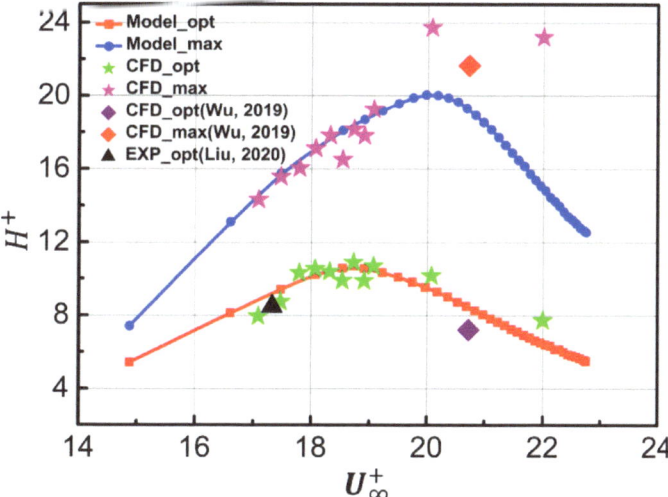

Figure 21. The dimensionless depths predicted by the model versus dimensionless inflow velocities.

5.2. Verification of the Quasi-Analytical Solution

In the previous section, we have established a theoretical model and solved the quasi-analytical solution of groove depths. In this section, firstly, the quasi-analytical solution of this model is verified by comparing the previous research and CFD results in the present with the model prediction results. Secondly, we redesign three grooved plates with different depths based on the quasi-analytical solution and compare the drag reduction effect of the grooved plates to further verify the correctness of the model.

Figure 21 shows the dimensionless depths predicted by the model versus the dimensionless inflow velocities. The results show that the predicted optimal depths are in good agreement with those from the present numerical simulations. However, for the prediction of the maximum depths, the model accuracy decreases with the increase of U_∞^+. Table 5 shows the details of the prediction accuracy of this model. For the prediction of the optimal depth, the average error of the model is less than 4.35% when $U_\infty^+ < 20.07$ (Re < 2.18×10^5). For the prediction of the maximum depth, the average error of the model is less than 4.09%

when $U_\infty^+ < 19.07$ (Re < 1.31×10^5). The main reason for the error may be that the changes of h and Re in the CFD cases are discrete, so the presentation of the optimal or maximum h in the CFD cases is not an accurate value (more details shown in Table 4).

Table 5. Comparison between model results and numerical results.

Re ($\times 10^5$)	Mean	0.436	0.544	0.653	0.762	0.870	0.980	1.09	1.20	1.31	2.18×10^5
U_∞^+	–	17.09	17.47	17.80	18.07	18.32	18.53	18.73	18.91	19.07	20.07
Model_opt(h)	–	0.54	0.47	0.43	0.39	0.35	0.32	0.29	0.27	0.24	0.14
CFD_opt(h)	–	0.50	0.45	0.45	0.40	0.35	0.30	0.30	0.25	0.25	0.15
Model_opt(H^+)	–	8.63	9.21	9.84	10.24	10.48	10.64	10.61	10.56	10.47	9.43
CFD_opt(H^+)	–	7.97	8.76	10.33	10.54	10.40	9.92	10.90	9.90	10.70	10.17
Relative error	4.35%	−7.7%	−4.9%	4.9%	2.9%	−0.7%	−6.8%	2.7%	−6.3%	2.2%	7.9%
Model_max(h)	–	0.93	0.81	0.72	0.65	0.59	0.54	0.50	0.47	0.44	
CFD_max(h)	–	0.90	0.80	0.70	0.65	0.60	0.50	0.50	0.45	0.45	
Model_max(H^+)	–	14.74	15.78	16.52	17.14	17.54	17.91	18.25	18.54	18.87	
CFD_max(H^+)	–	14.34	15.58	16.06	17.13	17.84	16.53	18.17	17.82	19.27	
Relative error	4.09%	−2.8%	−1.2%	−2.8%	0.0%	1.7%	−7.8%	−0.4%	−3.9%	2.1%	

Table 6 shows the details of the comparison between the predicted results of the model with the numerical results of Wu et al. [5] and the experimental results of Liu et al. [32]. Wu et al. [5] optimized the groove depth on the airfoil with a Reynolds number of 299,444 ($U_\infty^+ = 20.72$). The results show that when H^+ is 7.225, the drag reduction rate of the groove is the largest, while when H^+ is 21.675, the drag reduction rate of the groove approaches 0. On the premise of consistent dimensionless velocity, the prediction results of our model for the optimal and maximum depth are 8.475 and 19.290, respectively, which is close to the work of Wu et al. [5]. The inconsistency of the application object may have been the reason for the minor errors. Meanwhile, the optimal dimensionless groove depth for pipelines, $H_{opt}^+ = 8.488$, was observed by the water tunnel experiment of Liu et al. [32] with a Reynolds number of 50,000 ($U_\infty^+ = 17.326$). In this case, the optimal dimensionless groove depth predicted by the model is 9.215, and the relative error between the model results and the experimental results is 7.8%, which qualitatively proves the correctness of the model. The inconsistency of the medium may have been the reason for the minor errors.

Table 6. Comparison between model results and previous results.

		Medium	U (m/s)	Characteristic Length (m)	Re	u_τ	U_∞^+	H^+ (opt)	H^+ (Max)
Wu et al.	CFD	air	22	0.2	299,444	1.062	20.722	7.225	21.676
Present	model	air	27.5	0.16	299,444	1.327	20.723	8.475	19.290
Liu et al.	EXP	water	1.478	0.034	50,000	0.085	17.326	8.488	
Present	model	air	4.6	0.16	50,089	0.265	17.320	9.215	

Figure 22, which is transformed from Figure 21, shows the relationship between the optimal and maximum groove depths and the local Reynolds numbers. Based on this model, three grooved plates have been designed at U = 4.6 m/s, as shown in Figure 23. A 160 mm smooth plate is placed in front of the grooved plate in each case. The grooved plate of Case 1 is divided into four sections, each with a length of 80 mm. According to the average local Reynolds numbers of each section, the corresponding optimal groove depths predicted by the model are 0.44, 0.35, 0.28, and 0.23 mm, respectively. On the contrary, the groove depths of Case 2 ($h \equiv 0.5$ mm) and Case 3 ($h \equiv 0.21$ mm) are calculated by the model according to the local Reynolds numbers before and after the grooved plate, respectively.

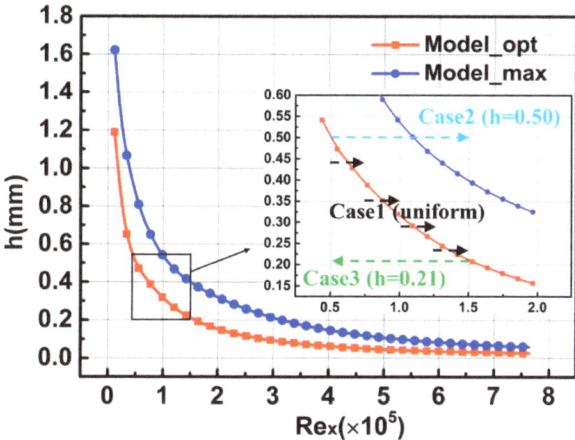

Figure 22. The groove depths predicted by the model versus local Reynolds numbers.

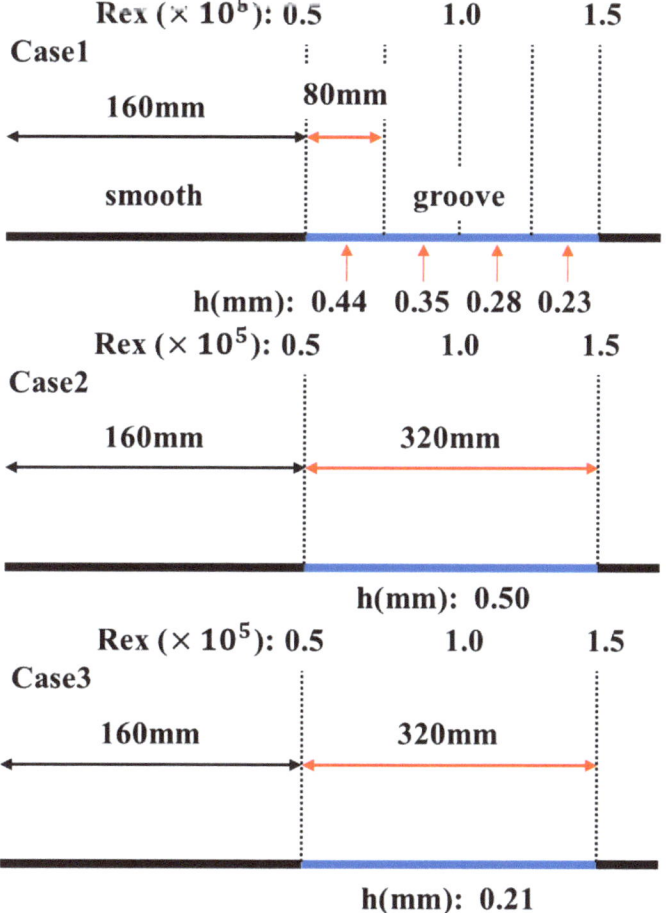

Figure 23. Schematic diagram of three grooved plates with different groove depths.

Table 7 shows that the drag reduction rate of the grooved plate (Case 1) according to the model design is the largest (12.25%) compared with other cases. On the contrary, the drag reduction rate of the grooved plate (Case 2) with the most deviation from the model design is the smallest—only 0.06%. The drag reduction rate of Case 3 is 7.83%, which is significantly less than Case 1. The reason why the drag reduction rate of Case 2 is almost zero is that the depth of the tail of the grooved plate has exceeded the predicted maximum depth (as shown in Figure 22), so these grooves at the tail cannot reduce the total drag, resulting in the decrease of the total drag reduction rate. Although the depth of the grooves in Case 3 do not meet the design of the optimal depths, it is less than the maximum groove depths predicted by the model, so Case 3 can slightly reduce the total drag.

Table 7. Comparison of the drag of three grooved plates.

	Viscous Drag (N)	η_v	Pressure Drag (N)	η_P	η
Baseline	0.01046	—	0	—	—
Case1	0.00756	−27.7%	0.00427	15.4%	−12.25%
Case2	0.00798	−23.7%	0.00557	23.6%	−0.06%
Case3	0.00839	−19.7%	0.00124	11.9%	−7.83%

Figure 24 compares the time-averaged entropy generation over different plates. The weaker the entropy generation, the smaller the total drag. Compared with the baseline plate, the entropy generation of the three grooved plates decreases gradually at the beginning of the groove area. From X = 0.16 to X = 0.48 in Case 1, the entropy generation at each place decreases significantly, which means that the grooves at each place play a role in reducing drag. However, in Case 2, the closer the groove is to the tail section of the grooved plate, the smaller the reduction of entropy generation, which means that the groove in the tail section may not reduce drag because the groove depth here exceeds the maximum depth shown in Figure 22. Compared with Case 2, the entropy generation of the front section (X = 0.32) of Case 3 decreases less, and the entropy generation of the tail section (X = 0.40~0.48) decreases more because the groove depth of the front section of Case 2 is closer to the optimal design depth and the groove depth of the tail section of Case 3 is closer to the optimal design depth, which further verifies the correctness of the model.

6. Conclusions

In this paper, we used the LES with the dynamic subgrid model to investigate the drag reduction characteristics of transverse V-grooves with different depths (h = 0.05~0.9 mm) at different Reynolds numbers ($1.09 \times 10^4 \sim 5.44 \times 10^5$). Based on the numerical results, the physical model describing the relationship between the dimensionless depth ($H^+ = Hu_\tau/v$) of the transverse groove, the dimensionless inflow velocity ($U_\infty^+ = U_\infty/u_\tau$), and the drag reduction rate (η) was established to quasi-analytically solve the optimal and maximum transverse groove depth according to different Reynolds numbers. The main conclusions are summarized as follows:

(1) The dimensionless groove depth (H^+) and dimensionless inflow velocity (U_∞^+) affect the magnitude of the slip velocity (U_s^+), thus driving the variation in the total viscous drag and pressure drag and thereby affecting the total drag. Therefore, the essence of balancing the benefit (the viscous drag reduction rate, η_v) and cost (the pressure drag-increase rate, η_p) to obtain the optimal drag reduction is to obtain the optimal slip velocity by matching the depth of the groove and the inflow velocity.

(2) The relationship between U_s^+ and η_v can be constructed by comparing the velocity profile of the grooved plate (slip) and baseline plate (no-slip), and the relationship between U_s^+ and η_p can be established by analyzing the effect of the slip velocity on the pressure distribution of the groove wall. When the value of $\eta_v + \eta_p$ reaches the maximum, the relationship between optimal depth H_{opt}^+ and U_∞^+ is obtained, and when this value is equal to 0, the relationship between maximum depth H_{max}^+ and U_∞^+ is obtained.

(3) The model results are consistent with the present numerical results and with previous data. For the solution of the optimal depth, the average error of the model is less than 4.35% when $U_\infty^+ < 20.07$ (Re < 2.18×10^5). For the solution of the maximum depth, the average error of the model is less than 4.09% when $U_\infty^+ < 19.07$ (Re <1.31×10^5).

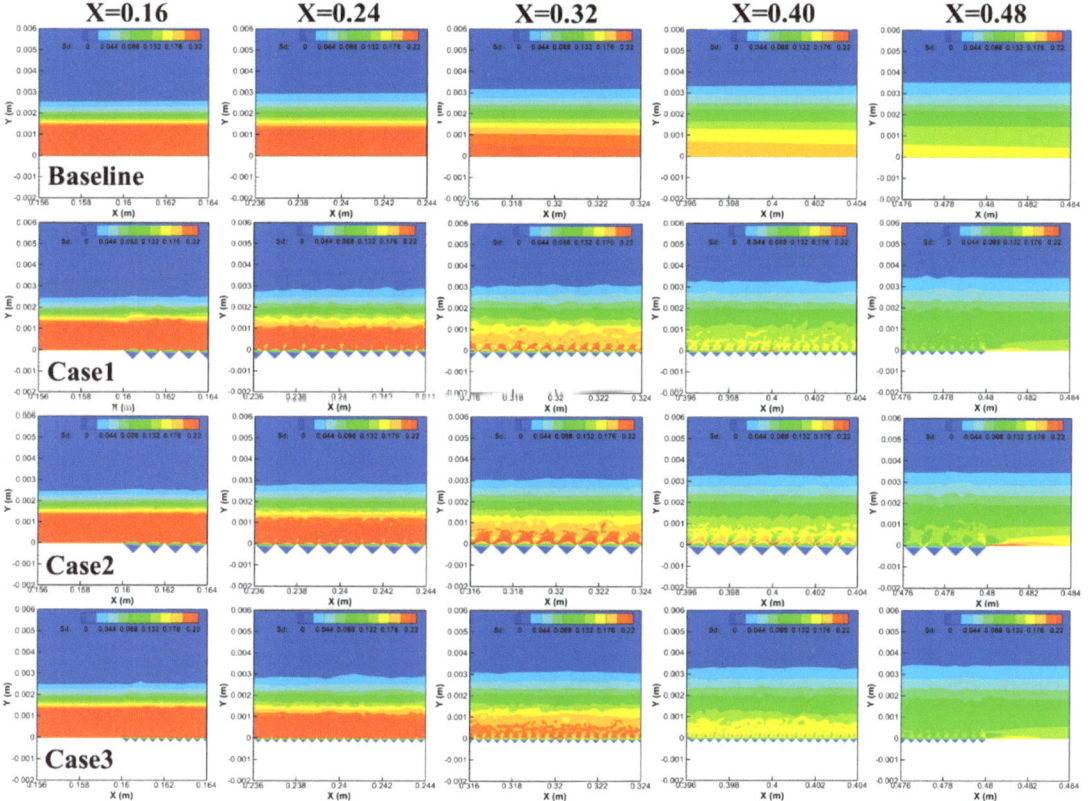

Figure 24. Time-averaged entropy generation over different plates.

Author Contributions: Among the authors in the list, Z.L. guided this research and completed the model establishment. L.H. completed model establishment, data analysis, and manuscript writing. Y.Z. completed the data calculation. All authors have read and agreed to the published version of the manuscript.

Funding: This work was funded by the National Natural Science Foundation of China (Nos. 52176032 and 51976005), National Science and Technology Major Project (2017-∥-0004-0016), and Aeronautics Power Foundation (No. 6141B090315).

Institutional Review Board Statement: Not applicable.

Informed Consent Statement: Not applicable.

Data Availability Statement: Not applicable.

Conflicts of Interest: The authors declare no conflict of interest.

References

1. Fu, Z.; Liang, X.; Zhang, K. Asymmetrical Velocity Distribution in the Drag-Reducing Channel Flow of Surfactant Solution Caused by an Injected Ultrathin Water Layer. *Symmetry* **2020**, *12*, 846. [CrossRef]
2. Yu, J.C.; Wahls, R.A.; Esker, B.M.; Lahey, L.T.; Akiyama, D.G.; Drake, M.L.; Christensen, D.P. Total Technolo-gy Readiness Level: Accelerating Technology Readiness for Aircraft Design. In Proceedings of the AIAA AVIATION 2021 FORUM, Online, 2–6 August 2021; p. 2454.
3. Zhang, Y.; Chen, H.; Fu, S.; Dong, W. Numerical study of an airfoil with riblets installed based on large eddy simulation. *Aerosp. Sci. Technol.* **2018**, *78*, 661–670. [CrossRef]
4. Zhang, Y.; Chongyang, Y.; Haixin, C.; Yuhui, Y. Study of riblet drag reduction for an infinite span wing with different sweep angles. *Chin. J. Aeronaut.* **2020**, *33*, 3125–3137. [CrossRef]
5. Wu, Z.; Li, S.; Liu, M.; Wang, S.; Yang, H.; Liang, X. Numerical research on the turbulent drag reduction mechanism of a transverse groove structure on an airfoil blade. *Eng. Appl. Comput. Fluid Mech.* **2019**, *13*, 1024–1035. [CrossRef]
6. Walsh, M.J.; Weinstein, L.M. Drag and Heat-Transfer Characteristics of Small Longitudinally Ribbed Surfaces. *AIAA J.* **1979**, *17*, 770–771. [CrossRef]
7. DeGroot, C.; Wang, C.; Floryan, J.M. Drag Reduction Due to Streamwise Grooves in Turbulent Channel Flow. *J. Fluids Eng.* **2016**, *138*, 121201. [CrossRef]
8. Wainwright, D.K.; Fish, F.E.; Ingersoll, S.; Williams, T.M.; Leger, J.S.; Smits, A.J.; Lauder, G.V. How smooth is a dolphin? The ridged skin of odontocetes. *Biol. Lett.* **2019**, *15*, 20190103. [CrossRef]
9. Lang, A.; Afroz, F.; Motta, P.; Wilroy, J.; Wahidi, R.; Elliott, C.; Habegger, M.L. Sharks, Dolphins and Butterflies: Micro-Sized Surfaces Have Macro Effects. In Proceedings of the Fluids Engineering Division Summer Meeting. American Society of Mechanical Engineers, Waikoloa, HI, USA, 30 July–3 August 2017; p. V01CT21A001.
10. Abdulbari, H.A.; Mahammed, H.D.; Hassan, Z.B.Y. Bio-Inspired Passive Drag Reduction Techniques: A Review. *ChemBioEng Rev.* **2015**, *2*, 185–203. [CrossRef]
11. Asadzadeh, H.; Moosavi, A.; Etemadi, A. Numerical simulation of drag reduction in microgrooved substrates using lattice-Boltzmann method. *J. Fluids Eng.* **2019**, *141*, 071111. [CrossRef]
12. Walsh, M.J. Riblets as a Viscous Drag Reduction Technique. *AIAA J.* **1983**, *21*, 485–486. [CrossRef]
13. Rastegari, A.; Akhavan, R. The common mechanism of turbulent skin-friction drag reduction with superhydrophobic longitudinal microgrooves and riblets. *J. Fluid Mech.* **2018**, *838*, 68–104. [CrossRef]
14. Ran, W.; Zare, A.; Jovanović, M.R. Model-based design of riblets for turbulent drag reduction. *J. Fluid Mech.* **2020**, *906*, A7. [CrossRef]
15. Peet, Y.; Sagaut, P.; Charron, Y. Pressure loss reduction in hydrogen pipelines by surface restructuring. *Int. J. Hydrogen Energy* **2009**, *34*, 8964–8973. [CrossRef]
16. Li, L.; Zhu, J.; Li, J.; Song, H.; Zeng, Z.; Wang, G.; Zhao, W.; Xue, Q. Effect of vortex frictional drag reduction on ordered microstructures. *Surf. Topogr. Metrol. Prop.* **2019**, *7*, 025008. [CrossRef]
17. Liu, M.; Li, S.; Wu, Z.; Zhang, K.; Wang, S.; Liang, X. Entropy generation analysis for grooved structure plate flow. *Eur. J. Mech.-B/Fluids* **2019**, *77*, 87–97. [CrossRef]
18. Wang, L.; Wang, C.; Wang, S.; Sun, G.; You, B. Design and analysis of micro-nano scale nested-grooved surface structure for drag reduction based on 'Vortex-Driven Design'. *Eur. J. Mech.-B/Fluids* **2020**, *85*, 335–350. [CrossRef]
19. Wood, D.H.; Antonia, R.A. Measurements in a Turbulent Boundary Layer over Ad-Type Surface Roughness. *J. Appl. Mech.* **1975**, *42*, 591–597. [CrossRef]
20. Choi, K.-S.; Fujisawa, N. Possibility of drag reduction using d-type roughness. *Flow Turbul. Combust.* **1993**, *50*, 315–324. [CrossRef]
21. Bushnell, D. Turbulent drag reduction for external flows. In Proceedings of the 21st Aerospace Sciences Meeting, Reno, NV, USA, 10–13 January 1983.
22. Cui, J.; Fu, Y. A numerical study on pressure drop in microchannel flow with different bionic micro-grooved surfaces. *J. Bionic Eng.* **2012**, *9*, 99–109. [CrossRef]
23. Wang, B.; Wang, J.; Zhou, G.; Chen, D. Drag Reduction by Microvortexes in Transverse Microgrooves. *Adv. Mech. Eng.* **2014**, *6*, 734012. [CrossRef]
24. Chen, H.; Gao, Y.; Stone, H.A.; Li, J. "Fluid bearing" effect of enclosed liquids in grooves on drag reduction in microchannels. *Phys. Rev. Fluids* **2016**, *1*, 083904. [CrossRef]
25. Wang, J.; Ma, W.; Li, X.; Bu, W.; Liu, C. Slip theory-based numerical investigation of the fluid transport behavior on a surface with a biomimetic microstructure. *Eng. Appl. Comput. Fluid Mech.* **2019**, *13*, 609–622. [CrossRef]
26. Davies, J.; Maynes, D.; Webb, B.W.; Woolford, B. Laminar flow in a microchannel with superhydrophobic walls exhibiting transverse ribs. *Phys. Fluids* **2006**, *18*, 087110. [CrossRef]
27. Rothstein, J.P. Slip on Superhydrophobic Surfaces. *Annu. Rev. Fluid Mech.* **2010**, *42*, 89–109. [CrossRef]
28. Wang, B.; Wang, J.; Chen, D. A prediction of drag reduction by entrapped gases in hydrophobic transverse grooves. *Sci. China Technol. Sci.* **2013**, *56*, 2973–2978. [CrossRef]
29. Seo, S.-H.; Nam, C.-D.; Han, J.-Y.; Hong, C.-H. Drag reduction of a bluff body by grooves laid out by design of experiment. *J. Fluids Eng.* **2013**, *135*, 111202. [CrossRef]
30. Lang, A.W.; Jones, E.M.; Afroz, F. Separation control over a grooved surface inspired by dolphin skin. *Bioinspir. Biomim.* **2017**, *12*, 026005. [CrossRef] [PubMed]

31. Mariotti, A.; Buresti, G.; Salvetti, M. Separation delay through contoured transverse grooves on a 2D boat-tailed bluff body: Effects on drag reduction and wake flow features. *Eur. J. Mech.-B/Fluids* **2018**, *74*, 351–362. [CrossRef]
32. Liu, W.; Ni, H.; Wang, P.; Zhou, Y. An investigation on the drag reduction performance of bioinspired pipeline surfaces with transverse microgrooves. *Beilstein J. Nanotechnol.* **2020**, *11*, 24–40. [CrossRef] [PubMed]
33. Breuer, M.; Jovičić, N. Separated flow around a flat plate at high incidence: An LES investigation. *J. Turbul.* **2001**, *2*, N18. [CrossRef]
34. Martin, S.; Bhushan, B. Fluid flow analysis of a shark-inspired microstructure. *J. Fluid Mech.* **2014**, *756*, 5–29. [CrossRef]
35. Hemida, H.; Gil, N.; Baker, C.; Baker, C. LES of the Slipstream of a Rotating Train. *J. Fluids Eng.* **2010**, *132*, 051103. [CrossRef]
36. Meneveau, C.; Lund, T.S.; Cabot, W.H. A Lagrangian dynamic subgrid-scale model of turbulence. *J. Fluid Mech.* **1996**, *319*, 353–385. [CrossRef]
37. Rhie, C.M.; Chow, W.L. Numerical study of the turbulent flow past an airfoil with trailing edge separation. *AIAA J.* **1983**, *21*, 1525–1532. [CrossRef]
38. Zhang, Z.; Zhang, M.; Cai, C.; Kang, K. A general model for riblets simulation in turbulent flows. *Int. J. Comput. Fluid Dyn.* **2020**, *34*, 333–345. [CrossRef]
39. Ahmadi-Baloutaki, M.; Carriveau, R.; Ting, D.S.-K. Effect of free-stream turbulence on flow characteristics over a transversely-grooved surface. *Exp. Therm. Fluid Sci.* **2013**, *51*, 56–70. [CrossRef]
40. Song, X.-W.; Zhang, M.-X.; Lin, P.-Z. Skin Friction Reduction Characteristics of Nonsmooth Surfaces Inspired by the Shapes of Barchan Dunes. *Math. Probl. Eng.* **2017**, *2017*, 6212605. [CrossRef]
41. Currie, I.G. *Fundamental Mechanics of Fluids*; CRC Press: Boca Raton, FL, USA, 2016.
42. Fukagata, K.; Kasagi, N.; Koumoutsakos, P. A theoretical prediction of friction drag reduction in turbulent flow by superhydrophobic surfaces. *Phys. Fluids* **2006**, *18*, 051703. [CrossRef]
43. Dean, R.B. Reynolds Number Dependence of Skin Friction and Other Bulk Flow Variables in Two-Dimensional Rectangular Duct Flow. *J. Fluids Eng.* **1978**, *100*, 215–223. [CrossRef]
44. Min, T.; Kim, J. Effects of hydrophobic surface on skin-friction drag. *Phys. Fluids* **2004**, *16*, L55–L58. [CrossRef]
45. Feng, B.; Chen, D.; Wang, J.; Yang, X. Bionic Research on Bird Feather for Drag Reduction. *Adv. Mech. Eng.* **2014**, *7*. [CrossRef]

MDPI AG
Grosspeteranlage 5
4052 Basel
Switzerland
Tel.: +41 61 683 77 34

Symmetry Editorial Office
E-mail: symmetry@mdpi.com
www.mdpi.com/journal/symmetry

Disclaimer/Publisher's Note: The statements, opinions and data contained in all publications are solely those of the individual author(s) and contributor(s) and not of MDPI and/or the editor(s). MDPI and/or the editor(s) disclaim responsibility for any injury to people or property resulting from any ideas, methods, instructions or products referred to in the content.